# Case studies in food product development

# Related titles:

*Food product development*
(ISBN 978-1-85573-468-5)
Product development, from refining an established product range to developing completely new products, is the lifeblood of the food industry. It is, however, a process fraught with risk, which often ends in failure. This book explains the means to making product development a success. Filled with examples and practical suggestions, and written by a distinguished team with unrivalled academic and industry expertise, *Food product development* is an essential guide for R&D and product development staff, and managers throughout the food industry concerned with this key issue.

*Consumer-led food product development*
(ISBN 978-1-84569-072-4)
No matter how innovative or technologically advanced a new or reformulated food product may be, it will only be a success if it gains widespread consumer acceptance. Ensuring that food product development strategies are consumer-led, therefore, is of importance to the food industry. *Consumer-led food product development* describes current state-of-art methods in this area. After an introductory section exploring reasons why food consumers purchase and consume certain products, chapters review methods to determine consumers' food-related attitudes and others issues in consumer-led food product development, such as opportunity identification and concept development, as well as techniques such as just-about-right scales and partial-least-squares methods.

*Making the most of HACCP: Learning from others' experience*
(ISBN 978-1-85573-504-0)
The Hazard Analysis and Critical Control Point (HACCP) system has now become generally accepted as the key safety management system for the food industry worldwide. Whilst there are numerous publications on its principles and methods of implementation, there are relatively few on the experience of those who have actually implemented HACCP systems in practice and what can be learnt from that experience. Edited by two leading authorities on the subject, and with an international team of contributors, *Making the most of HACCP* describes that experience and what it can teach about implementing and developing HACCP systems effectively.

Details of these books and a complete list of Woodhead titles can be obtained by:

- visiting our web site at www.woodheadpublishing.com
- contacting Customer Services (email: sales@woodhead-publishing.com; fax: +44 (0) 1223 893694; tel.: +44 (0) 1223 891358 ext. 130; address: Woodhead Publishing Limited, Abington Hall, Abington, Cambridge CB21 6AH, England)

# Case studies in food product development

### Edited by
### Mary Earle and Richard Earle

**CRC Press**
**Boca Raton Boston New York Washington, DC**

WOODHEAD PUBLISHING LIMITED
Cambridge England

Published by Woodhead Publishing Limited, Abington Hall, Abington, Cambridge CB21 6AH, England
www.woodheadpublishing.com

Published in North America by CRC Press LLC, 6000 Broken Sound Parkway, NW, Suite 300, Boca Raton, FL 33487, USA

First published 2008, Woodhead Publishing Limited and CRC Press LLC
© 2008, Woodhead Publishing Limited
The authors have asserted their moral rights.

British Library Cataloguing in Publication Data
A catalogue record for this book is available from the British Library.

Library of Congress Cataloging in Publication Data
A catalog record for this book is available from the Library of Congress.

Woodhead Publishing Limited ISBN 978-1-84569-260-5 (book)
Woodhead Publishing Limited ISBN 978-1-84569-413-5 (e-book)
CRC Press ISBN 978-1-4200-7204-4
CRC Press order number: WP7204

The publishers' policy is to use permanent paper from mills that operate a sustainable forestry policy, and which has been manufactured from pulp which is processed using acid-free and elementary chlorine-free practices. Furthermore, the publishers ensure that the text paper and cover board used have met acceptable environmental accreditation standards.

Project managed by Macfarlane Production Services, Dunstable, Bedfordshire, England (e-mail: macfarl@aol.com)
Typeset by Godiva Publishing Services Limited, Coventry, West Midlands, England
Printed by TJ International Limited, Padstow, Cornwall, England

# Contents

## Part IV    Technological development

# Contributor contact details and author biographies

(* = main contact)

### Chapter 1

Mary Earle trained as a process engineer then moved through technological research and teaching in food science and bakery technology, into food product development where she has worked for more than fifty years. Initially in large British multinational companies, she moved into cooperative product development research in the New Zealand meat industry, then into teaching and practice across the whole framework of New Product Development. From her work, generations of students have evolved successful careers, and a whole scheme of development methodology has been conceived and matured in lectures, articles, books, reports; and put into industrial practice in many countries. Now she is Professor Emeritus at Massey University.

Dick Earle trained as a process engineer, and moved through industrial and operational research into the meat industry. He then took up teaching and research in process strategies and development, working over forty years in a range of technologies to engineer, on a biological base, processes using novel methods and approaches. A major goal was to combine academic and industrial aspects of process research, teaching and development. Applications were particularly in food, pharmaceutical, and environmental areas. Experience over a wide field, led to lectures and publications including a basic food engineering text, which have had wide international exposure. Now he is Professor Emeritus at Massey University.

Mary Earle and Dick Earle
21 Montgomery Terrace,
Palmerston North,
New Zealand 4410
E-mail: rmearle@inspire.net.nz

## Chapter 2

Dick Inwood was an early Massey Food Technology graduate, who spent most of his career doing, managing or coordinating product development, at first for Unilever, in New Zealand, and then for the petfood operations of Mars, in Australia, in Germany, and finally in Europe, where he was responsible for the integration of research and development activities across a number of R&D centres in different countries. Social and economic pressures, and technological advance have caused many changes in New Product Development strategies in the past four decades, but current concerns about energy, transport, water conservation and other environmental factors, coupled with increasing consumer health concerns, portend an even greater opportunity for change in the design and supply of future food products.

Dick Inwood
Apartment 2906
6 Lorne Street
Auckland 1010
New Zealand
E-mail: dick@labrossadiere.com

## Chapter 3

Paiboon Thammarutwasik trained in food processing and quality management systems. He has been teaching and researching in food processing and product development for more than 30 years. His aim is to try to upgrade the quality of the product through product and process development research. He set up the research centre at the Faculty of Aro-Industry to assist small and medium sized enterprises in developing new products and processes. He is currently the Director of the Nutriceutical and Functional Food Research Centre at Prince of Songkla University.

Paiboon Thammarutwasik
Faculty of Agro-Industry
Prince of Songkla University
Hat Yai
Songkhla 90112
Thailand
E-mail: paiboon.th@gmail.com

## Chapter 4

Sam (Satya) Adapa received an MS in Food Science at CFTRI, Mysore, India and an MS in Clinical Nutrition at Columbia University, New York, followed by over 35 years in the US food industry. Most of his experience has been in product development, new venture technologies and innovations, technical aspects of NPD, QA and co-packer management and business functions. He has an exemplary record of new products at General Foods (Kraft), Rich's, Dole and Safeway with a broad variety of technologies for retail and food service products, juices, beverages, bakery, confectionery, dairy and non-dairy and fruit products as Director, Associate Director, Group/Programme Manager. He holds three US and international patents in food innovation. He has extensive experience in technology transfer and applications licensing work in the food industry in Europe, Japan, South Korea and India. He has been employed at Safeway since 1998 as Product Development Manager, Consumer Brands, and lives in the San Francisco Bay area of California.

Sam Adapa*
(Safeway, Inc.)
216 Shadybrook Court
Pittsburg Hills, CA 94565
USA
E-mail: *sam.adapa@safeway.com*

Kumaresh Chakraborty has been a Marketing, Sales, Business and Product Development executive in the Food, Ingredient and Beverage Industry for over 20 years, specialising in strategy, value-added marketing, new product and brand development and management. He developed and commercialised patented new products. Also, he is an adjunct faculty in the College of Business at Governors State University, Keller Graduate School of Management and University of Phoenix, where he teaches post-graduate and undergraduate courses in strategy, new product development, marketing, sales and international marketing, consumer and organisational behaviour, and health-wellness. He believes, innovation and new products are keys to industry growth; however, including voice of the customer while designing need-based breakthrough products that are new-to-the-world or new-to-the-market will ultimately drive success of the new products in the future.

Dr. Chakraborty received an MBA from the University of Chicago, a PhD from North Dakota State University, an MS from CFTRI, University of Mysore, India, and a BS Honours from J.N. Agricultural University, India. He has been cited in numerous editorials in trade publications, published several research papers in peer-reviewed journals, holds a patents and has made presentations at international meetings.

Kumaresh Chakraborty
Management Consultant
1412 Fran Lin Parkway
Munster, IN 46321
USA
E-mail: chakrabo@tsrcom.com

## Chapter 5

Bill Edwardson obtained his PhD in Food Science and Technology from Massey University in New Zealand, where he applied product development methodology in the design of food products in developing countries. He worked for almost twenty years at the International Development Research Centre (IDRC) of Canada, where he pioneered the application of product development and enterprise development methodologies in rural development projects in developing countries. He has worked in Asia and Africa but particularly in Latin America, when he was posted to IDRC's regional office in Bogotá, Colombia. Bill believes that commercially oriented methodologies, such as New Product Development, borrowed and adapted from the private sector, need to be more widely used in development projects if poor communities and farmers are to improve their incomes through participating in sustainable agrifood chains and enterprises. Bill is currently working as an independent consultant in international development projects from Ottawa, Canada.

Bill Edwardson*
10 Tunbridge Court
Ottawa
Ontario, K2E 7J8
Canada
E-mail: williamedwardson@hotmail.com

Rupert Best holds a PhD in Chemical Engineering from the University of Birmingham, UK. His expertise is in the area of rural enterprise development, with an emphasis on tropical products' processing and marketing. He worked for over 20 years at the Centro Internacional de Agricultura Tropical (CIAT) in Colombia and Uganda, where he managed the Rural Agroenterprise Development Project and the Cassava Programme. During this period he and his colleagues developed a 'territorial' approach to rural agroenterprise development aimed at improving the livelihoods of poor smallholder farmers. The approach incorporates elements of product development strategy. In 2004 he joined the Secretariat of the Global Forum on Agricultural Research (GFAR) in Rome where he specialised in research partnership programmes. He is now an independent consultant based in Bogotá, Colombia.

Rupert Best
Apartado Aereo 5813
Bogotá
Colombia
E-mail: rupertbest@hotmail.com

## Chapter 6

Dave Woodhams trained as a process engineer and joined the NZ Dairy Research Institute in 1963. After three years working on whey utilisation he then completed a doctorate on spray drying at the University of Wisconsin. Returning to NZDRI in 1970 he first set up the whey ultrafiltration project and later led the milk powder research team during a very active product and process development period. He has been technical officer for a dairy company and has managed a process equipment supply company. Since 1986, as a dairy process consultant, he has managed a number of cutting-edge, multidisciplinary R&D projects, often crossing the boundaries between biochemistry, process engineering and industrial microbiology. He is now mostly retired (occasional technical writing excepted) and lives in the country with a cat and a dog.

David Woodhams
460 Whatarangi Road
Palliser Bay R D 2
Featherston 5772
New Zealand
E-mail: woodhams@wise.net.nz

## Chapter 7

Brian Wilkinson trained as a biotechnologist and then worked in waste treatment research for the meat industry for four years before returning to university to gain the skills of a food product developer. After two years at university he once again joined the meat industry to work as a process and product developer. After four years the opportunity arose to join Mary Earle in a research centre that she had recently set up to carry out applied research for small to medium sized companies in the New Zealand food industry. After twelve years in the research centre he moved into teaching with special emphasis on the meat industry. He remained in that role for a further 10 years before moving into teaching product development. His main research efforts are focussed on the meat industry where he is currently involved in developing innovative products as well as developing processes to produce these products.

Brian Wilkinson
Institute of Food, Nutrition and Human Health
Massey University
PO Box 11 222
Palmerston North
New Zealand
E-mail: B.Wilkinson@massey.ac.nz

John Palamountain left school at 15 and worked on a number of farms in various roles for a period of seven years. He then applied for a position as a sales representative for a company making dog biscuits for domestic and farm dogs. After two years he and his partner bought this company which was based in a small rural North Island New Zealand town. He looked after plant operations and marketing of the company's products for a period of seven years before it was taken over by Goodman Fielder, a large Australasian company. He continued with Goodman Fielder for a further 10 years in the same roles before leaving and starting a new dog biscuit company in Wanganui in 1993. He and another partner sold this company to Heinz Watties and he continued to manage this plant for them until 2001 when he resigned to start his new company Vita Power Ltd.

John Palamountain
Vita Power Ltd
PO Box 448
120 Putiki Drive
Wanganui
New Zealand
E-mail: john@vitapower.co.nz

## Chapter 8

Liz Bowie (was Ashworth) was born in 1950 in Elgin, Morayshire, Scotland. She began cooking with her mother at the age of three and has not stopped since. Twelve years with Baxters Food Group started a long career in the food industry, firstly hands on and then working with food manufacturers to develop new products; many of which have won awards. She is a Member of the Guild of Food Writers of Great Britain, is passionately committed to promoting good healthy food and teaching would-be cooks of all ages. She is a working part of a 'cook school with a difference' on Westray, Orkney where all are welcome to enjoy real home cooking!

Liz (Ashworth) Bowie
NPD Consultant to the Food Industry
Member of the Guild of Food Writers of Great Britain
12 Kirkhill Drive,
Lhanbryde
Elgin
Morayshire, IV30 8QA
Scotland
E-mail: lizashworth@tiscali.co.uk

## Chapter 9

Buncha Ooraikul originally trained in agriculture, then moved to food science and technology. He worked briefly in the food industry as quality control and research and development manager before entering academia. During his 35 years service in several universities he taught quality control, process engineering, fruit and vegetable processing, and product and process development. His research, mainly in food processing and product development, ranged from potato processing, accelerated fermentation processes with immobilised microorganisms, modified atmosphere packaging of food, minimal processing of fruits and vegetables, to nutriceutical and functional food development. He is now Professor Emeritus at the University of Alberta.

Buncha Ooraikul, PhD
Professor Emeritus
University of Alberta
Edmonton
Alberta
Canada
E-mail: buncha.ooraikul@gmail.com

## Chapter 10

Rex Perreau studied Food Technology which included short stints in the dairy industry, and the fruit and vegetable canning and freezing industry. He worked in the confectionery industry in New Zealand as chief chemist, and then moved to the United Kingdom where he filled a number of roles in Research and Development. These included product and process development, process instrumentation, process engineering, and quality management. He led the UK division of a multinational confectionery company through the process of achieving accreditation to the ISO 9000 Quality management system. He is retired and lives in Birmingham UK.

Rex Perreau
112 Langleys Road
Selly Oak
Birmingham B29 6HT
UK
E-mail: Joeperreau@aol.com

## Chapter 11

Andrew Russell trained as a biochemical engineer in New Zealand and the UK. The majority of his career has been based within the Foods R&D organisation of Unilever in the UK – where he has led a number of large projects concerned with the development of both process and ingredient technologies. He has also held positions as a university lecturer and as a process developer within the New Zealand dairy industry. His main interest lies at the boundary between scientific understanding and product development – where new insights in aspects such as food microstructure or food chemistry can lead to radically new technologies and, ultimately, to greatly improved products in the marketplace.

Andrew B. Russell
Unilever Foods R&D
Colworth Science Park
Sharnbrook
Bedford MK44 1LQ
UK
E-mail: andrew.russell@unilever.com

## Chapter 12

Torben Sorensen graduated with a degree in Food Technology, after which he worked as a research scientist in the fish processing industry, both in Denmark and in New Zealand. This was followed by the establishment of Sorensen Laboratories, an independent food technology laboratory that has offered a diverse range of services to the food industry in New Zealand and in the Asia Pacific region for the past 30 years. Torben has developed a wide range of new products, and undertaken various assignments for the meat, fish and poultry industries, as well as for bakeries and a host of small companies.

Torben Sorenson
57A Krippner Road
Puhoi
Rodney District
New Zealand
E-mail: torben@sorensenlaboratories.co.nz

## Chapter 13

Choon Nghee Gwee studied Food Technology at Massey University, and was awarded her Bachelor of Technology degree in 1978. Keen to have a broader understanding of business practices, she attended MBA courses at UC Berkeley in 1993. She joined Cold Storage (Singapore) as Food Technologist on leaving Massey and after a thorough grounding in all aspects of beverages, dairy and bakery goods manufacture, she then switched to plant trials and process development with Alfa Laval. Her successful work in aseptic coconut cream enabled her to join PT Pulau Sambu (Kara Brand) where she headed the product development and quality control functions. In 1994 Gwee decided to try her hand at her own product concepts; the subject of this chapter.

Choon Nghee Gwee
Pacifica Food Consultancy Sdn Bhd
24-2 Medan Setia 2
Plaza Damansara
Damansara Heights
50490 Kuala Lumpur
Malaysia
E-mail: cngwee@gmail.com

## Chapter 14

Anne Goldman received her food science education from the Universities of London and Leeds before embarking on a career in product development for the food industry in the UK and New Zealand. Her experiences in the food industry, academia and regulatory bodies formed the groundwork for her subsequent consulting career that has always focused on meeting the business needs of R&D and marketing management. She is a principal and co-founder in 1986 of ACCE, a full service Canadian-owned market research company specialising in consumer guidance sensory research for clients in the consumer packaged goods and food service industries.

Elizabeth O'Neil has extensive experience in teaching/training and consulting in the areas of marketing, consumer behaviour and integrated marketing communications. As a qualitative market researcher for more than 20 years, she has expertise in the consumer packaged goods and food categories. Her research ranges from product development initiatives to all aspects of marketing communication including package/label design, advertising strategies, retail management and brand equity assessment. Elizabeth is a graduate of both the University of Guelph and the University of Toronto.

Anne Goldman* and Elizabeth O'Neil
2575B Dunwin Drive
Mississauga
Ontario, L5L 3N9
Canada
E-mail: agoldman@acceintl.com; eoneil@acceintl.com

## Chapter 15

Joe Bogue is a Senior Lecturer with the Department of Food Business and Development at University College Cork. His research focuses on market-oriented New Product Development (NPD) and research interests include: the management of NPD; increasing the competitiveness of the Irish food sector through NPD; consumer behaviour and the marketing of novel foods, such as health-enhancing and functional foods. One of his main NPD interests is encouraging Irish food firms to integrate the consumer more closely with the NPD process, and thus to develop a more market-oriented food industry. Prior to joining University College Cork he worked in strategic marketing and management within the food sector, and has qualifications from both science and business programmes.

Douglas Sorenson is a qualified food technologist, and holds an MSc in Food Business and a PhD in Food Science and Technology (Food Marketing). He has over eight years consumer and sensory research experience having worked in academia at University College Cork, and agency-side with Eolas International. His main research area focuses on consumer knowledge management in the food innovation process with over 20 publications and communications on consumer food choice, health and wellness, and new food product development.

Joe Bogue* and Douglas Sorenson
Department of Food Business and Development
University College Cork
Western Road
Cork
Ireland
E-mail: J.Bogue@ucc.ie

## Chapter 16

Helle Alsted Søndergaard holds a PhD in Innovation Management with focus on market-oriented product development in the Danish food industry. Most of her work lies in the interface between consumer behaviour and new product development with emphasis on how information can be used in the innovation process. Her teaching centres on New Product Development and Innovation Management from a marketing perspective. She holds a position as Associate Professor at MAPP – Centre for customer relations in the food sector, University of Aarhus.

Helle Alsted Søndergaard*
MAPP
Aarhus School of Business
University of Aarhus
Haslegaardsvej 10
DK-8210 Aarhus V
Denmark
E-mail: hals@asb.dk

Merete Edelenbos holds a PhD in Food Science and Technology. She has vegetable food quality experience having worked in academia at the Danish Institute of Agricultural Sciences in cooperation with the Danish food industry on chemical and sensory quality aspects of vegetables and fruits. Her work lies in the interface between molecular and sensory quality and the role of chemical constituents for flavour and taste and human health. She holds a position as senior scientist at Faculty of Agricultural Sciences.

Merete Edelenbos
Faculty of Agricultural Sciences
Department of Food Science
University of Aarhus
Kirstinebjergvej 10
DK-5792 Aarslev
Denmark
E-mail: Merete.Edelenbos@agrsci.dk

## Chapter 17

Andrea Currie's plans for a promising career in veterinary science were forestalled by a less than acceptable chemistry result in her final high school exams. Andrea reverted to her back up plan – Food Technology – prompted by an interesting careers brochure and, almost certainly, by her mother's love of trying new and appealing ways to tempt her large family to eat.

Her final year product development thesis, to make cottage cheese with added egg, (one could be forgiven for asking 'why?', as it does terrible things to starter cultures and curd formation) was a complete failure, but ignited a passion for product development that persists to this day.

Product development positions in food ingredients (flavour application, salt) allowed Andrea's knowledge base to expand across a wide range of food systems and led to her appointment as a food technologist with one of Australia's largest retailers, Coles Group Ltd, overseeing housebrand food development.

Fourteen years later, retailing is in her blood. She leads a team of technologists responsible for the entire packaged grocery offering in Coles' supermarket chain, within a housebrand programme she considers is second to none in Australia – and she'll tell anyone who'll listen about it too!

Andrea Currie
C/- 800 Toorak Rd
Tooronga, VIC 3146
Australia
E-mail: Andrea.Currie@coles.com.au

## Chapter 18

Ed Neff is a food technology graduate of Massey University, NZ. His career, in Australia, apart from some early years in the dairy industry, has been with large multinational companies including HJ Heinz and Nestle. His experience covers research and development, quality management, factory management and operations management across a wide range of products and processes. He holds a special interest in food product development where he sees a need to establish and maintain an effective Product Development Process, involving cross functional procedures, whilst encouraging a culture of creativity and innovation. His current activities involve some consulting together with voluntary work.

Ed Neff
7 Oak Grove
Malvern East, VIC 3145
Australia
E-mail: nefflot@tpgi.com.au

## Chapter 19

Gordon Fuller gained experience in product development as an extension professor, as a research chemist with multinational companies, and with an American research institute and a UK research association. He was VP of technical services for an international food company and spent several years in international consulting for government and industry, interspersed with guest lecturing on product development, agricultural economics and communication. Product experience includes fruit, vegetable and meat products, and dried, fermented, refrigerated, frozen, and canned products. He has several published papers, four books on food and product development and has contributed to a food encyclopaedia. He harbours a healthy scepticism of market research technology.

Gordon Fuller
71 Pinnacle Crescent
Guelph
Ontario, N1K 1P6
Canada
E-mail: gwfuller@sympatico.ca

## Chapter 20

Vichai Haruthaithanasan is one of the finest food scientists in the world today. His contribution to the field of product development is well recognised by his peers. He has developed as an outstanding product and process developer through experience in Kasetsart University and in the Thai food industry. His achievements as a researcher, especially in the field of peanut post-harvest handling and processing are noteworthy. His was head of the Department of Product Development at Kasetsart University for 12 years. Now he is Director of Kasetsart Agricultural and Agro-Industrial Product Improvement Institute.

Penkwan Chompreeda trained as a food scientist. She has undertaken teaching and research posts in food product development, working for more than twenty years in health food product development, sensory and consumer testing and nutrition. Now she is Associate Professor at Kasetsart University, Bangkok, Thailand.

Hathairat Rimkeeree is one of the many that Professor Mary Earle trained at Massey. She studied at Massey University, Palmerston North, New Zealand for her Master and PhD in product development from 1989 to 1994 then went back home to teach at the Department of Product Development, Kasetsart University, Bangkok. Her research efforts are applied to using natural raw materials in the development of consumer products especially in cosmetic products. Now she is Associate Professor at Kasetsart University.

Thongchai Suwonsichon trained as a product developer. His interest is in sensory and texture evaluation especially in non-destructive measurement using near infrared spectroscopy (NIR). Now he is Associate Professor at Kasetsat University and Head of Kasetsart University Sensory and Consumer Research Unit.

Hathairat Rimkeeree*, Vichai Haruthaithanasan, Penkwan Chompreeda and Thongchai Suwonsichon
Faculty of Agro-Industry
Kasetsart University
Chatuchak
Bangkok, 10900
Thailand
E-mail: fagihru@ku.ac.th; fagivch@ku.ac.th; fagipkc@ku.ac.th; fagitcs@ku.ac.th

# Preface

This book is dedicated to Jack Savage, pioneer in food product development, and Head of Food Product Development, Unilever Ltd, Colworth House, Sharnbrook, Bedfordshire from 1953 to 1974. Jack converted product development from an empirical art into a rigorous discipline, combining technology, marketing, strategy and management but keeping its creativity. He encouraged innovation and was never content with simple product improvement. So he would be interested in this book, which not only records innovative product developments in the food industry but also acknowledges the work of product developers in building companies and sustaining them in the long term.

There are wide variations in the book of product type, company size, company knowledge, company philosophy and therefore in product development but the underlying New Product Development (NPD) structure remains the same. NPD is an adaptable discipline, which changes not only with the focus of the project but also with the knowledge available. Because of lack of knowledge, there can be a need for basic scientific, consumer and market research in the project. As some case studies show, this can need years of research and large resources before the product can be launched into production and marketing. Smaller companies, on the other hand, are shown in this book to take risks without knowledge. Some case studies and also the discussion chapters are written by people with many years of experience and some case studies by young people starting their careers in product development. This gives the reader a variety of viewpoints but surprisingly all indicating the need to change as NPD techniques are developed and also as the focus of the companies changes with changes in society.

Linked with its companion book, *Food Product Development* by Earle, Earle and Anderson, they together illustrate the discipline and its application. They demonstrate the essential unity of the subject, emphasised by the brief italicised editorial introductions added at the beginning of each chapter. We are confident that Jack would have found this book both fulfilling and, inevitably for him, worthy of criticism.

We should like to thank all the authors for their cooperation in writing this book, and the companies who agreed to publication. Also thank you to the people in Woodhead Publishing Ltd, to Francis Dodds for agreeing to support this unusual publication, Sarah Whitworth for her suggestions on the form of the book, Lynsey Gathercole for keeping in contact with the many authors so that the book could be formed into a whole, and Laura Pugh who readied the book for publication.

We hope that the readers will enjoy this book; for the top managers in the companies, may it lead them towards new products; for the academics, may it lead to increased understanding of the knowledge and skills of product developers; and for the product developers, may they recognise the success and the failure of the work of other product developers, but most importantly be proud of the discipline of NPD and of its achievements.

Mary and Dick Earle

# Part I

# Introduction

# 1

# New Product Development: systematic industrial technology

**Mary and Richard Earle, New Zealand**

*New product development is essentially about creativity and ideas, but commercially these have to be implemented as a series of events through an organisation. Repeated successful developments have demonstrated a common pattern in these events. This pattern recurs and has been refined into the classical New Product Development (NPD). In essence NPD has grown to an industrial technology, a systematising of a whole series of patterns that fuse together into a larger common framework which transcends detail.*

*The classical outline of NPD is the basis for development of new food products. There is a start with a project aim and product ideas, which develop through the Product Development Process (PD Process) to a product concept and product design specifications, prototype products, a final product and production method, a marketing plan and launching on the market, and a final evaluation after launching. At certain stages in the PD Process there are critical points where the project is evaluated and go no-go decisions made. The framework is important both to the novice and to the experienced practitioner; it is a help and not a hindrance to new projects, a framework and not a fetter, a cost benefit and not a cost burden. It demonstrates its validity by maximising success and reducing risks.*

*Because of its origins, NPD is fundamentally empirical and does vary from project to project. This means that the product strategy is different, the activities in the development stages are different, and yet the overall structures are the same. NPD varies because companies differ in:*

- *size, the small company with 2–3 people to the large multinational company with many thousands of people;*

- *knowledge, new in food technology to 100 years of experience and knowledge;*
- *financial resources, a few thousand dollars to multimillion dollars for a project;*
- *risk taking, low risk to high risk.*

*Some companies depend on the tacit knowledge in the company and others have extensive knowledge sources and also organise research for new knowledge, in other words some companies may work with minimal knowledge and others seek as much knowledge as possible. Therefore the management of product development can vary a great deal; although basically it is managing people, finance, resources such as raw materials and processing plant, and of course time. NPD is based on people, their skills, knowledge, creativity, coordination, problem solving and decision making.*

*NPD can never be perfect. Perfection always eludes but the percentage of hits can go on increasing. This is accomplished by the summation of experiences; practical case studies build the data from which the improvements come. Case studies show the variability of NPD but also the similarities and demonstrate what works. Case studies are the essence of the testing and improvement of NPD.*

*Under commercial circumstances, the available base of NPD case studies is small and individuals and organisations see only their own experience, shared in-house but carefully guarded against the competition, real, potential or imagined. So on balance, case studies are seldom or never spread around. This works heavily against the building of a technology depending for its validation on demonstrable success, and for its improvement on detecting and eliminating flaws. Everyone gains from breadth and diversity, and from identification of problems that they did not think they had but now can see. Case studies show and tell.*

*This book,* Case Studies in Food Product Development *is a companion to* Food Product Development *by Earle, Earle and Anderson, which described and analysed classic New Product Development, and this chapter particularly relates to pages 1–41 and 317–347.*

## 1.1    Introduction

It is through case studies that one recognises and understands the variety of food product development. In this book, product development is described as it is practised in large and small companies, in universities/research centres, in rural development and small company start-up. The case studies span the whole food chain, and include primary, industrial and consumer products. These varied case studies illustrate the variations in food product development and also the similarities, so that the classical system of New Product Development can be made more realistic.

## 1.2   Food products in the food chain

The food chain encompasses everything from farm production or fish catching to the consumer's plate, and product development can therefore accordingly take in the total food chain or just part of it. The product can be for an industrial company, a small craftsman, a chef or the final consumer. It can be a cheap commodity, a designed ingredient, an everyday food, an expensive gourmet food. It can be a fundamental food necessary for health or it can be a 'fun' food. It can be an innovation or an improvement, or similar to a product already on the market but new to the company.

### 1.2.1   Primary products

Foods have a biological basis; therefore all have an agricultural or marine origin. Thence the food raw material has either to be grown or caught. Today, with the consumers' and therefore the food regulators' demands to know what is in the food, the traceabililty of the ingredients has become important. The product developer has to identify not only the country of origin but also the farming practices – is it GM free, reduced pesticides, 'organic'? The next generation of product developers will need knowledge of the total food system, including agricultural production, fish farming and fish catching, although they may never actually use fresh farm produce but only processed ingredients.

Development of new primary products is a long procedure, often taking years. Chapter 5 describes the development of a cassava industry in Colombia, South America, sponsored by international development agencies. This included firstly the development of suitable cassava tubers and of farming practices to ensure yields and quality, then drying procedures for cassava chips to be sold to animal feed companies, and also production methods for a cassava flour to be used as an ingredient in bread making in Colombia. This case study shows the importance of coordination of the agricultural research with the product, processing and market research needed in development of primary products.

Today with the increasing interest in nutriceutical, functional and organic foods, there is an increase in product development in primary production. Chapter 6 describes an early nutriceutical product, a specialised milk product produced by hyperimmunised cows that was claimed to provide relief from the symptoms of pain and swelling due to arthritic conditions. The chapter describes the complex and extensive research over many years that needed to be done in farm, laboratory and processing plant and also in clinical testing of the product to substantiate the claims being made during marketing.

Some product development can be very long term; for example, to develop a new apple variety may take years (see *Food Product Development* by Earle, Earle and Anderson, pp. 319–326). Firstly different genetic stock are crossed to identify the tree that produces apples with the colour, flavour and texture wanted by the consumers. Then several years are needed to develop the growing conditions for the chosen variety and to grow enough trees to develop the orchards. This is very long-term product development, which needs to recognise throughout the project the market and the consumers who will eventually buy the product. Even more long term is the development of a new fruit such as the original green kiwifruit, which took many years from the arrival of the wild plant in New Zealand to the time the kiwifruit was marketed.

Even what may be regarded as simpler crops to develop, such as wheat and oilseeds, because of the industry's 100 years of experience and extensive databases, still take years to develop to the final processed product. Fish farming also takes similar long development times to recognise the type of fish and the feeding regimes necessary for growth, as well as the best methods of handling on maturity.

Although not quite so long term as farming, catching new varieties of fish can also take years to develop the acceptable consumer product. The catching and handling methods are researched to give the optimum quality, and market research is needed to discover the consumer acceptability of the fish and also how to advertise and market it. In Chapter 12, the disaster of marketing a new product with unacceptable fish quality is described.

Product development in primary production can be aimed at improved sensory properties, nutrition, health, and increased storage life. It was an area where in the past there was little connection between the agricultural or marine research, and the processing, product and market research. But this has now changed with the development of products for specific markets and consumers.

### 1.2.2   Industrial products

Food ingredients have become a very fruitful area for product developers in recent years. The change from commodity product to food ingredient can increase the value markedly. The dairy industry is a good example where the development of specialised milk proteins has greatly increased the value of milk and whey powders (see *Food Product Development* by Earle, Earle and Anderson, pp. 332–339).

The aim is to develop the basic raw material into ingredients with specific reproducible properties. The ingredients can provide improved flavour: enhanced natural food flavours, non-calorie sweeteners; improved textures and stability: better emulsifiers, stabilisers; health benefits: essential minerals, vitamins, fibre, improved calcium absorption, a low glycemic index, cholesterol reduction. The list is endless. A typical change from commodity to industrial products is illustrated by milk – once sold as dried whole and skim milks, now as specialised protein, lactose and fat ingredients.

Industrial products can be marketed to large food companies, food service companies, supermarkets for their bakeries, smaller food processing companies, craftsmen and restaurants. Processing qualities for ingredients are availability, ease of use, variability, costs; and the important product qualities are composition, safety and sensory properties. Their main customers are the technical people in the company and the finance department, so they are being marketed on technical qualities and price.

### 1.2.3   Consumer products

The largest number of products is designed for groups of consumers. These are designed for the retailer, the buyer and the consumer – sometimes with technical qualities but often on aesthetic and emotional qualities. They are certainly very varied from gourmet foods to the everyday bulk food; from nutriceutical, functional and nutritional foods to the empty 'fun' foods; from basic ingredients to complete meals, from one-hour life to 3–4 years.

The activities in the PD Process vary for each of these product categories – primary, industrial and consumer, and accordingly the strategy and management also vary. Developing a new fresh fruit or vegetable takes a long time, so the product development strategy is always long term; a new canned fruit product can be developed quickly to compete with another new product so it is more fluid and short term. An important difference is between the launching of a new product platform and the addition of another product to the present product platform. The first is long term and can change the company and the market, the other is a quiet ripple!

## 1.3   Strategy and management in different situations

Just as there is variety in products so there is variety in product development strategy and management. Major issues centre on knowledge, systems and overall company strategy.

### 1.3.1   Knowledge

There are three basic sources of active food technology knowledge: multinational food companies, multinational supermarket chains and universities/research centres. There are many people and companies relying on these sources as either objective or tacit knowledge. Knowledge can also come from observation in the market place. An important source is from people, either group communication or one-to-one communication. Passive information can be in textbooks, journals, on the Internet. But how is this information and acquired knowledge turned into the knowledge and skills necessary to bring that new product to market? That is what NPD strategy and management is providing.

To the large multinational food company, all sources of information are available. They can choose to do basic research in product/processing technology or in consumer/market research or in both areas; or use their extensive databases, look for a suitable innovative company to buy, contract research to a university or to a private consultant. The choice of this basic research strategy depends on the overall company strategy – whether to build up a solid base of knowledge within the company or only seek knowledge when it is needed, believing that this reduces the costs of research.

The small company often has insufficient knowledge for the activities in product development and has to seek it. Often they turn to universities, research centres or consultants and this presents a serious strategy and management problem (Chapter 3). The small company has to bring outside researchers, perhaps from a university, into the project team. Several chapters describe how this can cause problems with technology transfer and sometimes failure in projects (Chapter 9). But it should always be remembered that there are small and medium companies working in very specific areas of technology and marketing who have great depth of knowledge in their own area and can effectively combine with the research centre (Chapter 7).

The large supermarket chain also has a major pool of knowledge, especially product and market knowledge (see Chapters 4 and 17). Because of the data they collect in checkout records and also through customer cards, they likely have the most extensive knowledge on consumer buying habits in the food industry. They are also developing new products for their own-brand products, which often started as cheaper versions of food company products, but in recent years have developed into higher priced and more original products. In the early years, they mostly contracted out the processing, but nowadays often have their own processing facilities, so they are moving into the total PD Process. Their product strategy is based on the product mix in the supermarket, how it needs to develop, on what is the best blend of their own-brand products and the branded products from the food companies.

### 1.3.2    Product development systems

As well as knowledge, the product development strategy and management is based on the product development system in the company and the people and financial resources that support it. In all companies, there needs to be a structured PD Process with stages, activities, critical analysis or evaluation points, and times specified for go/no-go decisions. This needs to coordinate the multidiscipline activities, particularly product and process technology and marketing. Systems vary from company to company as can be recognised in many of the chapters, from the large company with advanced planning techniques (Chapter 11) to the little company that makes the decisions when the problems arise (Chapter 7).

### 1.3.3    Company strategy

The product development strategy is also dependent on the overall company strategy (Chapter 2). The large multinational company may decide to have research and development (R&D) units in each country or they may centralise the R&D in one place. These two alternatives affect the strategy and management of product development as well as the types of products to be produced. A small company starting up usually has a specific product idea, often a specific market, and sometimes a specific process. So it has a very narrow company strategy, set by the owner, and starts to develop a company strategy as the project proceeds. There are, of course, different company strategies between these two extremes – some technology based and some market driven.

The other important part of the company strategy is the innovation strategy: simplifying the product mix, product improvements and addition to existing product lines, moving products into a new market, introduction of a new product platform and introduction of a complete product and process innovation. The large company usually has a mixture of these; for the small company 'starting up' is a complete innovation, introducing a product, process and marketing.

These three important factors – knowledge, PD system and company strategy together with the resources available, determine the product strategy and management. But, of course, there are intangible factors such as company philosophy, consumers' and customers' preferences and inclinations, and also tangible factors of economic, political and social changes and of competitive actions, all of which require flexibility in NPD. Chapter 2 indicates how the political changes in Eastern Europe and China have changed strategies for NPD in multinational companies.

Lastly, can product development strategy and management be used in non-commercial development projects such as those organised by national and international agencies in agro-industry development. This is not easy, as described in Chapter 5, because of the many organisations often involved and the political influences that occur. But it could give a rigour to the development, especially the use of critical points and go/no-go decisions to control the project, and an understanding about the multidisciplinary nature of development to give the balance of technical and market research.

## 1.4   Product Development Processes

The PD Process is the basis around which any NPD project is built. A company can start with little knowledge of the PD Process but by trial and error ends up by developing the PD Process. There are classical ones in the literature and commercial ones designed by individuals and software companies, but the basis of all are the stages, critical points and evaluations, go/no-go decisions and the activities needed within each stage. Activities change from project to project, sometimes activities are dropped because of either low risk in dropping them, or lack of people and resources, or lack of knowledge. There is also sometimes a blending of the stages, because circumstances change and there has to be recycling to earlier stages, either to find more knowledge or to change the focus of the project. For example, a competitor may have launched a similar product and the decision has to be made to either launch the present product or go back to the earlier stages of the PD Process and change the product, or of course drop it altogether and start with another product concept.

In Part III, there are four very different product development projects: the first developing a nutriceutical milk product showing the whole PD Process from farm to consumer; then two chapters showing how an invention/idea by one person can develop to a marketable product; and the final chapter describing cooperation between a university and a company in developing a product. These chapters show how the PD Process changes in different situations but still retains the same basic structure.

### 1.4.1    Product Development Process from farm to consumer

The PD Process is based on the total development project from farm to market. This is true not only when the changes are occurring on the farm or the fishery, but also today because the need to identify sources of ingredients is becoming necessary for many new products. The farm development of a new product is a long, complex and expensive NPD project and the early stages can be in government-funded research centres but in the end, it is a commercial NPD project. These farm to consumer projects are usually developed by large companies such as farmers' cooperatives or commercial farm companies who have the facilities and resources to sustain this long-term research. Development on the farm has become more important because of the pressure to have new 'fresh' products, organic products and nutriceutical products.

Although the main activity in the early stages is plant breeding or animal development, the other multidisciplinary activities have to be included. For example in Chapter 6, there was a need to not only research the vaccination of the dairy cows but also to test the milk to identify if it had the desired effects on the health of the consumers. It was a long time before the farm production and process research could be started. Where a new fresh plant product is developed, there is a great deal of plant breeding research to identify the most suitable variety. This is judged not only for production qualities such as production difficulties/ease, disease resistance and yields, and on handling difficulties such as deterioration rates after harvesting and on storage, but also on the consumer attributes, particularly sensory properties, nutritional composition, safety and use.

The PD Process still has, after the identification of the project aim, objectives and constraints, four identifiable stages: product strategy; product design and process development; product commercialisation; product launch and evaluation. In the first stage, product strategy, there is a need, for plants to identify possible varieties and do some growing trials, or if animals, identify breeds, feeding regimes and running trials over one or two generations. It is important at this stage to also bring in some market and consumer research so that the selection of different plants or animals to develop is dependent on market and consumer needs. In the case study on the nutriceutical milk, the product strategy, product design and animal experimentation were completed before the final stages of production and process experimentation, market study and the launching of the product. It had taken many years to develop the vaccination method and the testing of the product with consumers. Time is long in on-farm development, even for the most straightforward products such as new fresh lettuce.

### 1.4.2    The individual inventor in the Product Development Process

In Chapters 7 and 8 are ideas of individuals that were developed and launched on the market. The pet food supplement in Chapter 7 developed from marketing knowledge; a need for extra nutrition by farm dogs was identified when marketing dog biscuits that had limited acceptability to the dogs. In Chapter 8, the idea came from kitchen experimentation and a family need; the need for a nutritional snack based on oatmeal to give hungry boys after school. The first inventor had sufficient money to build a small company to develop the product and launch it on the market. The other inventor did not have money to do this and had to look for and persuade a company to develop, produce and market the product.

There are always new products coming onto the markets based on ideas of individuals or groups of people, some are successful and often end up by being bought by large companies, and of course quite a few which are failures because of lack of knowledge and resources, particularly finance. Surprisingly in the pet food project, where the owner's expertise was marketing, the company failed with the original product for farmers' dogs because they did not identify the farmers' acceptable price range before launching. Luckily, because the company was small, they could quickly move to niche markets where the higher price was acceptable. There was a lack of systematic NPD in this project that did cause recycling backwards into the early stages. This often happens with small companies, which just jump the early activities of NPD and do not have adequate knowledge on the product and marketing.

A middle-sized company working in the specialised field of baked oatmeal products took up the oatmeal snack invention. There was a small, knowledge-able management team who saw the possibilities of a new product platform. They had the knowledge and resources to design and build a new processing line and also develop the market. They followed the classical PD Process and were knowledgeable and skilled in many of the technical and marketing techniques.

### 1.4.3   The university and research centre in the Product Development Process

There are various ways that university staff can become involved in NPD. Very often today there are government funds available for applied research in specific areas. The funding organisation identifies research areas that they think are necessary for future developments and accepts proposals for specific projects in the area. In the area of food science and technology, they may have identified a need for research into a biological material so that it can be used in the food industry, or a specific consumer need such as nutrition, obesity or safety, or a processing innovation to increase productivity. Very often this research can lead to new products and processes but then the problem is to transfer the new knowledge into the industry.

There are published papers, patents, lectures and workshops, which likely fulfil the needs of the research funding organisation, but may not lead to industrial interest. The transfer of the knowledge into an active NPD project in one company is difficult and there is no agreed roadmap for success. It is complex, as the company tries to fit this knowledge into its general product strategy. There is usually extensive knowledge for Stage 2: product and process development, but this has to be fitted into the facilities and resources of the company. Of course Stages 3 and 4 have to follow as usual. If it is patentable, the product and process can be protected, but there is still a great deal of development work in the company.

The other avenues are that the company asks the university either for NPD ideas to solve a problem as in Chapter 9, or to undertake a specific part of the project, for example developing a formulation and a process as in Chapter 7, or to provide early consumer research to identify the product concept as in Chapter 16. In all cases, the university staff are part of the project team. There can be problems with the knowledge generated; the university staff wish to publish the research results in peer-reviewed journals but the company feels this information is commercially sensitive.

Where the university staff are carrying the major part of the NPD, there can be problems when they want to pass the project to the company who may not have developed the knowledge and skills and especially the enthusiasm for the project. The extreme is described in Chapter 9 where the company passed the ownership of the project to the university at the stage of commercialisation. The university seldom has the commercial position or expertise or finances to start up a food company. This problem can be overcome by involving the company staff in 'hands-on' activities during the development of the project.

## 1.5    Technological development

Technological development involves the product, process and distribution. As described earlier, technical research needs involvement with marketing in the early stages to develop the product concept, the product design specifications and the product prototypes, which form the basis for the next stage of product and process development. Sometimes the major innovation comes from some research in process development, which may take a long time and only when it is successful does marketing come into the project. In Chapter 11, the aim was to develop a new process for continuous ice cream production, and it was only towards the end of the development that it emerged that the process could produce an attractive low fat ice cream. In the beginning, marketing saw it as a creamier ice cream and only later realised the huge growing market for low fat ice cream. In Chapter 13, a powdered drink for the Chinese market started from knowledge gained in process development and building of a new plant for coconut milk. Process development continues for the total life of the product, changing from batch to continuous production, and then into larger production outputs as described in Chapter 10.

### 1.5.1    Building the production

In the initial stages of developing the process, the total process is divided into unit operations and each is studied as both physical processes such as heat transfer, drying, emulsification, grinding, cutting and also chemical and biological transformations such as nutrient loss, growth and destruction of bacteria. For example, in pasteurisation and sterilisation of milk the heat transfer and its destruction of microorganisms are studied to determine the optimum conditions of temperature and time required for destroying the microorganisms. Chapters 10 and 11 show how extensive the research on the unit operation can be when developing radical new processes; these developments took a long time and a great deal of resources.

Usually, the principal unit operation is developed, and then fitted into the process line. When the unit operations are combined in an existing production line, there can often be upsets. The new unit operation either has a faster output or the new plant is a different configuration from the old plant, and chaos can ensue without understanding of the complete process from modelling and active study. A new production line or a new plant for a new product is often a major change involving design and construction. When designing a process, the total process environment has to be considered. In Chapter 12, the air conditioning needed to keep the air temperature within certain limits was not considered and installed so that the successful pies, launched in the winter, fell apart in the summer and sales dropped rapidly!

Fundamental in developing production of a new product are raw materials, their quality and their availability. A new product platform was halted for many months because some of the ingredients had to be imported and permission to import had not been obtained. With a new fish product in Chapter 12, a large Japanese market was destroyed for a long time because there was a lack of knowledge about the raw material and the effect of its variation on the product quality.

Small companies are vulnerable in their ability to develop a process. In Chapters 7 and 13, the two companies developed successful processes, but in Chapter 12, there is a description of a small company that was partially processing in its own premises and then transporting for packing by a contracting company – with disastrous results for contamination of the product.

### 1.5.2  Building the distribution

There are two parts of distribution: transport and storage, and through both the product quality has to be protected. Packaging is important and this needs to be tested for adequacy before proceeding to the aesthetic design. For example, in Chapter 7, the product was packed in polythene bottles without any storage tests and it was discovered that the oil was deteriorating quickly during the marketing. It had to be withdrawn from the market and repackaged. In other cases, the market for the product has been destroyed for a number of years because the package did not protect from moisture and the product picked up water, so browning occurred. Storage tests do take time at ambient temperatures and often impatient marketing departments feel they cannot wait this time before launching the product. Higher temperatures can be used to hasten the storage tests, but the results may not be directly predictable for ambient temperatures because important deterioration modes may be different.

Friability of the oatbakes during handling and transport became important in Chapter 8, where the packaging had to be changed from the oatcake cartons. Everyone on the NPD team must be aware that distribution testing is essential before the final design of the packaging can be signed off, and this can be critical in the timing of the project.

## 1.6    Consumer and market research

Market driven and consumer driven NPD are sometimes suggested to be missing in food companies. But in product line extension, packaging change and product improvement, marketing is the dominating activity and therefore consumer and market research are primarily important. The packaging and advertising designs are the major activities and usually the product design is minor unless there are any processing changes involved.

For innovation driven by the market, there is a need for study of long-term trends in society, politics and economics. This is wider research than inter-viewing consumers and studying the present or past market sales. Market innovation and, in particular, the introduction of new product platforms into the market takes time, just like the technological innovation.

### 1.6.1    In the early stages

In building the product concept and, in particular, identifying the product attributes, there is a need for consumer research. Focus groups of small numbers of people are often used to collect knowledge on what consumers' attitudes, needs and wants are in the products at present on the market and what they would desire in new products. Once the product attributes are identified, more quantitative techniques, as described in Chapters 15 and 16, can be used to identify the important product attributes so that the product concept for the new product can be developed.

A larger group of people can examine the product concept for its suitability, but it is more rewarding to develop two or three product prototypes based on the product concept and test them, perhaps with the written description of the product concept. In Chapter 16, testing the product concept descriptions with photographs of products did not add more information than was obtained in the first part of the study, so it appears that allowing consumers to see the actual product prototypes and indeed tasting them could be more useful. The con-sumers can make judgements more easily when they see the actual product. This certainly leads into decisions on whether they would buy and eat the product.

The next stage in building the product design specifications does need to include consumer testing so that detailed and reasonably accurate information can be used in design of the product. Product design specifications are commonly developed in other industries but the rigour of doing this is often ignored in the food industry. In many development laboratories, it is a sequence of make it and taste it, which can take a long time and may lead in the wrong direction. So it is usually more efficient to take time and do preliminary consumer and technical research on which design specifications can be built.

Sensory testing has been part of many food development laboratories for over fifty years, sometimes by consultants who were contracted on a narrow basis. The product may be acceptable to eat, but it could have other attributes that would inhibit purchase. In Chapter 14, not only sensory testing but also the study of the broader aspects of the product is seen to be necessary. There is a need to identify consumers who are in the target market and to have them comment generally on the product prototypes.

By the end of the early stages of the NPD, the consumer should have specified the product concept, the product attributes, the method of buying, storing, cooking and eating. Further, the size of the market, its location, competition and prices should have come from the market research. The best source of this data is supermarket records as described in Chapter 17, but this is not always available for food companies and they will instead use commercial market research companies. For the small companies who cannot afford to buy research, they can use government statistics and reports, and do their own mini survey in the retail outlets.

### 1.6.2   In product design and process development

Consumer testing of the products as they are being developed is desirable but time and finance resources may restrict it to critical decision-making points. The product design specifications need, when possible, to be translated into objective tests such as texture measurement and nutritional composition, which can be used as the product and the process are being developed. For some attributes such as flavour there is no physical or chemical test and sensory testing has to be used, but this is not for acceptability but for quantitative product attributes by a trained sensory panel, if necessary using a standard sample.

At the critical points, there is a need to go back to the consumers and to test the acceptability of the total product as well as individual attributes. Gradually the numbers in these consumer tests are increased until, after the production trial, the product is tested in a test market or in a large consumer test in the identified markets. As Chapter 17 says, consumer-centred product development begins and ends with the consumer.

### 1.6.3   In launching the product

In planning the launch, there is a need for consumer input into the building of the packaging and the advertising. A great deal of information has been built up during the development of the product, which can be used as the basis of the design. There is a need to confirm and build on this and it is very important to work from this knowledge and not to start from scratch. There is nothing worse than marketing writing advertising material that does not relate to the product as developed, for example, saying that a pet food is rich in liver when it does not contain any liver, or a product is labelled fat reduced when it has the same fat content as other products on the market. In Chapter 16, the consumer research results, used when building up the product concept, helped the development of the advertising.

Very often, focus groups are used to develop ideas for packaging and advertising, but quantitative techniques, similar to those in Chapters 15 and 16, are also used to find more quantitative consumer data. At this market development stage, it is very important to have reasonably accurate data, so it is important to be doing general market research not only to predict sales but also to identify the customers – retail or industrial – to whom the product is first sold. Their needs for the product are crucial at this stage; if they do not particularly want the product, they will not be active in selling it in retail! The industrial product which may have been designed to fit with the technologists and their process, now has to be sold to ingredient buyers who are interested in price, delivery and quantities.

In the present book, there is not a case study that includes the product launch but in the companion book *Food Product Development* there are three such case studies. There is a description of the launch of an apple variety (pp. 324–326), with customer demonstrations and tasting in different countries. From this consumer data, they finally settled on a markets 40% Asia, 30% North America and 30% Europe. The great problem with a new variety of fruit is balancing production and marketing. An industrial product, whey proteins, was launched at the (US) Institute of Food Technologists' Annual Conference (see pp. 338–339) where the product was demonstrated in soft drinks. Before this, possible customers in the market screened the product prototypes as they were developed. For a consumer product, a range of sauces, design consultants selected the brand name and, prior to launch, the sauces were presented to the retailers (pp. 343–345).

## 1.7    Managing the Product Development Process

Very often in industry, NPD is in separate parts, usually closely related to the two main operating sections of the company: production and marketing, but with connections into finance, planning and management. This separation can cause endless problems in the larger companies, because of the natural wish by particular departments to be dominant and have the major power. In the smaller companies, there are only a few people who form a team and have decision making either through discussion or directly by the owner.

The separate activities of marketing and production in large companies have led some of them to sell the factories and contract out their production. Only marketing as an operational activity is left in the parent company. This of course led to marketing's complete domination of product development and a tendency to follow the 'fashion' in the market rather than looking for an innovation that would give the company unique strengths.

### 1.7.1    Technical development

In Chapter 18, in the discussion of a road map for the technologist in product design, process development and manufacturing, there is a plea for cooperation with, not dominance by, marketing. During product design, there needs to be a cooperative effort so that the needs and wants of the consumers and the markets emerge as the basis of the design specifications. But right through the product and process development, there is a need for product testing at critical points in the development, feeding in of market information such as size of market and competitive products, and also environmental shifts such as regulation changes, political changes, industry changes. A new regulation can change, for example the final composition of the product or the conditions of pasteurisation. As described in Chapter 5, freeing or tightening of import regulations and duties can change the economics of the marketplace with cheap imported products destroying the market for the local, more expensive, ones.

### 1.7.2    Market development

There are many market changes that can affect product development as discussed in Chapter 19, and they need to be monitored both before the project starts and during the project. Often, there can be prediction of these changes at the beginning of the project but it is illuminating to survey them again later and find how far the predictions match to the reality. As discussed in Chapter 2, the fall of communism in the Eastern European states caused an influx of European and American products and eventually the production of these products in Eastern Europe. These were new and large markets that were content with a short range of products. It caused a change in company philosophy about new products and their proliferation.

### 1.7.3    Controlling the Product Development Process

There are critical points in the PD Process where not only the progression of the project is analysed but also the changes in the environment sought out and identified. It is very expensive to cancel a project just before launching, or even worse after it is launched.

In Chapter 18, there is the use of defined stages, each of which is followed by the project evaluation and a go/no-go decision. The project can be divided into a number of stages, usually 3, 4 or 5, and the evaluation method between the stages is identified. So there are review meetings, firstly amongst the NPD team and then between team managers and senior executives. In the latter meeting, the decision is made to provide the resources for the project to continue or to stop the project.

Chapter 19 also shows that there are critical points in the project that needs to be identified when planning the project. These critical points can be changed for different projects because all NPD projects are different, some only marginally, others extensively. For example, in a simple product extension, the packaging design and advertising design can be critical and there needs to be careful control of these designs so that they are not taking too long. In developing a new process for a new product, there are many critical points, sometimes major and sometimes minor. These have to be carefully monitored because sometimes the minor critical points become major, for example a change in raw material, a lack of stainless steel for the plant. The minor and the major critical points have to be identified and monitored.

## 1.8   Educating the next generation of product developers

NPD needs skilled and knowledgeable professional practitioners. They can arise through experience, but a more efficient and effective way is to select and educate them for the purpose. The width and diversity of the knowledge needed by product developers make this a substantial undertaking.

The education has to cover all the major aspects of NPD, and at a sufficient depth to make the contributing disciplines accessible for further acquisition of knowledge. This is a big specification. It means a multidisciplinary degree course, incorporating the chemical, biological, physical, mathematical and social sciences and combining them in technological subjects such as process engineering, consumer science, market research, product quality measurement, quality assurance. These different technological subjects are brought together in specific product development subjects, which might be named as stages of NPD over a 3–4 year course, and joined together in the final year with NPD management.

The courses would be rounded out with practical projects, starting with a simple project and leading in the final year to an industrial NPD project sponsored by an industrial company. This combination of knowledge and techniques is highly effective in building abilities in problem solving, which is a skill that is very often needed in the food industry. The combination of the sciences and the social sciences means also that the students are blending the consumer research into their projects and there should not be criticism in the future of lack of consideration of consumer needs and wants. An essential component is a course on ethics, because of the vital role of food in population health, and showing how the community, because of a lack of ethics, so often imposed food regulations in the past. It is most critical that the teaching is related to the food industry, to large and small companies, to consumer, industrial and food service product development.

Quantitative aspects of both technology and marketing must be emphasised because of their power and precision, and increasingly they can be used in NPD projects because of the computer software available. Today, there is more and more quantitative knowledge, for example chemical change reactions in food processing and consumer reactions to food products. Also in the PD Process, there are many quantitative techniques such as experimental design, which can be used to model not only the chemical and microbiological changes but also accompanying sensory changes in the food, and lead to optimum product quality. This is going to need increased research in the future of rather a different character than in the past and similar to the research in Chapter 15, where consumer research is combined quantitatively with process and product research in a multidisciplinary research team.

Chapter 20 shows how in Kasetsart University, Thailand, this education was accomplished with not only in an integrated undergraduate degree, but also postgraduate degrees, in NPD. The example was in a rapidly developing country in a specific environment, but the need for product development education is global. The Thai success suggests that this could show the route towards an accessible solution of an international problem with wide application

## 1.9   New Product Development: a technology

NPD has gradually matured over the years from the operational sections of the company, production and marketing, with contributions from finance and management. It is now a multidisciplinary technology combining the results of technological research with those of social research. It combines design of product and process with design of aesthetics and marketing, all based on consumer needs and wants, into a systematic procedure usually called the Product Development Process. It brings together food chemistry, food microbiology, food engineering, consumer science and market research. This is a very wide area of knowledge. But there are people in the food industry who have this multidisciplinary knowledge, partially from education but mainly from experience. There are of course also specialists in the various areas, found often in academic research, who increase basic knowledge and also create new techniques and approaches.

We think it is now possible, and profitable, to acknowledge the creation of new food products as an area of study and industrial application, a technology in its own right, New Food Product Development. Within overall NPD, it covers a very major specialisation, the creation of new foods. It is a holistic technology. Holistic is defined in the Concise Oxford English Dictionary as a 'tendency in nature to form wholes that are more than the sum of the parts by creative evolution'. This fits NPD of foods very nicely. There is a need for someone in the future to write the complete NPD textbook with methods and techniques, so that students and product developers in industry can have the quantitative, multidisciplinary knowledge needed for NPD.

The following case studies exemplify NPD practice and could be part of the basis for a new textbook as they illuminate its accomplishments and failures, and show its possibilities.

# Part II

# Product development strategy and management

# 2

# The multinational food business – strategic, organizational and management issues for product development

Dick Inwood, Australia and Europe

*This chapter explores overall strategies for new product development and how experience can be examined, systematized, and incorporated. One of the excitements of working in New Product Development (NPD) is that each project is different. It's the reason that case studies, rather than study of a methodology, are peculiarly useful to those seeking to use others' experience to improve the efficiency with which new products are identified and launched. Identifying social and political changes, technological innovations, changes in the structure of the food system, market place changes, and then developing a strategy and programme for a company is exciting and mind refreshing. The NPD strategy is developed in the context of the company – its capabilities, technological knowledge, market availability and its resources, as well as the social and economic environment. Company strategy evolves. It shapes, and is shaped by a NPD programme that gives a 'pipeline' of new products for the future.*

*Organizing and supervising this programme is a NPD executive who is often the research and development manager, or perhaps a new product manager in marketing.*

*The chapter particularly relates to pages 27–37, 45–78, in* Food Product Development *by Earle, Earle and Anderson.*

## 2.1  Introduction

### 2.1.1  The management of New Product Development in the 21st century – learning from the past

Management of New Product Development (NPD) in a multinational food business is complex and wide-reaching. The considerations that the person managing this function needs to address include cultural and social as well as the technical and economic. The operations organized and managed by the R&D are in many countries and different types of organizations, and they vary. The aim is to relate the experience of one R&D manager in a multinational food company to the PD strategy development in the literature. The case I wish to study, then, is how to achieve this alignment, obtain the commitment, and organize the NPD function to use this. The 'case' studied is that of the management of a NPD function within a company that operates across international borders.

## 2.2  Multinational food businesses

### 2.2.1  Who are they?

The largest multinational food businesses are shown in Table 2.1. These firms are quite diverse in their history and background, and have a diverse range of issues with which to grapple, as they seek to find organizational structures appropriate to the 21st century. Bartlett and Ghoshal (1989) reviewed the administrative heritages of a number of companies, and classified them as multinational, international, and global. They point the way to an organization they term 'transnational'.

An exemplar of the *multinational* company was Unilever – a European-based company that developed before WWII, and, partly because of the dislocation that caused, and the slower communication of that era, became managed as a series of national companies, locally managed, with some central control of plans, capital budgets, but with maximum local flexibility to exploit local market opportunities. Companies of this type usually had nationally based R&D functions, although they typically had some core centralized research, always under fire from the provinces.

Their exemplar of the *international* business was Procter & Gamble, a US-based company whose international expansion was predominantly post WWII. National subsidiaries had far less autonomy. Strategic control was never relinquished, and the central control worked actively to rationalize manufacturing, centralize R&D, and coordinate marketing. The majority of the big multinational food businesses are of this type.

*Global* companies in their classification were more typically Japanese. Partly conditioned by language, these companies are much more centrally controlled, with local companies basically sales offices. These firms start with the concept of a unitary global market.

They define the *transnational* company as one that has a new management mentality, which allowed for the development of global competitiveness,

**Table 2.1**  Multinational food businesses

| Rank | Company | Revenue (billion) | Origin | Fortune 500 |
|------|---------|-------------------|--------|-------------|
| 1 | Nestle SA | $82 | Swiss | 53 |
| 2 | Archer Daniels Midland | $36.2 | US | 157 |
| 3 | Kraft Foods | $31.0 | US | |
| 4 | Unilever | $30.0 | UK/Net | 106 |
| 5 | PepsiCo | $29.3 | US | 175 |
| 6 | Cargill | $27.3 | US | |
| 7 | Bunge | $24.2 | US | |
| 8 | Tyson Foods | $24.5 | US | 231 |
| 9 | Coca Cola | $21 | US | |
| 10 | Mars | $17 | US | 267 |
| 11 | Groupe Danone | $14.9 | France | |
| 12 | Con Agra | $14.5 | US | 434 |
| 13 | General Mills | $11.1 | US | |
| 14 | Sara Lee | $10.7 | US | |
| 15 | Swift | $9.9 | US | |
| 16 | Dean Foods | $8.9 | US | |
| 17 | Kellogg | $8.8 | US | |
| 18 | Heinz | $8.4 | US | |

The Fortune 500 columns are July 2006, while the revenue numbers, which are from 2005, represent food and beverage figures, but don't include brewers and distillers.
Other important multinationals are the food service group – McDonalds ($51.3 billion), Burger King ($11 billion), Compass ($20 billion), Sodexho ($13.9 billion) and Yum! ($27.9 billion).
Note – these figures include Franchisee figures.

multinational flexibility and worldwide learning simultaneously, and suggest that both Procter & Gamble, and Unilever are moving in this direction.

### 2.2.2   How do they work?

At the heart of a multinational will be its strategic plan, typically a five-year plan into which a great deal of energy is expended each year, reviewing and revising its relevance. This plan will be an amalgam of the ambitions and constraints of the various functions within the business, and the vision of the senior management team of the future estate into which the business might grow. Probably, at the highest level the plan will be a synthesis of the plans of the various operations or regions, to which international management may well apply discount factors, reflecting their views of the optimism or conservatism of the various plans.

Multinational (or international or intranational or transnational or global) businesses are complex entities, run by management teams with different functional responsibilities. The strategic plan enables these functions to establish clear objectives, to which they are committed and against which they are measured.

Typically this plan will show a growth in sales in the later years of the plan to be achieved by products not yet developed, and in earlier years by products

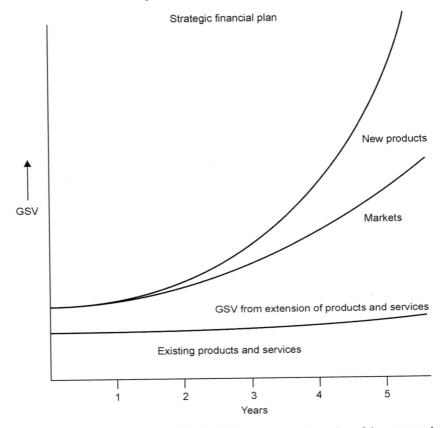

**Fig. 2.1**    Strategic financial plan. (Note: GSV is the gross sales value of the company's goods and services.)

already in the development pipeline (Fig. 2.1). They will have someone in the business whose job is to build that pipeline, and ensure new products come out of it. Now, if this person, with that responsibility (and I shall call this person 'him' in future, fully realizing that it may well be 'her') goes to the literature to gain advice from those who have studied the subject, he will invariably find two strictures:

- The NPD programme should be aligned with and part of the strategic objectives of the company.
- The product development programme needs 'top management commitment' to be truly successful.

A significant managerial problem is the extent to which the manager should expend his energies ensuring that these conditions exist, as distinct from his prime role in ensuring the 'pipeline' is in good condition, and is delivering as planned. This alignment, and this commitment doesn't come automatically.

The need for a business to have a strategic plan is unquestionable. But it is usually chastening to review these plans five years later, and realize how naïve the forecasts of five years prior were. There's always a reason, which the planners often term an 'externality', meaning that the conditions forecast in the plan are thrown into disarray by an event – perhaps opportune – that could not have been reasonably forecast. And the 'externalities' have pushed MFBs' diverse histories into a common set of conditions, which has shaped their attitudes to innovation and new products.

## 2.3   Recent history shaping the behaviour of multinational food businesses

Multinational food businesses (MFBs) are essentially oligopolies. Oligopolies traditionally have tried to compete in three ways:

- Increasing market share with a marketing and new product programme. The way of the 1970s and 1980s.
- By buying up domestic competitors. Often the 'fit' isn't good. Many – some studies say most – mergers fail.
- By expanding internationally.

Coming off the high inflation of the late 1970s and early 1980s when their innovative efforts had been driven by the need to minimize assets, most MFBs had large, flexible, highly productive factories, concentrated in the then stable economies of Northern America, Western Europe, and Australia. They sought to achieve efficiencies in administrative costs, and lever their size and experience, not only in manufacturing and technical knowledge, but also in marketing and especially in advertising. Television was increasingly expensive and increasingly international, especially in Europe, so the branding, which had often developed erratically in differing markets, had to be internationalized and harmonized. Companies had to ensure that brands represented the same thing in the differing markets, particularly with the arrival of the EU, and the ability of the supermarket trade to purchase from different source countries. So, for example Marathon™ in the UK became Snickers™, as in the US, and Wiskas™ in Germany became Whiskas™ as in the UK, and in companies operating internationally there was much harmonization of logo, image, and product position to achieve more power from the brand. Marketing was the nature of the business.

That the value was not the brand, but customer loyalty to the brand was soon recognized, and brand loyalty was integral to the business strategies of most companies. These were the opening rounds of the key long-term battle for customer loyalty, between the manufacturer and the retailer.

One result was a large increase in 'pseudo' varieties, as manufacturers sought to lever brands, and acquire shelf space in the supermarkets. NPD was directed at finding technological pegs on which to hang new news, and product differentiation. But it got more complicated.

In 1989, the Berlin Wall came down. The markets of Eastern Europe, the Baltic and Balkan States, and of the Ukraine and Russia became 'available'. There was a scramble to establish a presence in these hitherto unsupplied markets. Initially, this took the form of supply issues, with production from existing factories, using brands that had been advertised in the West. Factories and supply routes became the battle lines for MFBs, supplying the new markets with products and establishing brand footprints before the East, too, devolved into fairly static markets, fought over with advertising and promotion budgets, as was the case in the West. This was followed by the establishment of factories local to these markets, and the development of local raw material sources. These new plants, often producing a simplified product range but replicating products using streamlined processes, and staffed with lower paid but skilled and motivated labour forces, became a preferred (or at least cheaper) supply source to the 'home' markets. A further development was the purchase of local brands in the newer markets, as some firms sought to increase their market and market share.

Then, just as this was settling down, the Chinese economic reform opened the market of China, to be rapidly followed by India – both markets having more and different restrictions, but both offering low labour costs, and huge potential market expansion. Again, MFBs were concentrating on supply issues, people and organization and reorganization issues. Of course, new products were not ignored, but they certainly slipped down the priority list in many a management agenda. This continued through the 1990s into the new century. Importantly, NPD structures were thinned of people who were redeployed to assist this growth.

The opportunity for NPD in the older markets was seized by others, particularly the supermarket chains, who developed or commissioned development of new products, aided by their particular knowledge of the changing, localized, purchasing patterns, acquired in real time at the cash register.

Few of these events were forecast in the business strategic plan. When business decisions, especially those involving new geographies, were involved, the R&D and NPD issues were very much a secondary issue, and often properly so.

## 2.4   Current state of innovation in multinational food businesses

While MFBs are often considered as a particular group, they are quite heterogeneous. While profitability and growth can be correlated with innovation, the reverse is not necessarily true. Alfranca, Rama and von Tunzelmann (2003) argue 'that a small nucleus of innovative MFBs chiefly rooted in home laboratories, heads the world's production of food technology'. Not all MFBs are innovative, however. Indeed, they classify just 22% of the MFBs they review as persistent innovators. They also suggest, based on their study of 100 MFBs:

- In contrast to other multinational businesses, MFBs locate a larger percentage of their R&D abroad.
- That although MFBs remain the most important players in food technology, they are losing ground to other innovators. This contrasts with non-food multinationals. The authors suggest that this is because MFBs have been competing on other grounds than innovation.
- Many MFBs innovate only sporadically.
- Particularly in US firms, 'the pace of innovation has lost momentum'.
- In terms of R&D priorities, MFBs devote most of their efforts to food-related technology with the aim of controlling the supply of innovations needed to produce food and drinks.
- There is increasing use of a complementary mix of technical and design patents in MFBs (design patents usually relate to packaging and package format).
- MFBs innovate at a slower pace than other innovators in food and food-related fields. One nation firms, smaller food multinationals, universities and research centres are innovating at a faster pace than the top MFBs.

Multinational companies, however, have some great strengths. They offer a range of exposures to skills, cultures, experience, commerce, technology and science, and an opportunity to acquire knowledge and experience from a concentrated collection of experts in the array of necessary disciplines. Their size usually ensures they have specialist sections and established procedures in the many disciplines that expand the processes of NPD. In the development and introduction of new food products, they have been for many years the dominating influence in the market places of the world. They have investment in, or access to fundamental research, on which new products are based. They are natural prime targets of those who develop the 'upstream' industries – packaging machinery and materials, ingredients, flavours, control systems – which can unlock new products. Above all, they have access to markets, test markets, and skills in launching that enable them to invest in the economies of scale that enable the critical product launch phase to sail through that often troubled period. And they usually have purses robust enough to absorb failures.

## 2.5   The roles of the research and development manager

Into this environment, steps the person who has responsibility for NPD, usually an R&D director, (or vice president). Often, because R&D can cover a wide range of problems from the environment, technical intellectual property, recall and laboratories, there will be one or more NPD managers reporting to 'him', as well as those responsible for more incremental 'service' R&D. Businesses have played around with other structures – e.g., NPD within a marketing function. Each market segment may have a team – a product manager, a food technologist with market research resources as its core NPD unit.

### 2.5.1   The need to be a manager and a businessman

In addition to the technical and operational responsibilities, for which his training and experience has fitted him, a NPD manager first and foremost has to be a businessman. If he wishes to be part of the business strategy forming team, and to champion the cause of a vigorous PD programme and the activities that develop it, he has above all to retain his credibility as a member of the management team. This team is presented with a wide range of causes in which it might invest, short term and long term, geographic, technological, structural, emotional, anticipatory and reactive.

The NPD manager's role is to keep the NPD programme on the company agenda, its resources appropriate to its agreed objectives, the strategy alive over the relative timeframes, and ensure the opportunities for investment in NPD are properly represented when business strategy is considered.

Thus, there is need for a managerial as well as an operational plan, and for that plan to be communicated properly to those both within the function, and across the senior management team. Ball (1998), in a thoughtful review of the literature on R&D management states in his conclusion, 'The message for R&D senior management is to more clearly communicate the place of R&D within the organization'. He also wisely warns that senior managers 'should examine whether or not their own views are polarized by the business environment that prevailed when they were achieving their early successes'.

Reading the literature – and it's extensive – always helps, but Ball describes the writing of practitioners as 'largely anecdotal' and the bulk of the literature as 'being by academics and read almost exclusively by other academics'. He concludes that 'the publications do more for the careers of academics than for the careers of R&D managers they set out to inform'. Techniques of idea generation, selecting technology platforms, surveying potential markets, evaluating project potential, resource planning, building a project portfolio, piloting projects, establishing project teams, product development, launch and post-launch management have been much studied and reviewed. Consultants queue to assist, and sometimes do.

Such study exists because the rewards for increasing efficiency of the R&D process are great. Management and control systems – usually today some form of stage/gate approval review – have been recommended, introduced, refined and pursued, and the basic elements are similar across a wide range of industries.

### 2.5.2   Counsel for research and development managers seeking involvement in company management

What you need to do to become an important counsel:

- To be invited into business council, it's necessary to be able to speak the language of business. For a start, this means fluency in finance and financial models, their virtues and limitations.
- To be involved in business strategy you need to be involved in the right meetings. To do this you have to get your credibility high and keep it high.

- To be effective in these meetings it's necessary to understand how decisions are made within the firm, in particular the firm's ethos, both that published, and that pervasive.
- And to understand what motivates the key decision maker, what are his problems, issues, pressures, what promises has he made that he must keep. And what pleases his ego and pricks his interest.

You have to get your subject onto the strategic agenda, because if it doesn't get into the agenda, it won't get into the budget, and if it doesn't get into the budget. ...

You need to keep your credibility high by delivering on your promises, and your welcome high, by helping your peer group deliver their promises. You ensure others know what you believe in and your technical and operational constraints. You need to be consistent and truthful, and reward others who are truthful.

It's necessary to be clear when decisions are made that are inconsistent with the strategic plan, to be honest when decisions have to be made because they are expedient and bring to the table any 'externality' which might invalidate the plan. It's all about trust, really, establishing yourself, and your function, as a brand.


## 2.6    Organization of research and development

Businesses become multinational for a variety of reasons and from a variety of histories. They grow internationally through takeover, aggregation, expansion into new territories, through joint ventures. Many authors (e.g. Pearce & Papanastassiou (1999), or Gassmann and von Zedwitz (1998) noted that R&D is increasingly globalized, and we have the analysis of Alfranca *et al.* (2003) that suggests that food industry leads this. In the food industry, there is always the especial push to adapt international products to local markets, local raw material supplies or localized processing options.

He who leads the R&D function within an MFB need develop a very clear vision of how he wishes to develop the function – it's too easy to allow the growth by adopting expedient solutions, or continued extensions of some past model. But to have an organization development plan, with coherent and cogent argument for it, is an essential part of the R&D leader's toolkit. Gassmann and von Zedwitz (1998) suggest that 'the conventional organization of firms is inadequate for the requirements of modern global R&D because it is constrained by hierarchical and regional barriers'. They present a number of global R&D organizational developments in technology-based companies.


### 2.6.1    Changing a research and development organization
MFBs are constantly evolving in structure and organization. There is much to do when merging or introducing a new region into an R&D organization. Some of the issues in changing the NPD structure are:

- How much to centralize R&D, how much to regionalize?
- How close should the R&D be to centres of excellence – how are relationships with such centres to be built?
- Should the hierarchy be organized regionally, functionally, or by product group?
- How much control to exert, through systems, budgets, procedures – how much to harmonize, and how much should be shed?
- How much 'R' how much 'D'? How many scientists versus engineers, and how do we manage the liaison between them?
- How much knowledge creation versus knowledge transformation?
- How much IT?
- How much do you want or expect people to travel? Or, relocate?
- What is the overall level of language skill available, and how is language to be managed?
- How much outsourcing of 'R', of 'D', of market research and other services, such as analytical work, microbiology?
- How are library, records and information services managed?
- Project management – especially when pilot production, test market production or launch takes place in a region separate from the intended market.
- Career planning in R&D.

The structure and organization will depend a lot on the nature of the 'parent' firm, its evolution and the rate of evolution. The rate of change in R&D needs to be consistent with that of the rest of the business if industrial tensions are not to be induced, and if the need is to accept such problems as a necessary part of organizational change, then preparation for handling such problems needs to be planned. As well as language barriers there are legal and cultural barriers. It's a lot easier to make people redundant in some countries (US, UK) than in others (France, Germany).

A necessity is a positive plan for the design of the organization you wish to run. Most companies have first flirted with some sort of matrix design in their R&D, where the formal hierarchy is across national boundaries, but local conditions apply for support services, as people have to comply with national, hence local, conditions of employment This works, sort of, when the matrix is of two dimensions – country and function – but the usual problems of matrix organizations are still readily apparent. Where local history is involved, local loyalties and the informal network of alliances that are so important in a big organization, nearly always remained local. When a third dimension, such as a project management structure, gets introduced, the problems are increased exponentially. Management becomes turgid, and matrix organization discredited. Recent history has involved the leaner, meaner, flatter organization, often with painful redundancy, particularly in the older parts of the multinationals, and it would appear that this has enabled a reappraisal of the matrix. Some personalities don't work well in a matrix and where one is imposed, some seek to destroy or exploit, hoping the change might be reversed. But structural changes should be swift, and rare.

People resist change that does not confer personal benefit and minor issues can be major sticking points. Simple needs for harmonization can involve much argument, loss of face and time.

- The keeping of records and filing systems. The merging or acquisition of a business that had not previously been integrated, probably means a change in one company's systems for experimental recording.
- Recipe writing. Considerable energy is used harmonizing such a fundamental as recipe writing, or testing of products. One group of people are going to do 'non-productive' work while losing a familiarity with a system and a knowledge base they have learned. Generally resistance is high and skilful.
- IT systems need to be harmonized, and while IT and the communications revolution is a great enabler, it can cause heat, as well as shed light.
- Language is a big problem. Great progress has been made in the internationalization of English, but apparent fluency can be so deceptive. Face to face it's possible to glean meaning in a new language, but a day's work in a strange language is exhausting. To write in one, requires a quite different dimension in knowledge and skill. Scientists/technologists are proud people, trained to be precise and the grammatical sleight of hand (or mouth), tolerable in a spoken foreign tongue, is not tolerable in print. Idioms must be avoided.
- Travel, relocation, especially temporary transfers of specialist staff to aid start-up, product relocation, new project work, also pose peculiar strains. The days of single career families, where families were ready to relocate for both the opportunity to experience new cultures and advance the career of one partner have changed, and the trust that staff have in the potential of the 'home' unit to reabsorb or even promote a returning transferee has also suffered in recent years. Even short-term relocation puts stresses on the lives of those most likely to be asked.

So, as well as the production of the operational plan for the activities of the NPD and R&D function, the R&D manager needs to be able to answer the following questions about his managerial plan:

- How much of your business growth over the next five years is going to come from NPD within existing markets, and how much from new markets?
- On what evolving science and technology are you relying for the development of new products?
- From where do you expect to get this knowledge? Which special knowledge and special skills are held within the organization?
- How do you study the market, and how are you doing to acquire knowledge of consumer motivations five years hence?
- How do you see the shape of R&D organization evolving over the next five years?

- What variations do you see in organization of New Product Development in different countries? How do you overcome language problems?

### 2.6.2   Specialist staff for research and development managers

Managing and directing an international R&D function requires a specialist staff, the nature of which immediately challenges the maturity of organization design and commitment to a matrix.

*Travel agent*

This may seem facetious. It's nonetheless critical. The manager will need a skilled travel agent. There is such a need to make the diverse locations feel part of a group, and to represent the corporation and energize loyalty to it. This is best done with the feet on the local ground.

*Financial analysis*

The function will need someone to maintain records and perform analyses on the benefit of different projects and programmes, to coordinate budgetary control and forecasting. These skills are often best found amongst those with an industrial engineering background, rather than those who are financial analysts, because of their knowledge and skills in overall operational analysis.

*IT Support*

R&D, especially an internationally dispersed one, is a heavy user of IT, particularly communications resources. Indeed, a matrix organization is unlikely to work without active IT communications support, and many of the early failures of such structures were as a result of IT inadequacies. Long-term success is predicated on the increasing flexibility of evolving IT. Projects which are managed at diverse locations, the use of shared and open lab books, the keeping of records for intellectual property reasons (particularly for the challenge to and defence of patents), information science, librarianship, and video conferencing facilities are good investments in efficiency. The technical training of much of the R&D staff tends to make them both critical and sophisticated users of IT, and energy spent in seeking their input, and getting good investment in a high quality IT infrastructure pays one of the biggest benefits in R&D integration, its efficiency, and its morale.

*Operations research/mathematics/database management and analysis*

Access to these skills is essential to the R&D toolkit. Collection of hard data and its analysis for evolving trends is the basis of much experimental design and research. The availability of international data is a prime advantage of an MFB.

*Market research*

Many functions within the business can claim to be the natural home of market research. Marketing, of course, but they are often using it to develop and monitor the latest campaign, advertising programme, or annual launch. Financial planning wants to assess potential market size, and the potential of alternative investments. Local or national markets want to study local markets (and to prove 'our market is different and we need a special product because' and sometimes they're right). R&D desperately needs to measure comparative product quality.

Where the work is quantitative, big rewards are to be gained by harmonizing data internationally. Harmonizing of systems, with its implications of change, and often implications of one system being the 'better', is difficult anyway; doing this when the process is 'owned' by several functions requires immense persuasive energy. Modern techniques of analyzing trends, patterns and detecting movements need quantifiable and consistent data to be effective.

But much of this is short-term operational market research. Market research or trend research is vital feedstock for a R&D function, especially when selecting the investment in longer-term research that precedes NPD. Two years ago it would have been rare to see 'Carbon Tax' or water conservation in the consideration of company strategy, and the consequent long-term research programme.

*Legislation and ER and patent issues*
This is a fast evolving work area demanding detailed knowledge and attention.

*A secretary*
Anyone involved with a lot of travel and a lot of meetings will need a good secretarial backup, preferably bilingual.

The quandary is whether or not these 'staff' functions should report directly to the R&D head. Many of these skills have their natural or traditional repository within other functions. This cameos the problem exhibited by all matrix management structures, in that the R&D head needs someone who is an advocate of R&D, needs available resources in IT, operations research, market research or financial analysis. If these positions report to other functions, there is danger, or at least suspicion that the work is aimed *to* R&D, not *for* it. So you may get IT solutions that suit the efficiency, timing and vision of an IT function, not those that are best fitted to the needs of R&D, or cost analysis performed by those who report to a financial controller who has perhaps differing objectives than NPD advocacy. Politics exist within business – no bad thing, it's the way of mankind. The R&D manager in an MFB does need a staff function to cover some of these areas.

## 2.7   Marketing of research and development

It has been suggested that the prime role of the R&D head is to be part of the business management team, but it's a close second role to communicate where R&D fits within the organization. In short, to market R&D and more particularly the research into technologies that unlock new products, or revitalize brands.

The functions that show clarity of management, understanding of markets and the pressures that will shape them are those to which the company will turn for growth. The disciplining of data, the ability to extract information from data and to be able to manage and make intelligible technical information contributes much to the impression and impact of the overall function. It also helps to unify a function spread across many regions and countries, to give it a central core, and technical gravitas.

For R&D needs to market itself. The case history of MFB management over recent years has been a series of acronyms (TQM, JIT, zero base budgeting, business process reengineering), all of which have been addressing that which has been termed by some commentators, denominator management, the increasing of return on investment by concentrating on the costs side of the business. This has been against a background and often a reward system based on MBO (management by objectives). While purporting to give clear focus to each management task, this often results in fragmentation of the tasks, the opposite of the real goal of senior management, which is surely the integration of objectives. New product R&D is a long-term investment. The R&D budget and the advertising budget are often the targets when short-term action needs are to save or generate cash, and the advertising budget has, as *its* advocates, marketing professionals!

The marketing of a function is not unlike the marketing of anything else. In its simplest form, it's the four Ps – product, promotion, pricing and placement supported by a unique selling proposition (USP). Much energy can be used discussing what *is* a product, and what is a *new* one. In the broadest terms if you have a customer you need a product.

Who is your customer? – essentially the other functions within the business.

- Manufacturing – simpler processes, greater process flexibility, improved yield, reduced wastage, better asset utilization, quality management.
- Supplies – extended raw material shelf life, raw material alternatives.
- Product management – new varieties, events or claims on which to hang new things to communicate, product improvements, USP reinforcement, new packaging, quality assurance.
- Distribution – shelf life extension, product protection.
- Finance – planning surety.

It's in pursuit of these needs that the management by objectives approach sometimes fails. For each of these functions have these short-term objectives. It's only as a group, that they need the growth and invigoration that new product innovation can provide.

A product needs **promotion**. Normally and properly the NPD development has good and close relations with the marketing department of a business. To other departments they often just bring change, hence problems. R&D should have regular and formal reviews with the other major internal customers. These should be joint sessions, a review of joint activities, business objectives, anticipated difficulties, and the need for the R&D skills to bring solutions, 'new products', to the potential customers of R&D.

A product needs **pricing**. Much of that which was written on the need to have some financial analysis skills, presentation and project management skills was to do with this. Pricing, or clarity of budgeting, and clarity of the status of projects go a long way to demystifying R&D.

To discuss **place** under such a heading is just a little contrived. The classic debate is whether or not R&D is better performed centrally or regionally. There is no obvious general case. There is no doubt that the assembly of a group of specialists together and the scale of much scientific equipment point toward centralization. Certainly, for specific areas, scale is necessary. But this is to neglect one of the very strengths of being a multinational. In terms of the 'marketing' of R&D, and the objective of making it systemic to the business, regional R&D is significantly the more powerful choice. The 'mystique' of a R&D campus, though often impressive, is losing its impact as a marketing tool, and widespread business involvement is powerful in a multinational company. Centralization is becoming less necessary with modern IT. Cold winds often blow through ivory towers.

## 2.8  Research and development and its contribution to corporate growth

The best USP I have seen for a R&D function is that suggested in an EIRMA 1996 report – R&D as 'the prime mover of internal corporate growth'. This report defines the three main factors of R&D's contribution:

- Maintenance of technical capabilities supporting present products – *value preservation.*
- Development and improvement of key technological capabilities to extend the present base – *value extension.*
- The acquisition or development of new technologies to open new markets – *new value creation.*

Most products need continual maintenance if their value is to be **preserved**. The variability of food raw materials with season, sourcing, economics and availability requires that continual small adjustments need be made to recipe and process as an ongoing operation to keep the products at their intended position. This is not NPD, but is often a demand on its resources.

**Value extension** is much bound up with existing brands, and the marketing function. A MFB will have a mix of international, internationally regional, and 'local' (usually national) brands. Brands are a form of promise to the customer, and much of the promise is of performance consistency. But to market brands, the marketing people need to produce 'news' about a brand and **extend** its value. So we have the brand relaunch, the new and improved, new flavour, new packaging, new variety, new communication, new sizes.

The strategic planning of brand management needs to have technical pegs, or developed and tested product changes, or quantified claims on which to base such relaunches. This is the stuff of brand management, who need to present to the sales people, often at a sales conference, a programme of events that will keep a brand fresh and relevant while consistent with its core values. The identification and securing of the technological pegs, and the guiding and monitoring of many of these projects through to the market place is the responsibility of R&D, although when these projects are large and complex, the formation of special project teams is practised. Most companies will be managing a portfolio of programmes and projects through some control programme which covers the next two to three years of operations. These, in turn, need to be based on technical skills that have been or are being acquired, and whose success is probable.

These first elements, **value preservation** and **value extension** are concerned with the existing business and its extension, and these are of great importance. But any company interested in new products and new markets, or who forecast a sales or activity gap in their strategic plan, needs to have activity in place to fill it. If this is not to be filled by acquisition or joint venture, then it is very much the focus of R&D, those 'prime movers of internal growth'. This is **value creation**, the pure form of NPD.

## 2.9    The New Product Development programme – establishing the investment in the 'R' of R&D

Experience would suggest that keeping the development process as short as possible is highly desirable. To keep this short, the business draws on its fundamental strengths, the essential knowledge of the business, its customers and both its commercial and technical opportunities. These are its core competencies.

*Knowing the core competencies*
An existing multinational business will possess a certain number of core competencies. They are the reason the business exists. This core competency may be grounded in a technical cometency, but it may be a commercial, organizational or managerial competence, or a mix of all or any of these. The business management need to be clear what these are.

*Establishing a market vision – 'The proper study of mankind is man' (Pope)*
Early on the need for access to or even control of market research is emphasized. A systematic view of the pressures and issues that will shape markets over the longer term should be available to stimulate the imagination of those responsible for new technology. The future condition of the market shapes the new product need.

To make judgement of the consumer wants and needs five years hence, and, thus predict technical competencies a business needs, so that it might have commercial advantage in their supply, obviously justifies investment. Yet this is the weakest area of that mix of skills that come under the heading of market research. While techniques of long-term forecasting do exist, the longer-term trends in a market have always been hard to assess – indeed, a company strategy based on staying close to short-term trends is certainly a valid one.

But it's more suited to businesses managed nationally or regionally than a MFB. MFBs need to gain product advantage and brand power based on unique intellectual property, on science and skills in managing the supply of goods. Worryingly, for the MFBs, the best data on the way customers behave is held by their prime competitors for customer loyalty, the supermarket chains. Increasingly this is used as the basis for NPD.

A good review, published in the *Financial Times* magazine in 2006, discusses the use to which Tesco puts the data generated by its Clubcard, and the sophistication that is used to analyse the 'five billion pieces of data generated by the weekly shop of 13 million customers'.

With the analytical power of current data mining and trend analysis techniques, and the acquisition of ten years of data, Tesco is now able to get instant information on responses to price, quality, variety changes, to new launches, to new advertising and promotion change – for all elements of the marketing mix. They are able to detect early the realities of changes and trends within and between market segments. True, Tesco sells this information to major suppliers, but it's they who moderate how much and what information is sold. From this, suppliers can not only identify potential new segments, the supermarkets can directly commission new products from 'own label' suppliers.

It's so, too, with other chains. Fast food franchises are able to gather point of sale data on which products, combinations, and what commercial skills affect the products they sell, and identify the customer segments (age groups, lifestyles) to which they sell them. Yet the quotation in the heading which leads this segment remains valid. For market vision needs to study customer behaviour first, not technology. Market vision is about customer behaviour. Technology should follow.

R&D are, however, charged with having the **technical vision**.

- defining and developing the key technological capabilities supporting business core competencies
- identifying and securing new technological competencies
- securing and managing projected budgets
- finding universities and research institutes to work with to acquire the identified competences
- finding potential commercial partners.


## 2.10   Bundling

The management act required of the market and R&D functions is sometimes termed 'bundling' (Fig. 2.2). Market opportunity is non-linear. Competitive activity, capacity shortfall, new markets, opinion shifts, not to speak of

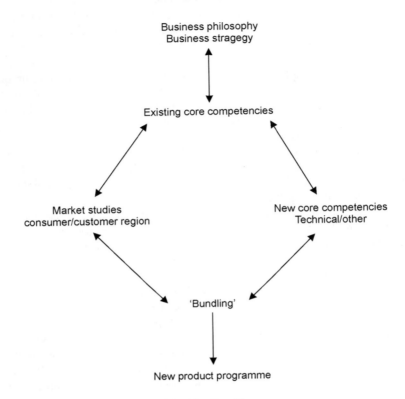

Fig. 2.2    Bundling.

catastrophes such as BSE, crop failure, political barriers, all change, rarely predictably. Research into new technology or new skills results in variable time scales, with success targetable, but not guaranteed.

One approach is to review the stock of technological capabilities and competences, and choose when to meld them with market opportunities as they become identified. Such a process gives a flexibility of response to changing conditions, keeps the Product Development process (PD Process) short, responsive, and hopefully opportune. The emphasis is keeping the 'library' of technical competences well and currently stocked, and keeping the market under continuous review. The approach is to review the state of technical advance, the business need, the market condition and 'bundle' the various elements into a series of new product launches. A business may choose to save all its opportunities for one major new product, or dribble the advances into a range of smaller product improvements. Such an approach allows for continuous adjustment to the business need, the market, and the rate of new products desirable or necessary. So, a **bundle** is the agglomeration of one or more elements of the technical programme, with one or more of the elements of the market opportunity, wrapped together to form a product extension, a new variety or new claim, or a new brand.

It is part of the management process to decide when a research programme has been of sufficient progress to introduce it to the market. For example, a research programme may have potential to reduce the calorie content of a chocolate bar by 100%. To be successful may take five years, but a 50% reduction is possible now and an 80% reduction guaranteeable within five years. A similar packaging research programme is developing a compostable package system, whose rate of compostability will also improve the longer it is left with research. Also there is a flavour development programme in hand which shows potential, whose progress is sound, but will improve in time, and a shelf life programme that will extend the length of time the chocolate bar stays fresh. When these elements should be combined, and launched becomes a management issue. Do they justify a new brand, should they be tested in one region, should the packaging 'benefit' be saved for a later launch? The different elements of the increasing technical competence can be 'bundled' to form a continuous stream of 'new products'.

Figure 2.3 suggests some of the market conditions or technical research that may be 'bundled'. Such an approach favours incremental NPD rather than big launches of new products. This, however, has insurances. The major launch of a new product is often dominated by marketing rather than technical costs.

Momentum, in NPD, often overcomes many of the inevitable problems, especially bureaucratic ones, that NPD brings. Enthusiasm, important to NPD management, can never be continuously sustained. Short lead times in NPD reduce the opportunity to be blindsided with competitive activity.

This process has the advantage of having long research goals, but the potential of shortening the development process. Momentum, fixing and fast realization of design parameters are a virtue in NPD. The procedures lend themselves to management review and control through a series of management stages, as suggested in Fig. 2.4.

## 2.11    The research and development role

If the management of NPD is seen as the timely bundling together of market opportunity and business core competency, then the business of R&D strategic planning is to ensure that procedures exist for acquiring the necessary skills, and that the technical skills, or competencies exist to support the core competencies on which the business is based. The core competencies of a business are not only technical ones (Table 2.2). But it's the R&D role to supply those that are.

### 2.11.1    Market vision
One part of the bundle is the evolution of the market – the 'demand' side of NPD. Earlier, it was stressed that R&D need access to customer research, and

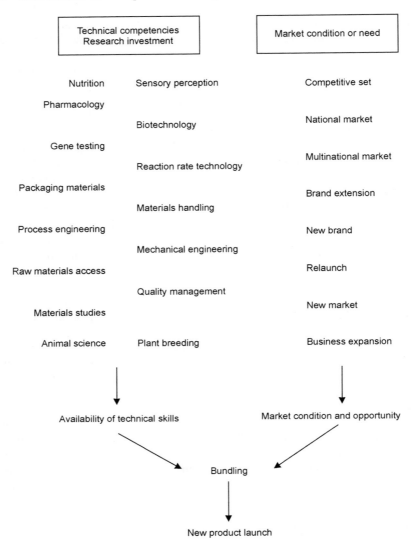

**Fig. 2.3**   Bundling technical competencies and market conditions.

preferably direct control of some of it. Typical aspects of 'demand' driven influences on product development:

- *Demographics*. Recently, a greying of the west, where most MFBs have their locus, smaller portion sizes, concerns with the health problems of the aging (diabetes vitamin/mineral supplementation, fibre content, low sodium, low cholesterol foods, low fat, the nature of fat) MSG or sugar level, 'refined' flours, extracted soy, responses to declining chemosensory perception, energy requirements.

**Controlling the bundle**

Stages and gates

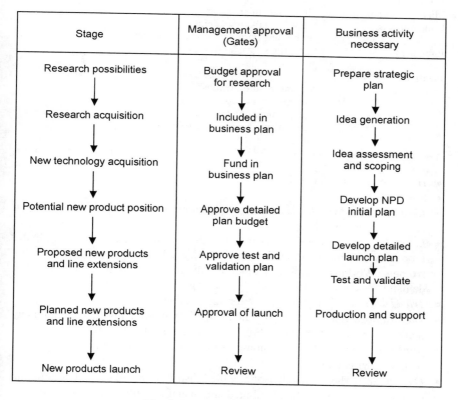

| Stage | Management approval (Gates) | Business activity necessary |
|---|---|---|
| Research possibilities ↓ | Budget approval for research ↓ | Prepare strategic plan ↓ |
| Research acquisition ↓ | Included in business plan ↓ | Idea generation ↓ |
| New technology acquisition ↓ | Fund in business plan ↓ | Idea assessment and scoping ↓ |
| Potential new product position ↓ | Approve detailed plan budget ↓ | Develop NPD initial plan ↓ |
| Proposed new products and line extensions ↓ | Approve test and validation plan ↓ | Develop detailed launch plan ↓ Test and validate ↓ |
| Planned new products and line extensions ↓ | Approval of launch ↓ | Production and support ↓ |
| New products launch | Review | Review |

**Fig. 2.4**  Controlling the bundle.

**Table 2.2**  Examples of core technical competencies

| Business | Core competency | Technical competency |
|---|---|---|
| Pet food | Small animal nutrition<br>Animal behaviour<br>Human/companion animal bond | Nutrition, biochemistry<br>Behaviourists<br>Psychology |
| Soft drinks | Water quality management | Chemistry |
| Frozen vegetables | Crop maturity prediction | Botany |
| Chocolate bars | Cocoa production | Agronomy |

**Table 2.3**    Demand driven NPD – some current health issues

| Negatives | |
|---|---|
| Bacterial | Salmonella, clostridia, E coli |
| Viral/prions | Avian flu, BSE |
| Micro-contaminants | Phthalates, micro-oestrogens |
| Contaminants | Heavy metals, insecticides |
| | antibiotics, hormone residues, nitrates |
| Ingredients | Salts, trans fats, sugars, high GI starch |
| | |
| **Positives** | |
| Inclusions | Natural, nature identical |
| | Vitamins, minerals, nutriceuticals |
| | antioxidants, omega-3 fats, |
| | free radical scavengers, probiotics |
| | prebiotics, artery softeners |

- *Health concerns*. See Table 2.3.
- *Food safety*
- *Religion* – products for specialist ethnic communities – Kosher food, Hindu or Muslim communities.
- *Lifestyle changes*. Food not as fuel, but as a treat; the desire to enjoy better food. Aspirational food, caviar for the citizen. Single households. Working women. Working couples without children (DINKS). Pressure on time, demand for convenience, snacking, grazing.
- *Travel by taste*. The demand for foreign cuisines.
- *The disappearance of seasons*. The availability of so many foods in so many markets throughout the year. Fresh raspberries for Christmas.
- *Appliances for all*. In every home, a microwave, fridge freezer, an electric wok, a slow cooker.
- *Distribution*. The availability of a chilled distribution and control system that enables chilled food to be distributed and managed with reasonable economy. The relatively low cost of transport generally, coupled with the increasing concern about its environmental impact, does seem to be one of the instabilities in a modern world. In the longer term, addressing the carbon energy in producing and transporting food, the waste produced and the water used is surely going to be a requisite, potentially a driver of innovation.

### 2.11.2    Technical vision

In making the assessment of core competencies and core technologies, there is the equal difficulty of technology forecasting – **technical vision**. This is at least as difficult as market place forecasting, technology in many ways moving faster than markets. But it was not ever thus. Beer, bread, cheese, wine and much else was accidental biotechnology and much food practice has preceded the understanding of the underlying science. In which food science should a food research group invest? How can any person keep up with advances on all fronts or judge their relative potential? One of the truly exciting things that faces the food industry, and in which the MFBs have the best opportunity to lead, is the transfer of the food business from a science-*uncovered* business to a science-*led* business. The last forty years has been spent trying to understand the bases of 'traditional' foods. Much of what technology and science has done is to conserve the image of an established brand, while making it cheaper, less seasonal and more available. We do stand upon the threshold of making new foodstuffs, and the challenge will be to market new products, which may be produced in a world which needs a food production system which demands less energy and less water, less waste to dispose, yet will contribute positively to health, feed a burgeoning population, and hopefully delight the senses.

The disciplines of biochemistry, physics, robotics, mathematics and computer science, pharmacy and nutriceuticals, gene technology and microbiology will be increasingly interwoven with the technologies of packaging, process or materials research, with nutrition, flavour research, food chemistry, toxicology and food texture such that barriers are merged and intertwined. It is the R&D functions of the MFBs that can achieve the overview of the varied sciences and technologies to address these issues. The challenge will be to communicate to the market place such new products – mankind is conservative where food is concerned.

Linear analysis of potential problems is not straightforward. The need to build a flexible production line for chocolate bar production in a new market could depend on having the biotechnology to enable the detection of minute levels of peanut residue on process equipment. Or the expansion of retort pouch technology might depend on high speed video optics, necessary to ensure seal integrity equal or superior to that established by the canning industry, and high speed weighing systems necessary to dispense precise weights of difficult to handle materials. Medical technology, to support some health benefit claims, or process control to facilitate aseptic processing, are examples of the skills a MFB may need. An MFB ought to have available to it and access to:

- *Mathematics and computer science*. The integration of statistics with the computer has resulted in immense power to extract information from complex data systems. The predictive power of neural nets, fuzzy logic, expert systems, their use in many aspects in science from market research to process control, from the electronic nose or to the analysis of mouth feel makes a competence in such processes a high priority.
- *Biotechnology*. Probably the most significant area of modern food research. Much work is directed at the manufacture and expression of functional proteins, which are being developed to control texture, replace the functions of fats, and control water binding. The use of biotechnology to produce natural but expensive proteins (such as that which can control ice crystal formation and prevent ice crystal growth through freeze-thaw cycles) is one application. Its use in rapid diagnostic techniques, using DNA assays, or for rapid microbiological assessment, as well as the development of enzymic methods of food processing justifies much research.

- *Reaction rate technology*. The increasing application of reaction rate technology to the food industry is a further sign of its maturation. The ability to target and predict flavour, texture, colour from application of knowledge about the composition and chemistry of materials and the control of process conditions becomes more successful and more important as data and skills build. The ability to precisely predict and control conditions to achieve product stability and safety are another potential benefit. It is not long since expert systems were being employed to deconstruct the art of the chef. We move towards constructed design by the understanding of reactions and process conditions – this is the hallmark of science.
- *Controlled or modified atmosphere packaging*. The key to much of the viability of the now huge chilled food segment.
- *Packaging technology*. The integration of packaging and process has resulted in many 'new' products. The mechanical development, robotics, electronic control of packaging machinery, together with the development of new sealing materials, new structures, crystallized polyethylene terephthalate, and sensor technology have produced products for microwave ovens, chilled cabinets, the basis of ready meals or meal components. The integration of machine, product and material has become a very sophisticated mix of technologies, and the addressing of this is a complex management problem for a the food industry. 'Clean room' and low temperature/low humidity packaging rooms, and the hygiene control systems, and the issues and opportunities these present are something that need to be part of a technology forecast.
- *Nutrition/pharmacology*. Ability to make nutritional claims about a product is clearly prized, and the research necessary to both deliver and justify claims has much potential. Typical is the research into flavonols/phenolic phyto-chemicals that are present in chocolate as a result of modified cocoa processing and chocolate manufacture. These result in claims for the presence of food products with natural antioxidants and free radical scavengers, with the potential to lower blood pressure, cholesterol, reduce diabetes and heart disease. A report by Lang *et al.* (2006) examined the claims by many of the world's largest MFBs to have health as part of their research programmes, ranging from products which are 'organic' and 'natural' (additive free) through to low calorie, to specific health claims. And maybe irradiation, or dehydrofreezing, or dehydration will surprise us again if energy and water get on the agenda. One prediction that is already showing as a trend is the merging of the pharmaceutical and food industries to generate new products, new fruits, vegetables, new claims.

## 2.12    Organizing and acquiring knowledge

So much depends on from where you start. If the MFB has an efficient and centralized research facility and function, or extensive pilot plants, or specialist facilities, then it is easy to 'clip on' an extra project. It's a major, and often politics-heavy decision if R&D is dispersed across many sites. Sources of fundamental work can be:

- universities
- equipment or packaging suppliers
- research associations
- specialist consultants
- in-house research
- market research/advertising agencies
- customers.

The easy answer is to maintain a network of contracts and alliances, in which the MFB sits in the middle of a web, adjusting the tension on the various strings in the net. It's not easy to achieve, however, and requires continual management.

*Universities*
In many ways the most cost efficient. Commissioning work through a university has many advantages, with access to a lot of specialist equipment (though often not the most up to date), with post graduate and post doctoral workers, generally relatively inexpensive, supervised by specialists in their respective fields, with access to good library facilities, with colleagues with whom they can hold collegiate discussions.

Contracts can be terminated under clearly defined conditions, if they are not working out, potential company employees can be vetted, the prestige of the university can be used in marketing claims and generally, as the R&D staff probably went to a university, they know where to start. Universities usually like industrial contracts.

But they do want to publish their work, they can be slow to find appropriate workers, they are getting precious about intellectual property rights, they can be leaky with confidentiality. Working with universities is an excellent way of getting involved with new markets, such as Eastern Europe and the Far East. They generally like to build relationships with MFBs. Many universities have spun off 'development' sections, or 'incubators' for the nurturing of innovation. Experience does suggest that university staff often attribute to the representatives of the MFBs, degrees of freedom that their representatives just don't have.

*Suppliers of ingredients, equipment, packaging*
Building of alliances with suppliers can be the most efficient way of working, but can also be the most tense. Partnerships are difficult in so many ways. Suppliers, particularly flavour houses, have specialist equipment and an experiential background that can be tremendously powerful. Equipment and packaging suppliers are often essential in a new product role. The commercial negotiations can often be extremely difficult, as relative contribution to both a success and failure, or a dispute as to the completion or not of a brief can be the basis of long-term distrust. This is particularly so if the development is an extended one,

and the technology though successful, is outflanked by changed market conditions. Sharing risk across partners is a commercial hazard.

MFBs, with their potential expansion into big markets, have an advantageous position with the upstream suppliers of technology, vis-à-vis nationally based food manufacturers.

- *Customers.* The reverse is true of customers, particularly if they are providing you with consumer feedback. Customers, especially the supermarket trade, would generally prefer to work with national organizations.
- *Research associations.* Research organizations, too, often receive much national funding. The technical ability of many national research organizations is excellent. In recent years there has been significant scaling back of these facilities.
- *Developing in-house skills.* This has obvious attractions. While histories vary, many MFBs have cycled through the phases of internal growth of R&D, realignment, downsizing or rationalization. This has happened over some years. It might be the result of mergers, of plant closures, as a result of the expansion into new markets, or the desire to rebalance the organization into the newer territories. It is also true that creativity, the energy to create change or to pursue new ideas, does not correlate well with middle age and comfort, and a department can be reenergized by such review. It can also be a painful process, and result in the loss of hard won experience.

This is pursued in the attempts to leverage efficiency of management, to ensure focus and is in accord with the leaner, flatter, organizational structure demanded of all functions. Many papers (e.g. Bonner and Jalajas, 1999) have been written which suggest that staff say they can't be creative unless they are comfortable and secure and unworried. History would, however, show that society is often most creative when stimulated by crisis, and most comfortable prior to a catastrophe.

## 2.13   Multinational research and development organization

Whether or not the business owns the research, it needs to own and retain the ability to assess potential value rapidly and to test and feed back its results. The paradigm suggested is a virtual network of alliances, which the MFB coordinates. Companies ought be quite clear on the competencies they must own, and technologies they must encourage.

A multinational company should also play to its strengths, and distribute its research effort geographically. It's good politics, it's good market research, and it's a first step in breaking down the national loyalties and solving organizational problems that will occur.

The picture painted is of access to in-depth science and skills with regular fusion of these into energetically and rapidly deploying new products, with the core axis of marketing and R&D presenting to the various regional structures a portfolio of recommended projects. The objective should be for the gestation phase to be as rapid a possible, as the market condition against which they are proposed is only transiently valid. Development is about following market opportunity.

All this suggests the classic divide, twixt research and development. The R&D leader, he who was earlier postulated, has to manage across the matrix of regional affiliation, fitting for administrative purposes in a local or national structure, across language barriers, across distributed research locations and yet hierarchically part of a global R&D function through which he can achieve some unity. Above all, he needs to infuse the group with the enthusiasm that begets creativity. He has to build the trust between the 'R' and the 'D' group that ensures belief in each other's work, its energy, direction and value. He has also to comply with the other demands of a modern business, of budgets, of rules, of staff development, as well as cope with the competitive strain that processes such as management by objectives induce across the business.

To restate their conclusion to a paper in R&D Management, Gassmann and von Zedwitz (1998) assert that, 'the conventional organization of firms is inadequate for the requirements of modern global R&D because it is constrained by regional and hierarchical barriers'. They state that, 'organizing industrial R&D for global efficiency is an oft neglected task', and go on to suggested overlaying structures of personal and informal networks and project structures that provide part of the 'glue' that holds a R&D network together. Part of that glue also has to be provided by the energy and attitudes of the manager.

The use of matrix management structures has often been criticized as leading to lack of clarity, divided loyalty, and as demanding too high a level of sophistication for an organization to be effective. Many firms flirted with such structures in the 1980s and 1990s, only to revert to either tighter regional, or tighter functional hierarchical control, as problems arose. But a form of matrix management is inevitable.

It's easy to forget how far and fast the world has globalized and how fast has been the development of global communications. E-mail didn't really become common till 1990. From 1998–2000 the IT people were obsessed with the Millennium bug, and if you pressed further for help, they said 'bandwidth', which meant 'go away'. Even six years ago, video conferences were a study in stilted conversations on equipment of marginal reliability. Today a video conference centre is properly convenient. Intranet, shared lab books are normal, and this has happened in fifteen years in which time businesses have absorbed Eastern Europe, China and India. People are adapting and quickly.

As younger managers comfortable with such networks, as the language of science trends toward English, as comfort with the changes in communication increases, and as the harmonizing of employment conditions and practices follows globalization, so management skills will evolve towards the sophistication and culture required.

Figure 2.5, adapted from Gassman and von Zedwitz (1998), suggests some of the complexities of a MFB R&D organization, with its matrix of global, local,

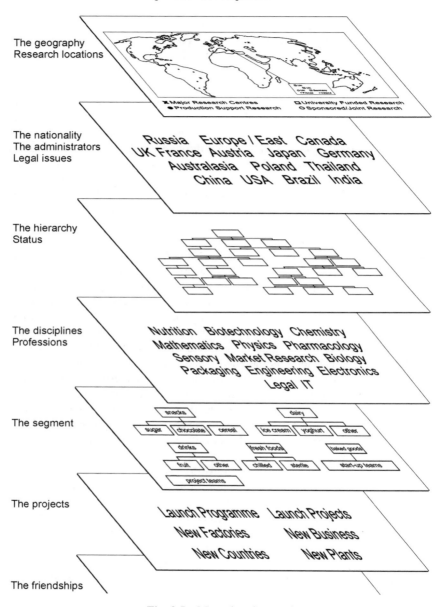

**Fig. 2.5**   Managing the matrix.

hierarchical, professional and market segment relationships. Figure 2.5 shows the various layers of the matrix of management in a MFB. As the world shrinks and globalizes, the role of senior management is to facilitate the inevitability of such a structure and to coach the staff through the inherited loyalties that give such organizations problems. The alternative in recent years has been constant reorganization which is worse. It requires true leadership, and a focus on the strengths and advantages a MFB has.

## 2.14    Problems met and lessons learnt

The R&D manager needs to:

- Believe that business growth can be achieved through the creative application of technology. He has to market that belief both in his function, and across the business.
- Have a clear idea of the organization that is necessary to provide that growth, and a plan to achieve the organization.
- Provoke a programme of understanding the business strategy, and exposing the opportunity that the strategy will present.
- Develop a programme of acquisition of knowledge that will nurture the business strategy.
- Use these two programmes to create an ongoing list of activities that become the NPD plan for the business.

But the R&D manager also has to create the right environment by:

- Communicating that he expects creativity.
- Supporting it with resources.
- Rewarding it.
- Accepting the chaos it sometimes causes.
- Accepting the failures.
- Rewarding candour and honesty.
- Leading by suggestion, and enthusiasm.
- Having a flexible management style with personality, and interest, and being conscious always of the degree of evolution of management in the different areas and regions of the business.

## 2.15    References and further reading

ALFRANCA O., RAMA R. and VON TUNZELLMANN N. (2003) *Innovation in Food and Beverage Multinationals* (UK: Haworth Press).

BALL D. (1998) The needs of R&D professionals in their first and second managerial appointments: are they being met? *R&D Management*, 28(3), 139–145.

BARTLETT C.A. and GHOSHAL S. (1989) *Managing across Borders: The Transnational Solution* (Boston: Harvard Business School Press).

BONNER M. and JALAJAS J. (1999) The threat of industrial downsizing, *R&D Management*, 29(1), 27–34.

CHRISTENSEN J.L., RAMA R. and VON TUNZELLMANN N. (1996) *Innovation in the European Food Products and Beverage Industry*, EIMS Publication 35 (European Commission).

EUROPEAN INDUSTRIAL RESEARCH MANAGEMENT ASSOCIATION (1996) *Core competencies and R&D*, EIRMA Workshop, w307 (EIRMA).

*FINANCIAL TIMES* WEEKEND MAGAZINE (2006) *Tesco is Watching you. How Clubcard took off.* FT Magazine Technology Special, 11–12 November.

GASSMANN O. and VON ZEDWITZ M. (1998) Organization of industrial R&D on a global scale. *R&D Management*, 28(3), 147–161.

LANG T., RAYNER G. and KAELIN E. (2006) *The Food Industry, Diet, Physical Activity and Health: A Review of Reported Commitments and Practice of 25 of the World's Largest Food Companies* (London, Centre for Food Policy, City University).

PEARCE R. and PAPANASTASSIOU M. (1999) Overseas R&D and the strategic evolution of MNEs: evidence from laboratories in the UK. *Research Policy*, 28(1), 23–41.

VON ZEDWITZ M., GASSMANN O. and BOULETTIER R. (2004) Organising global R&D: challenges and dilemmas. *Journal of International Management*, 10(1), 21–49.

# 3

# The food research centre – assisting small and medium sized industry

Paiboon Thammarutwasik, Thailand

*This chapter shows the place of a food research centre as a generic contributor to food product development knowledge. Food research centres have created knowledge in food science, food technology and food engineering for over 100 years. In the USA, they were often connected to agricultural research in universities in providing the research beyond the farm gate, and became renowned in the areas of food chemistry, food microbiology, sensory testing and food engineering. In Australia, food research was a separate centre in the government research organisation, CSIRO. There were often separate research centres for particular industries, such as the fishing industry and meat industry. These were important to the food industry, until the growth of large multinational food companies, supermarket chains and agricultural companies who had the resources and people for their own research. The research centres turned then more towards the smaller companies who did not have the resources for research and development.*

*The research centres started to use the systems and techniques of the large multinational companies, in particular adopting the Product Development Process (PD Process). They found that there was a major problem in organising the product development between the company and the research centre. The knowledge in the two groups can be widely different, the company having marketing and commercial knowledge and the research centre scientific knowledge. The two types of knowledge need to be combined and then used for successful product development. A two way transfer of knowledge – both explicit and tacit – needs to occur throughout the New Product Development*

*(NPD) project. There is a continuous knowledge conversion from tacit to tacit, tacit to explicit, explicit to explicit, explicit to tacit, which can raise communication problems and therefore barriers to successful completion of the project.*

*There needs to be an understanding of the PD Process. The specific PD Process for the project has to be designed by all the parties together, so that an understanding of the stages, the planned activities, the outcomes and decisions is agreed in the initial stages of the project. The PD Process has a similar structure to the classical description in the textbooks, but there are often not the financial and people resources to do everything, and choices have to be made and risks recognised. A lack of good communication and understanding can lead to the failures experienced in some research centres.*

The chapter particularly relates to pages *161–191* in Food Product Development *by Earle, Earle and Anderson.*

## 3.1    Introduction

Food product development research was conducted in Thai research centres and universities, but few results ever reached commercialisation. Most food product development projects were based on the researchers' interest with disregard for the markets. For small and medium sized enterprises (SMEs), there is a lack of funds and resources to conduct research and development compared to the larger, often multinational, companies. Hence, the Agro-Industry Development Centre for Export (ADCET) was established in the Faculty of Agro-Industry, Prince of Songkla University, Hatyai, southern Thailand and it aimed at helping small and medium sized enterprises develop new products and processes.

ADCET is the country's first centre in southern Thailand, established in 1997. This technology transfer unit offers academic and business expertise to companies including small and medium enterprises (SMEs) looking to develop new healthier products and to expand their market. It also acts as the bridge to span the gap between industry and academia, from the Faculty of Agro-Industry and others. The unit also helps the SMEs to gain access to new technology, provide training, consultancy, research in food product development and shelf life extension.

Policy: To provide the knowledge and technical know-how to the industry in order to enhance the competitive capability.

Aim: To assist the small and medium size food industry in the creation of technical know-how and its application in the development of food products.

## 3.2    Product development in the research centre

The food research has to have the capabilities of creating knowledge and applying this knowledge in the development of food products.

### 3.2.1   Creation of technical know-how

The food industry is a core activity in Thailand's economy, involving many sectors, producer through to processing, export, import and the domestic retailing sectors. Thailand exports agricultural commodity products and processed foods throughout the world, with an export value at US$42.5 billion in 2004. Thailand is ranked fifteenth in terms of food exporting in the world market. However, the food production pattern has been changed drastically by many factors, e.g. climate, innovative processing, tariff and non-tariff barriers to trade, as well as labour, costs and skills.

Thus research and development (R&D) is needed in order to create knowledge to support food industry development. Nonaka and Takeuchi (1995) explained the word 'knowledge' as a dynamic human process of justifying personal belief toward the truth. Knowledge creation is a continuous, self-transcending process by means of which one transcends the boundary of the old self into a new self by acquiring a new context, a new view of the world and new knowledge. In short, it is a journey 'from being to becoming' (Prigogine, 1980).

There are two types of knowledge: explicit knowledge and tacit knowledge. NPD is an example of the successful conversion of tacit knowledge into explicit knowledge.

Personnel working for the ADCET are skilled in different disciplines such as food technology, food product development, food engineering, quality control and safety, food packaging systems, and food industry management. Besides teaching and research, these specialised experts and technicians also carry out some community research projects, especially product quality improvement, process development and food safety aspects.

When the problems are brought by customers, the ADCET manager will identify an appropriate expert in the Faculty who has experience in this problem. The experiments are designed to solve the customers' needs. The technological know-how would be created by these experts. However, the expert has to work to compete against the limited time and resources available within the company.

### 3.2.2   Application in development of food products

Food product development is becoming more important nowadays due to the changing lifestyle of consumers, the high quality needed in products, including nutritional value and safety, as well as the trade barriers. Food product development in the Thai food industry can be classified as follows (Suwannaporn and Speece, 1998):

- New product line: This refers to new products that are produced for the first time by the company. They may be new to the Thai market or already available from another company.
- Additions to existing product line: The company uses existing or minor modification of existing facilities to produce a new product, which may or

may not be new to the market. This new product is an extension of the product platform, offering a distinctive claim, feature and market position relative to competition.

- Modification of existing product: The company adds more varieties of taste, aroma, form, content and packaging to their existing products. The purpose of this type of new product is to maintain market image, particularly by satisfying the desire for variety.

Although research and development will create new food products with more value added, research and development capability in the Thai food industry is very poor. For Thailand, the budget invested in food research was about 0.26% of GDP comparing to 2% of GDP for the well developed countries. The UK Office of Science and Technology has outlined the main research requirements for the food and drink industry. Most important among their research results for the food industry are productivity and quality improvement. Australia has recently begun to invest in more research related to food product development. The quantum of research dollars has not reduced; there seems to be less spent on basic research and more on the marketing side of product development or adaptation of products developed overseas to Australian conditions (Agriculture, Fisheries, and Forestry, Australia, 2002).

Existing tacit and explicit knowledge from the ADCET and experts will be used for creating new food products for the clients who come to the ADCET with an idea. If any new technological know-how is needed, the experiments will be designed. When the kitchen experiments are completed, the scale-up trials are set-up. Usually, ADCET provides the scaling-up facilities for the clients for the marketing test.

## 3.3   Guidelines for technology transfer in the Agro-Industry Development Centre for Export (ADCET)

Technology transfer plays an important part in industrial development. Transfer of technology and knowledge contributes to learning and development of the capability of the industry. Before discussing the mechanism of technology transfer, Gee (1993) defines technology as a set of knowledge contained in technical ideas, information and data, personal technical skills and expertise, equipment, prototypes, designs or computer codes. Thus, technology is tangible, reproducible and scientific. It can be products, processes, or related findings.

Technology transfer is the mechanism by which the accumulated knowledge developed by a specific entity is transferred wholly or partially to another one to allow the receiver to benefit from such knowledge (UNIDO, 2004). In another sense, technology transfer is to improve the technological capability of business enterprises in developing countries. The advantages of technology transfer could be (Bennett, 2002):

- A production process or part of a process which improves production efficiency, reduces costs, improves quality control and/or reduces environmental pollution.
- A product which is of better quality, has greater functionality, better appearance, less damaging of the environment in its use; or
- A combination of process and product as production of a better product often requires changes in the process.

The ADCET delivers the technology to the clients by different means such as seminar, demonstration and workshop. Demonstration and workshop were more effective methods. In practice, the ADCET prefers the clients working closely with the expert or researcher. They can directly learn from the expert.

### 3.3.1   Product design
The food industry, nowadays, has become more conscious of customer needs. Modern retailers in Thailand, some owned by foreigners, have products that are standardised, high quality, long shelf life and better packaging. Therefore the design of products is needed in order to conform to customers' requirements. The concept of food product development should focus on healthy, safe and convenient products.

The innovation of product design can be transferred to the private sector in form of patent, copyright and trademark. The ownership of this property right will be shared between the company and the researcher if it is contracted research and development. For the SMEs, the new product design knowledge is transferred to the employees by mean of training for commercialisation. The researcher sometimes also provides consultancy to the small and medium enterprises till this new product is sustainable in the market.

### 3.3.2   Process development
Process development is another strategy for the private sector in order to boost their competitiveness. For product diversity, the process design and development should aim for high flexibility so that the process can be performed under different raw material conditions. An internal or external R&D unit can carry out process development. SMEs choose to contract out R&D because they lack the necessary facilities and expertise. However, some large companies also choose to contract out because the companies avoid investment in facilities used only occasionally for short-term projects.

For the technology transfer process, the researcher (supplier) will directly transfer the technology to the SMEs. From our experiences, particularly with the SMEs, the technical know-how resulting from R&D will be transferred through the training (including workshop) procedure. The transfer process will be a long-term process depending on the absorptive capacity of the firm for new knowledge. Sometimes, it needs written instructions, design drawing or prototypes or the combination of these. For large companies, transfer of such knowledge and technology can be done through purchasing the patent, licensing or trademark.

### 3.3.3    Pilot scale production

The R&D personnel are responsible for developing a new food product prototype. The prototype recipe may be adjusted to meet customers' specifications, then this is the end of research and development tasks.

The next step is the pilot scale production, which is useful in determining how the new food products and processes perform under industrial operating conditions. The pilot scale production is often used for the production of large samples used for storage and marketing tests.

For our ADCET's experience, we also provide pilot plant production testing to clients who seek facilities for the big mass production for a marketing test. The ADCET and client have an agreement to keep the work confidential. The clients can also rent the pilot plant facilities for the production of products when they start their business. The reason is that most SMEs lack funds for establishing a plant equipped with production facilities. This is also one of the strategies of ADCET to make sure the contract research has been extended to the pilot scale production and its success.

## 3.4    Cost estimation and return on investment

### 3.4.1    Costs

Once the new product is accepted by the market, the cost of product, potential sales and profits must be analysed. Costs are divided into fixed costs and variable costs. The fixed costs are mainly plant and equipment, while the variable costs are those that vary with the level of output; they are mainly direct labour and materials.

The analysis of the break-even point is necessary in order to estimate the profits of the firm under different operating conditions or output levels. Product costing can be accomplished by either job costing or process costing.

ADCET usually estimates the manufacturing cost or cost of goods produced for the clients. Besides the material costs, labour costs and operational over-heads, the cost of sales and administrative costs must be added in order to know the total product costs.

### 3.4.2    Return on investment

There are many techniques that can be used for evaluating the financial suitability of the investment projects. The basic techniques involve estimating the relationship between the investment required and the benefits to be gained. Payback period is the most popular technique often used to justify investment but it is unsuitable for long-term investment because it ignores profits expected beyond the point of payback time and does not consider the time value of money (Babcock and Morse, 2002). The other techniques are: net present value, average rate of return and initial internal rate of return.

The clients from the small and medium enterprises have been taught the concept and how to estimate the return on investment in order that they can decide whether or not to invest in the project.

## 3.5    Intellectual property

Research and development is conducted to develop and improve technological processes and products. Usually, researchers and/or companies will have rights in a discovery. These rights are a form of intellectual property rights.

### 3.5.1    Protection
The research findings can be protected through patents, copyrights, trade secrets and trademarks.

A patent is a grant from a government to exclude others from making, having made, using, selling or offering to sell in or import into their territorial jurisdiction the claimed invention (UNIDO, 2004). There are three classifications of patents: (1) utility, (2) design, and (3) plant (Babcock and Morse, 2002).

Copyrights are used to protect the writing of an author from being copied. Copyrights are widely used for protection of computer software, in particular outside the United States of America (UNIDO, 2004).

Trade secrets, or confidential technological and commercial information, are the most important assets of many businesses. Trade secrets have no precise definition, but to be protected by courts it may be a formula, process, know-how, specifications or often business information (Babcock and Morse, 2002).

Trademarks are used to indicate the origin of goods or services and to distinguish them from others. They are very powerful tools. Some trademarks are registered, others are not (UNIDO, 2004).

In Thailand, new product and process development which satisfies some specific criteria, can be patentable. Either the inventor or the university can apply for the patent from the Department of Intellectual Property, Ministry of Commerce. However, there are very few new product and process developments that are patented because the process is more complex.

### 3.5.2    Using intellectual property
From our experiences in transferring the research findings or technology from the Centre to the private sector, it needs good technology transfer people. They should have clear discipline and the skill of negotiation to transfer the property rights and licenses to SME manufacturers. The intellectual property rights agreement should be established firstly and the agreement needs to be looked at

from both angles. The agreement should include royalty and royalty payment, measures to protect the patent licensor, measures to protect the patent licensee (Tanasugarn, 2000).

In practice, the royalty charged to the SME by the Centre is a very small amount, although the royalty is normally based on the valuation of the technology or research findings. This encourages the transfer of technology or invention to the private sectors.

## 3.6    Communication with clients

It is important that the technical staff have the manner for good coordination and communication with the client and expert. The communication takes place at three levels, as shown in the Fig. 3.1.

Formal communications are normally among the client, researcher and technical staff (ADCET, contractor) for any issue occurring during the project. ADCET also often has a senior specialist providing supervision to the project as well as to ensure that correct information has been transferred between the researcher and client.

Two case studies have been selected to illustrate the work and the achievements of the Centre.

## 3.7    Case Study 1 in ADCET related to food product development

**Problem identification:** One of the SMEs in the south of Thailand wanted to extend the shelf life of fresh Sator beans for at lease 2–3 months. Sator beans (*Parkia speciosa, Hassk*) are popular in the South-east Asia countries, particularly Indonesia, Malaysia, The Philippines, and Thailand. When this produce is off-season, the price goes up, while it is very cheap when it is in season. The target market includes local markets and nearby countries, as well as

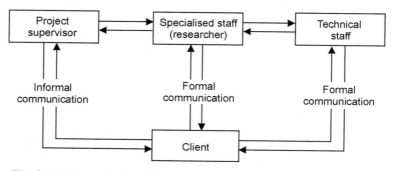

**Fig. 3.1**    Communication among the client, technical staff and researcher.
Source: Adapted from Babcock and Morse (2002).

America and Europe. It is of major importance to the company to extend the shelf life of the beans and therefore extend the time of marketing at high quality.

**Development of the project**: Firstly, the client consulted with the ADCET's technical staff to describe the problem and particularly the product idea. The technical staff identified an expert, who works on fruit and vegetable processing technology in the Faculty. The product concept and details were discussed between the client and the expert. When both agreed with the concept, the proposal including the budget was prepared.

**Product concept**: When the Sator beans are consumed, they should be fresh, crisp and green. Therefore, the product idea is that Sator beans should be packed in packaging with modified atmosphere conditions.

**Study design**: The mixed varieties of Sator beans were purchased from the same farm through out the experiments. The Sator beans were tested to determine the chemical properties (protein, fat, moisture, ash and fibre), the physical properties, especially crispness by texture analyser (TA-XT2) and colour by Munsell colour chart, and microbial count particularly aerobic and anaerobic microorganisms.

The Sator beans were graded according to the size and maturity and then soaked in salt solution (1%, 10 min) and dried in the cooled air. Then 100g Sator beans were placed in a foam tray ($8 \times 16$ cm$^2$) and packed in a flexible film bag (Nylon/LLDPE, $200 \times 240$ cm). The bag was filled with a gas mixture, either ($CO_2$ 20%, $N_2$ 77%, $O_2$ 3%) or ($CO_2$ 24%, $N_2$ 73%, $O_2$ 3%) and then sealed.

The whole package was kept at 2–5 °C. The chemical, physical and microbiological properties of the samples were analysed every 10 days. The sensory characteristics especially crispness and colour were also evaluated by using a nine-point hedonic scale.

**Conclusion** Overall the results showed that the shelf life of Sator beans packed in $CO_2$, 20, 24%, $N_2$ 77, 73% and $O_2$ 3% can be kept for 50 days at 2–5 °C without any growth of pathogenic bacteria and with acceptable colour and texture. But to the taste panel, the Sator beans, packed in ($CO_2$ 24%, $N_2$ 73% and $O_2$ 3%) were more acceptable than the other beans.

As the research was going on, the expert and the client worked together in order that the client could learn the technology as well as the product changes. At the end, the technology was transferred to the client through learning by doing. The academic expert also provided consultancy to the client till they could operate correctly by themselves. The intellectual property was also patented by the university and the client on agreement.

## 3.8 Case Study 2 in ADCET related to food process development

Fish cracker is a traditional popular snack for Thai people. It contains mainly carbohydrate and protein. These fish crackers are made by Muslim communities in Nara-thiwat province, the farthest southern province. Therefore, this product is called 'one district one product'. For the market, the product is popular for the local consumption and visitors and is also exported to Malaysia and Singapore. Snacking is increasing in popularity, so there is a strong push to increase production and improve the quality through improved processing technology.

**Problem identification**: The project was brought to the ADCET by the Department of Regional Industry. The problem was to improve the processing of fish crackers. ADCET set up a team composed of processing technologist, engineer and quality control. The weakness and strength of the steps in the process were analysed. Study of the process showed that the boiling and drying steps must be corrected (Fig. 3.2). The nutritive value and safety quality of the product would be improved.

**Development of the boiling project**: For the boiling step, the dough was formed into a cylindrical shape (60–80 cm length), then the dough was boiled in hot water (90 °C) for 40 min. The gelatinisation of the dough occurred when the temperature at the centre of the dough was 70 °C. The boiling caused nutritive loss and the dough absorbed more moisture.

**Boiling equipment design method**: The existing boiling kettle was modified to be a steamer (Fig. 3.3). The dough was steamed until the temperature at the centre was 70 °C, which took 30 min. The benefits of this modification were that the nutritive value of the product was maintained, the gelatinisation time was shortened, and the moisture content in the dough was reduced.

**Development of the drying project**: There are two steps of drying, which are: pre-drying the cooked dough and (2) drying the piece of fish cracker. Usually the cooked dough and piece of fish cracker were dried in the open air by

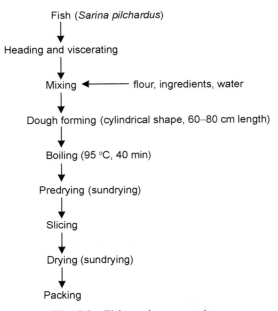

**Fig. 3.2**  Fish cracker processing.

**Fig. 3.3**   Modified steamer.

sun energy (Fig. 3.4). The existing process was highly weather dependent and caused an inferior quality product. Dust, flies, insects and mould contaminated the product when it took many days for drying. The fish cracker could not be produced during the rainy season.

**Drying design method**: (1) Sun dryer: The design concepts were based on material available locally and simple operation. The capacity of the sun dryer was 150 kg per lot of cooked dough or slices of fish cracker. The drying chamber

**Fig. 3.4**   Original sun drying of dough and fish cracker.

**Fig. 3.5**  Sun dryer.

size was $4 \times 10 \times 3$ m with plastic roof (2 mm thickness), while the wall of chamber was made of plastic net in order that the moisture could be easily removed. The whole dough piece or the slices of fish crackers were placed on bamboo trays (Fig. 3.5).

(2) Air dryer: The dryer was used during the rainy season. The dryer was composed of a drying chamber and a furnace for supplying heat. The drying

**Fig. 3.6**  Air dryer.

chamber size was $2 \times 3 \times 2$ m, with capacity 150 kg per lot of product. The trays were made of bamboo.

The furnace was made of ordinary bricks. Rice husk and small pieces of rubber wood were used as heat source. The cooled air was heated and brought into the chamber by two air blowing fans run by electric motors. The dryer was installed with chimneys and butterfly valves to control the air flow. The drying temperature was 60–70 °C (Fig. 3.6). The dough and fish cracker, dried by either sun dryer or air dryer, gave superior quality of product without contamination of dust, flies and mould.

**Conclusion:** The improved process was tested and then introduced to the producer. The fish cracker has smooth texture, and light colour. The steaming step for the dough reduces the nutrients loss and moisture content. The fish cracker, dried by either sun dryer or air dryer, will also have improved quality. The producer says that she can get very high price and the product is more acceptable to the consumers.

## 3.9    Problems met and lessons learnt

### 3.9.1    Problems
- When the ADCET was set up, it was hard to get the Faculty staff to join the activities. They preferred to work at their personal interests in the laboratory and to concentrate on teaching.
- Research culture of researchers, they want to spend more time carrying out a research project while the industry prefers to get the results within limited time.

### 3.9.2    Lessons learnt
- Most Faculty staff and technicians changed their attitude. They learned how to manage their research in order to compete against the time and how to communicate with the clients.
- The staff are willing to work in the community rather than in the laboratory. The research finding can be either transferred to commercialisation or patented. They work as a team and share their ideas on how to tackle the problem.

## 3.10    References

AGRICULTURE, FISHERIES AND FORESTRY – Australia 2002. *Australian Food Statistics* (Canberra, ABRAE).

BABCOCK, D.L. and MORSE, L.C. 2002. *Managing Engineering and Technology.* 3rd edn (Upper Saddle River, NJ: Prentice Hall).

BENNETT, D. 2002. *Innovative Technology Transfer Framework Linked to Trade for UNIDO Action* (Vienna: UNIDO).

GEE, R.E. 1993. Technology transfer effectiveness in university–industry cooperative research. *International Journal of Technology Management*, 8 (6–8): 652–668.

NONAKA, I. and TAKEUCHI, H. 1995. *The Knowledge Creating Company* (New York: Oxford University Press).

PRIGOGINE, I. 1980. *From Being to Becoming: Time and Complexity in the Physical Sciences* (San Francisco, CA: W.H. Freeman and Co.).

SUWANNAPORN, P. and SPEECE, M. 1998. Organization of new product development in Thailand Food Processing Industry. *International Food and Agribusiness Managemen Review*, 1 (2): 195–226.

TANASUGARN, L. 2000. *Guide on Patent License Agreement drafting: business to business.* (Thailand, Department of Intellectual Property, Ministry of Commerce).

UNIDO 2004. *Technology Transfer Operations, Inducing Agreement Formulation and Negotiation* (Vienna: UNIDO).

# 4

# The supermarket industry – private label brand development

Sam Adapa and Kumaresh Chakraborty, USA

*The supermarket chains are a major part of the food industry today and at least manage the controlled flow of new product introductions of both the national brands from the food companies and their own private brands. For consumer products, the initial step in food marketing of national brands is to procure shelf space in the supermarket, unless the company is selling to speciality stores. The private brands of the supermarkets are steadily increasing their share of the food market, often the processing is contracted to food processors but supermarkets also have their own manufacturing plants and of course food production in the supermarket such as baking, meat preparation, and delicatessen.*

*They are an important source of knowledge in the food industry, particularly of marketing. Their data capture and storage at the checkout is a major source of buying information, not only what foods are bought but also who is buying. Because of the spread of supermarkets, they have knowledge of buyer behaviour in countries, states and counties and individual cities and towns, down to the socio-economic groups. So they have extensive knowledge of consumers' buying behaviour in the present and how it has changed over the years. Already they have wide knowledge of storage and distribution. With further development of their manufacturing facilities, they will also increase their technological knowledge. So their major position in New Product Development (NPD) is evident at the present and can only increase in the future. They do not conduct basic research on processes and products but tend to use the available technological knowledge, so they do not have large R&D facilities like the multinational food companies.*

*Their product strategy is essentially market-led innovation and product improvement, but the development of their own manufacturing technology will also lead to technological innovation. Their product strategy for their private brands started as marketing low cost versions of the highest selling foods, but rapidly increased to developing the fresh products produced in the supermarket. Particularly with the introduction of bakeries, delicatessens and ready-to-eat meals, and the build-up of fruit and vegetables, the supermarket name became firmly attached to fresh foods. They rapidly became strong competition to the baking companies with their attractive, fresh loaves, rolls, buns and eventually cakes.*

*Today they often have several brands covering the range from the high-class foods to the everyday commodity foods. They are no longer aiming to achieve only the label of low cost foods but for quality and also for serving special needs such as nutritional foods. So their product strategy is wide and encompasses several aims. The more expensive products will give them improved margins, but they are also aiming for an enhancement of the shopping experience.*

*The chapter particularly relates to pages 50–59, 65–69, 96–117, in* Food Product Development *by Earle, Earle and Anderson.*

## 4.1    Introduction

The evolution of the supermarket industry has been phenomenal. In the beginning supermarkets limited their marketing efforts strictly to traditional retailing with attractive displays, clean appearance, ambience and sensible pricing. The advent of competition from among supermarkets as well as the dominance of multinational brands, resulted in the disappearance of pricing power and significant profit margins. Supermarkets in all the mature markets of Western Europe and the USA began revitalisation and product differentiation. They retained their loyal customer base by being innovative. They enhanced the quality and perception of their store brands/private label brands with a systematic focus. Most of the successful supermarkets have a challenging line of their own brands and products appealing to the consumers against the well established, aging multinational brands. This rejuvenation and revitalisation is bringing a new spirit to the supermarket industry for their ultimate survival and success. This chapter details these changes and strategies which are vital to the underpinning of the 21st-century industry. They continue to innovate and attract a broader base of consumers.

## 4.2    Development of store brands and brand equity

Most retailers attract customers to their stores to satisfy their needs in several ways such as type, variety or exclusive product or service offerings. Most supermarkets and retailers cater to a group of consumers in the locales where the stores are located, therefore the same retailers, when located elsewhere in different cities or states, may offer different product mixes.

#### 4.2.1   Private labels in the USA

Private labels have a long history, as far back as 100 years ago. Safeway Corporation, a supermarket chain from Pleasanton, California, recently celebrated the 100 years of their *Lucerne* dairy brand (website: Highbeam.com). In the 1970s, supermarkets in various regions of the United States offered their own products under the name 'generic' in white and black label at a discount price to attract consumers who otherwise might not buy name brand products at a higher price. Consumers had an established image of the generic brands and understood the product quality was either questionable or not quite satisfactory.

Consumer perceptions have changed since then. The supermarkets have upgraded the quality of their generic products by marketing them under the 'private-label brand' or 'store-brand' or 'controlled label brand' categories. The store-brand products are available only from that retailer.

Consumers these days have many choices. They can purchase nationally known brands (e.g., Campbell's soup, Dole pineapple or Kellogg's corn flakes), licensed brands (e.g., Crisco shortening), and private-label store-brands (e.g., Safeway, Kroger, Supervalu) to satisfy their needs and wants, within their budget.

In recent years, supermarkets have increased their size considerably due to the consolidation of companies and the diverse merchandise offerings. At the same time stores have become more sophisticated, and have reached a new height in their identity. Private labels have many obvious benefits to both retailers and consumers. As mentioned earlier, the exclusivity of a stronger store-brand improves consumer loyalty to the store, enhances store image and price discounts of 10–20% over comparable US national brands and 25% lower prices in Europe (Levy and Weitz, 2004). Also it has lower promotional and display costs, more control of the manufacturing, quality and distribution. This is resulting in higher gross margins to supermarket chains.

There are limitations to store-brands. Some consumers often still recognise them as low-priced alternatives rather than for their quality. Retailers are investing significantly in the product design, consumer awareness and image. Additionally, the retailers are beefing up training of staff in sales and marketing to compete against the national brands.

Store-brand provides 'value' to consumer and retailer, and thus influences buyers' decisions to make a purchase. The 'value' the brand offers is called 'brand equity'. Supermarkets are building brand equity of their store-brand products through creation of awareness by placing their products next to the national brands on the shelves, showing unit price comparison in order to get buyers' attention. The supermarket managers strongly believe that price is the single most important factor that influences consumers in purchase decisions. Additional awareness is brought through flyers, newspapers, magazines, store/ club card discounts and free samples at the point of purchase.

### 4.2.2    The growth of private labels internationally

Throughout Europe and America, private label is winning the loyalty of more and more consumers every day. According to a research study conducted for Private Label Manufacturers Association (PLMA) by MORI, a public opinion consultancy firm, private label has achieved unprecedented acceptance with shoppers, who say that they are now more aware of private label and plan to increase their purchases. The largest increases in the percentage of consumers who are more aware of private label were in the United Kingdom, Spain and France. Increased awareness is leading to more purchasing.

ACNeilsen (2007) reports in PLMA's International Private Label Yearbook sales of private label across Europe increase to record levels (Fig. 4.1). Private label is increasing its market share in Europe's largest and most mature retail markets as well as in markets with historically low private label penetration. For the first time, market share for private label has surpassed 40% in four countries – United Kingdom, Germany, Belgium and Switzerland. The market share for retailer brands has reached an all time high in two more of Europe's important retail markets, France and Spain, where private label now accounts for one in every three products sold.

Demand for private brand foods in Japan is strong and growing. Faced with growing competition for sales to increasingly discriminating consumers, Japanese food traders at all levels are seriously looking into privately labelled products as a solution to their declining profit margin. The Japanese retailing industry predicts a rapid rise in private brand merchandising and prospects are good for US food exports to Japan that can contribute to private brand sales. The vast majority of private brand imports are off-the-shelf products that can be sold in Japan with minimal adaptation to the Japanese market. Beyond off-

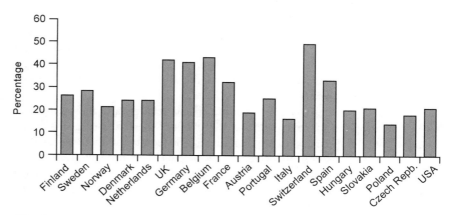

**Fig. 4.1**    Private label market share by country (unit volume). Source: PLMA World Private Label, 2007 (www.plmainternational.com).

the-shelf products requires producers to work with their Japanese buyer to create a product suitable to Japanese tastes. Medium-sized supermarkets rely on large wholesalers or joint buying cooperatives to provide them with imports. Three of the largest supermarket chains, Daiei, Ito-Yokado and Jusco have imported private brands for years. While private brands in Japan make up a smaller share of the total market than in the United States, they are a growing share of total sales. The percentage of private brands in total food sales is 5–7% for supermarkets, 2–3% for convenience stores and 6–7% for food service establishments. Although national brands continue to have prestigious image among consumers, retailers and distributors are more inclined toward private brands. The profit margins on national brands are about 15%, retailers expect roughly double that margin from private brands (AgExporter, 1996). At the same time, private brand retail prices run from 20% to 50% lower than the national brands.

### 4.2.3   Supermarkets in the USA

As with all retailing, the supermarket industry went through major transformation in the last thirty years. The change in the population demographics, urbanisation, and ethnic diversity in the United States brought competitive pressures on retailing. Supermarkets became highly innovative in their product offerings and services by including in-store banking, 24/7 store hours, membership cards, wider selection of produce, meats, floral, and pharmacy departments. Inflation and competitive pressures resulted in smaller profit margins. Productivity improvements through computerisation and cost cuttings offset the added costs of offering consumer choices, inventory control and distribution.

The major supermarkets in USA and their ranking in sales in US$ billions are shown in Table 4.1. Successful supermarket chains advanced their creativity by transforming store brands from the 'generic' or 'no-frills' brand into attractive value-added propositions to consumers. While the national brands such as Kraft (General Foods), Quaker, Ralston and ConAgra lost their innovative edge through consolidation and repetitive cost cuttings, the supermarket brands (private label) have seen a resurgence of interest and value in the consumers' shopping habits. Many national brands such as Borden, Chef Pierre, Keebler, Beatrice, and General Foods have been disappearing rapidly from the effects of acquisition and consolidation. It is conceivable that half of all the current national leadership brands in the US could disappear in the next 20 years. The consumer will likely reject antiquated marketing techniques without innovation. Mere slogans 'it's a Coke', are not going to be enough to entice consumers. Only brands with identity of innovation and leadership are expected to prevail on the supermarket shelves.

**Table 4.1**    Private brands in US supermarkets. The top supermarkets in the USA and their sales rank in US$ billions

| Rank | Company | Stores | Store brands | Annual food sales (US$ billions) |
|---|---|---|---|---|
| 1 | Wal-Mart and Sam's Club | 3850 | Great Value<br>Sam's Choice<br>Sam's<br>Equate<br>Specialty<br>Ol' Roy | 355.4 |
| 2 | Kroger | 2500 | Kroger<br>Ralph's<br>Smith's<br>King Soopers<br>Fry's<br>Payless<br>Food4Less<br>Foodco<br>Cost Cutter<br>Private Selection | 66.1 |
| 3 | Costco Wholesale | 422 | Kirkland<br>Kirkland Signature | 62.4 |
| 4 | Safeway | 1760 | Safeway Select<br>O Organic<br>Lucerne | 40.6 |
| 5 | SuperValu and Albertson's | 1750 | Acme<br>Albertson's<br>Bigg's<br>Bristol Farms<br>Cub Foods<br>Farm Fresh<br>Hornbacher's<br>Jewel-Osco<br>Lucky<br>Save A Lot<br>Scott's<br>Shaw's Star<br>Shop'n Save Shoppers<br>Sunflower<br>Osco<br>Savon<br>SuperValue<br>TLC<br>HomeBest<br>Shoppers Value<br>Favorite<br>Supervalu<br>Guaranteed | 36 |

**Table 4.1**    (*Continued*)

| Rank | Company | Stores | Store brands | Annual food sales (US$ billions) |
|------|---------|--------|--------------|----------------------------------|
| 6 | Publix | 880 | Publix<br>Publix Premium<br>Publix Upromise<br>Gift Card<br>Publix Apron's<br>Publix Green Tag | 21 |

Information extracted from multiple sources: Progressive Grocer Annual Publication, www.Yahoo.com/finance, www.msnmoney.com and individual company websites and annual reports.

## 4.3    Product Development Process for private brands

### 4.3.1    Stages and special activities

To enable a high degree of innovation and to combat the competitive advantages of the national brands, the supermarkets need to engage in the following ideal Product Developmental Process steps.

1.  Global market research and database search.
2.  Market research – focus groups.
3.  Product identification and positioning.
4.  Product concept and prototype development.
5.  Product evaluation, costing and approval.
6.  Formulation, label and package development.
7.  Scale-up testing and shelf life evaluation.
8.  Specifications for formula ingredients, finished products, package, process, QA specifications.
9.  New product commercialisation and distribution.

*Brand maintenance and new line extension products*
The following steps can be omitted: 1 Global market research and 2 Market research, but pursue steps 3–9. In general, the line extension of new products take about half the time needed as compared to developing novel or innovative new products. *Innovative technologies* require a feasibility study and involve complexity in formulation, ingredients, processing and establishing shelf life.

New products can be produced either in the supermarket's own manufacturing facilities or by co-packers. Line extensions provide self-manufacturing plants the cost-effective means of additional product volumes. Self-manufacturing also expands the new product opportunities. Co-packers are often unwilling to bear the brunt of capital costs involved in new processing and packaging equipment. Proprietary customisation is used by the innovative supermarket chains to provide attractive value-added product positioning.

### 4.3.2    Comparison with classical New Product Development processes

The traditional national brand companies historically have large structured, specialised research and development departments. They thrive on new product innovation and existing product line extensions. New Product Development is pivotal to market share growth, revenues and the ultimate survival of the company. Successful food companies historically budgeted about 1% pre-tax sales revenue for research and development expenditures to maintain the new product pipeline and growth.

During the last two decades, leading companies by-passed this strategic advantage by acquiring brands and market share in company take-overs, sacrificing the large research and development staff and programmes. It is evident the national brands are rapidly losing ground on innovation based on the supermarket shelf presence and tradeshow offerings. In the mean time, the supermarket chains are advancing their innovative designs in package and product. They are vigorously pursuing goals with thinly staffed R&D and using supply chain partnerships effectively.

By nature of their broad scope, supermarkets active in New Product Development usually employ veteran generalists in product development. Surprisingly, they can churn out a stream of new products in a 3–4 month time span. There is, however, a broad inconsistency among the supermarket leaders on how they develop and commercialise new products. Companies like Whole Foods, Costco, Trader Joe's, Target, and the upcoming Tesco are known to depend heavily on a small committee of tasters, especially product buyers. Wal-Mart and Sam's Club have limited store brand presence with only family/budget tier products. The more successful store brands leaders, such as Publix, Kroger, Safeway, and HEB, are known to use a broader team of food technology, marketing, sourcing and quality assurance professionals. The degree of evaluation and approval is contingent on the specific brand tier and revenue that the product is expected to generate. Not all products can go through this strict regimen and controlled testing, due to a large portfolio of new products in any given year. Safeway uses some external technical resources for consumer research, focus groups, and sensory analysis for major product offerings under the premium brands. Supermarkets apply business models differently, consistent with the product pipeline, available technical resources and degree of sophistication of local consumers for new products.

Most of the national brands' companies have their own dedicated multi-manufacturing plants to customise proprietary products. In the supermarket industry this varies widely. While the large chains such as Wal-Mart, Sam's Club and Costco do not have the extended flexibility of captive manufacturing plants, other leaders like Kroger and Safeway have the option of self-manufacturing as well as contracted external co-packers. At Safeway, almost 70% of self-branded items are co-packed. Safeway utilises its 31 plants in the US and Canada in dairy, grocery, bakery, beverages, pet foods and frozen desserts in addition to 900 or so co-packers for food and non-food items.

A smaller chain like Trader Joe's has become highly creative in utilising the strengths of aspiring entrepreneurs to create products under Trader Joe's brand. This provides an easy way through the otherwise difficult roadblocks most retailers have. Target is known to have a highly demanding buyer's approval/ rejection process for their Orchard Farms brand. Super Target has ambitious plans to improve the quality and expand the stores geographically.

## 4.4 New Product Development leadership culture

During the last two decades, there was a thrust for greater returns on invested capital (ROI) across the USA. Co-packing/contract manufacturing seemed to be a norm for an economical entry. This routinely resulted in the outsourcing of the manufacturing function to maximise the ROI by simply concentrating on marketing. Increasingly competitive environment and smaller profit margins made this strategy widely popular. Food industry consolidations and greater reluctance of retailers to commit capital to manufacturing chores, are causing further drain on available manufacturing capacity for co-packing. This also runs counter to paradigms of innovation. Co-packers relish the 'cookie cutter' approach resulting in a guarantee of larger business volumes. Innovative products often have questionable future success in advancing the interests of co-packers.

Supermarkets who exclusively deal with co-packers vary widely in the contractual arrangements. Contracts are generally 1–5 years with renegotiation clauses tied to certain finished product volumes and pricing. This offers a very attractive ROI in the short term; however, it takes away the versatility in the longer-term business planning. Product buyers in sourcing/purchasing and legal departments generally negotiate the contracts.

During the last five years Safeway Corporation reinvented itself. They conducted extensive reviews of their numerous brands. Based on consumer recognition and brand strategy, Safeway focused the resources to ten power brands in 2006. Safeway SELECT for Premium line, Eating Right in health and wellness, Primo Taglio in the delicatessen, Safeway across broad centre of store product lines, O Organic, Lucerne in dairy, Basic Red for no-frills economy, Go to Cola, Rancher's Reserve in meats and Priority in pet food brands are in current use (Safeway public domain).

Safeway Corporation re-energised the emphasis of fine balancing of brand maintenance – line extension for near-term gains, and innovative new products for long-term customer loyalty and business growth. Safeway is known to target its premium private brands to far exceed the quality in comparison to national brands and other competitive store brands. This is a great evolution in the private-label brand business model.

While the national brands are refocusing to maximise the efforts in innovation and speed-to-market agenda, the private-label store-brands are aiming to broaden an attractive portfolio of products with consumer appeal.

### 4.4.1 Suggested Product Development Processes

*Suggested process for New Product Development with contract manufacturers*
The supermarket and the co-packer have to work together in the PD Process but each group is responsible for the activities in different stages. The co-packer is

responsible for the development of the product and the process, and the supermarket for the product idea, plant audits, product evaluation and commercialisation approval.

| Function | Responsible party |
|----------|-------------------|
| Product ideation | Supermarket and co-packer interaction |
| Prototype concepts and improvement | Co-packer |
| Finished product, costing | Co-packer |
| Label specifications | Co-packer |
| Plant audit, approval | Supermarket QA |
| Product evaluation | Supermarket technical staff and marketing |
| Label reviews | Co-packer and supermarket |
| Commercialisation approvals | Supermarket teams |
| Product quality guarantee | Co-packer |
| Customer complaints | Co-packer, supermarket customer service |

*Suggested process for New Product Development for self-manufacturing option*
Similar to the PD Process for the co-packer, by virtue of self-manufacturing, the technical group and manufacturing operations generally have the added flexibility of designing the product with more consumer-desired attributes through formulation or process refinement. This kind of customisation may be either limited to premium products or a new product ascertained by sensory profile and cost parameters. Once the product testing passes through pilot plant scale-up evaluation, the formulation and process requisites are passed to corporate QA and plant QA for on-going product quality maintenance through plant audits, compliance of the State and Federal Regulatory requirements. Most companies in the supermarket industry conduct competitive product analysis of their store brands against the best selling national brands (NB) to keep their product quality and cost structure competitive.

Process development follows the regulatory and product safety for thermal process studies, acidification of product, packaging, distribution, warehousing requirements and shelf life criteria. It is imperative that product safety is primary and product quality and costs are secondary in consideration.

As a resource, an extensive database of ingredient suppliers with raw material specifications are maintained for future operational needs. This information is routinely reviewed when there is a need for an alternate ingredient supplier.

## 4.5   Future trends and evolution in food retailing

Most US supermarket chains are mobilising efforts into superstores like the Super Wal-Mart as well as into small neighbourhood stores like 7-Eleven and Trader Joe's to satisfy a broad variety of consumer needs.

Superstores have a broad array of merchandise and services on offer. Wal-Mart, Marks & Spencer, and Tesco are customising the store décor and products to suit the local consumer tastes. Globally, joint venture partnerships are becoming the mainstay of international staying power and survival. There are new developments in clean energy, recyclable green packaging, RFID for inventory control and GPS tracking for effective distribution. RFID is Radio Frequency Identification tagging Device and unlike the UPC barcodes, this provides extensive data on every piece of product tagged with automated data capture, real-time remote monitoring, inventory tracking, theft and diversion detection, demand planning and more.

Global sourcing of ingredients keeps the ingredient costs and availability under control, thus minimising the risk of inflation in Western countries. Global sourcing is assuring an abundance of supplies for the newly industrialised countries of Asia, Europe, Middle East and South America.

The new technologies and their implementation are pivotal to what global supermarkets and retailing could look like twenty years from now. According to the world future society, emergence of mega-malls will meet broader needs of consumers for one-stop shopping. To attract greater consumer traffic, Wal-Mart is seriously considering health clinics in the Super Wal-Mart locations.

In the USA, store brands are on the verge of Europe-style popularity. The private label manufacturers, retailers and industry professionals are trying to drive private label to the next level. Private-label unit market share is at nearly 21%, many retailers have set goals of 25%, 30% or more for their brands (website: press update 2007 plma.com). The private-label brands in the USA have risen from $39 to $65 billion, as shown in Fig. 4.2.

Some 41% of Americans now describe themselves as frequent store-brand purchasers, up from 36% just five years ago, and a major change from 15 years ago when it was 12%. Nearly one in three consumers say that they are more likely now than a year ago to purchase private label (plma.com show info 2007). Almost seven out of ten agreed that the private-label products they buy are as good, if not better, than their national brand counterparts, a significant increase from five years ago when a little more than half of the respondents agreed. Private labels' greatest appeal is among consumers in the lowest income range; however, consumers at the highest and middle-income levels are significantly more likely to increase their purchase of private label in the upcoming year than those in the lowest income range. The supermarket industry will continue to

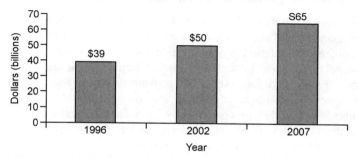

**Fig. 4.2**  US private-label industry growth. Source: *Supermarket News* 2007 (www.plma.com)

consolidate, making private label products an even stronger business than it is today (highbeam.com website).

Ipsos MORI, an international consumer polling organisation, UK, surveyed 1,017 people in a representative sample across the USA of consumers who are responsible for their households' main grocery shopping, including food, household cleaning products, personal care, and health and beauty items. The survey found that the popularity of grocery store brands has a 'halo' effect on non-grocery private-label products and offered proof that private-label products' level of acceptance in the grocery channel is making its presence felt in categories and channels far from food stores. About 20% of consumers reported that they already frequently buy private-label health and beauty products, home office products, household products, and home improvement products irrespective of the channel of trade in which they are sold. Some 75% said overall satisfaction with non-grocery stores' private-label products in the past is likely to encourage them to buy a larger number of the products in the year ahead. Nearly 67% said new and innovative store brands and a greater variety of store-brand products would spur them to purchase even more. While private label is gaining strength in grocery stores, profound changes are taking place in the way consumers shop in those stores. The frequency of regular store visits continues to decline and food dollars are more widely distributed over different types of stores than ever (website: press update 2007 plma.com).

The role of mega-retailers and specialty chains is growing. While 80% of consumers acknowledge that they regularly shop at supermarkets, a combined 45% of consumers said they also shop at supercentres and warehouse clubs. When making trips, one in six consumers said Wal-Mart is where they do their main grocery shopping, equal to the total of the next three chains. The penetration is growing for specialty food stores, such as Whole Foods and Trader Joe's. These changes are beneficial to private label. The majority of consumers shopping at multiple chains are looking for the best value and lowest price, so the need for the US retailers to build a unique image with consumers has never been greater (website: press update 2007 plma.com).

## 4.6    Pursuit of global food retailing strategies

A global research study by ACNielsen showed that retailer brands are growing to a dominant market share in the fast-growing fresh departments. The study, which covered 45 national and regional markets, found nine out of ten top private-label categories, in terms of market share, are in the fresh product categories. Retailers are focused on their consumers' needs, and some have even created specially designed in-store areas for fresh and prepared foods. Consumers can quickly and conveniently shop for the meal components they need for their families (PLMA e-scanner, website: plmainternational.com).

The pace of new product introductions is growing sharply, according to a new study by Mintel Global New Products Database. During 2006 about 182,000 new SKUs (stock keeping units) were introduced on a global basis, almost

105,000 of them food and drink items, about 300 products per day. This rate of product introductions is about 17% higher than in 2005, and more than double the growth rate between the years 2004 and 2005 (PLMA e-scanner 2007, website: plmainternational.com).

## 4.7    Problems met and lessons learnt

The current sales of store brands in the US are estimated to be over $65 billion annually. In the developed Western economies, especially in North America, food consumption has reached a plateau. In the 1970s what was offered as orange juice, now comes in low acid, regular, high pulp, no pulp, from concentrate, not from concentrate, fortified with calcium, omega-3, and heart healthy nutrients, frozen, chilled, and shelf stable packaging forms. Supermarkets have become a classical case of burgeoning variety of choices for consumers. This is an added burden against cost controls, inventory management and distribution, magnifying the overall business complexity. There is added pressure on the limited supermarket shelf space due to the proliferation of new products and flavour options. Some growing problems are:

- Supermarkets are trying to reach out to the growing share of global consumers and the under- represented ethnic consumers in the domestic markets.
- The global warming trade-offs and clean energy needs are adding further strain to the supermarket industry.
- Supermarkets face a critical balancing of the clout of established national brands against their own private-label brand promotions. Alienation of national brands is seen to be mutually detrimental to the economic health of the nation.
- Growing pressures of regulatory food safety laws from the local county, state, federal and international agencies add to the burdens of supermarket business.
- Successful supermarket chains have learned to have captive manufacturing as well as the flexibility of co-packer/contract manufacturing to maintain attractive profit margins and consumer choices.
- Above all this, consumers are known to be highly sensitive to any inflationary price increases on their staples. They are quick to abandon their loyalty and move to less-expensive choices in club stores. Cost controls and price controls play a significant role.
- Enticing consumer interest with a variety of choices, services and ambience is great, however, and the No Frills Club Store and Dollar Store businesses are cannibalising the existence of conventional supermarkets, as we have known them.

The future of the supermarket industry lies in further consolidation as seen in the recent past with other industries such as ingredient suppliers and multinational companies (website: highbeam.com). The ultimate survival and success of supermarkets would depend on leveraging effectively all the above factors from new products to cost controls.

## 4.8    References and bibliography

ACNielsen 2007. In Private Label Manufacturers Association. www.plmainternational.com.

AgExporter 1996. April–May. Demand for private brand foods in Japan is strong and growing. USDA Foreign Agricultural Services (FAS).

R. Blackwell, P. Miniard and J. Engel 2006. *Consumer Behaviour*. 10th edn. Thomson South-Western, The Thompson Corporation. Mason, Ohio, USA.

*Dairy Field*. 2004. April. Stagnito Communications, Inc. USA.

*Food Product Design, Research and Development*. 1994. Virgo Publishing. Phoenix, Arizona, USA.

J. Fraser 2004. *Private Label: A Challenging Growth Market*. Industrial Directions Inc. http://industrydirections.com.

Ipsos MORI 2007. UK Survey conducted for Private Label Manufacturers Association. www.plma.com.

M. Levy and B. Weitz 2004. *Retailing Management*, 5th edn. McGraw-Hill Irwin Inc. New York, USA.

Mintel Global New Products Database 2006. Market Research. International E-scanner, www.plmainternational.com.

MORI (Market and Openion Research International Ltd, UK) 2007. Research Study conducted for Private Label Manufacturers Association International Council. www.plmainternational.com.

*New Product Magazine*, 2006. March. Stagnito Communications, Inc. USA.

PLMA (Private Label Manufacturers Association). 2005. New PLMA Research Finds Gain in 'Frequent store brand shoppers. www.plma.com.

PLMAs 2005. International Private Label Yearbook. 2005. *A Statistical Guide to Market Share Trends*. Private Label Manufacturers Association. International Council, Amsterdam, Netherlands.

PLMAs 2006. Private Label Yearbook. 2006. *A Statistical Guide To Today's Store Brands*. Private Label Manufacturers Association. New York, USA.

PLMAs World of Private Label, Private Label Today. 2007. *Growth and Success*. www.plmainternational.com.

PLMAs 2007. Market update. www.plma.com.

PLMAs 2007. Private Label: Getting Bigger Every Day. www.plma.com.

PLMA 2007. Press update. 'Star Power' of Store Brands. www.plma.com.

PLMA 2007. Annual Meeting, Orlando, Florida, USA. www.plma.com.

PLMA e-Scanner Internationl. 2007. March. www.plmainternational.com.

*Supermarket News*. 2002. April. www.highbeam.com.

*Supermarket News*. 2004. November. www.highbeam.com.

*Reference websites*:
www.bloomberg.com
www.brandweek.com
www.msnmoney.com
www.progressivegrocer.com
www.supermarketnews.com
www.yahoo.com/Finance

# 5

# Rural agroenterprise – cassava development in Latin America

**Bill Edwardson and Rupert Best, Canada**

*This chapter shows how New Product Development (NPD) can be applied in aid programmes, that it can be an effective technique and that it might reduce failures. In many countries food science and technology research and development is often financed and led by government, especially with regard to strategic agrifood commodities. Apart from the multinationals, a high proportion of knowledge development is in government organisations, universities and international organisations. In developing countries, resources have been provided through bilateral and multilateral aid programmes for several decades to increase food supplies through agricultural improvements and most of this research and development work has been carried out in public institutions.*

*More recently a focus on rural agro-industry development has been seeking to identify new products that will improve incomes for small farmers, rather than solely addressing crop and livestock production and productivity. This has meant increasing attention to identifying markets and how they can be accessed, and how viable enterprises can be established that will benefit poor farmers and rural populations. The question is 'can the NPD developed in commercial companies be adapted to this type of agrifood product development?'*

*Developing a new product strategy in a company is very focussed and the final decisions are made by a small group of people inside the company. It is based on the company's business strategy and their future growth expectations. Is it possible to produce a focussed product strategy for a complex group of intertwining organisations with a general aim such as to help increase the small farmers' incomes? Are the product strategy and the Product Development Process (PD Process) adaptable enough to be used in countrywide development?*

*Developing a product strategy in an aid programme context needs coopera-tion between agencies and the development of a focussed outcome. While this is similar to the development of a product strategy in a company, in a developing country situation there is firstly a need to identify a small coordinated team from various agencies to lead the process. This team has to identify the product portfolio, the product development strategy and the product development pro-gramme and get agreement from the various partners involved. In addition, they have to deal often with the issues of establishing a viable new enterprise in a rural area with deficient business development services and infrastructure. This is not easy!*

*In recent years, there has been a move to follow an NPD path because of the many failures of aid programmes and R&D investments in producing products with good commercial prospects. This case describes the evolution and experience of a product development strategy in a developing country as a means to improve incomes for poor farmers producing a strategic crop, cassava.*

*The chapter particularly relates to pages 46–93, 153–157, 319–332 in* Food Product Development *by Earle, Earle and Anderson.*

## 5.1    Introduction

This case study describes the evolution of applied research approaches that incorporated product development strategy in pro-poor rural development projects, which were aimed at reducing poverty in cassava farming communities in a developing country in Latin America. The project took place in Colombia over a span of 12 years in a continually changing political, economic and social environment. The case illustrates the challenges of developing appropriate com-mercially oriented approaches in a public sector research and rural development context. It underlines that farmers must participate in the value chains and that technological innovations must be tailored to specific market needs if the farmers are to obtain improved and stable incomes for their crop and livestock raw materials.

## 5.2    Background

This case study begins in the early 1980s, at a time when typical agricultural development projects, funded by foreign donors, in poor countries focussed on technical improvements to productivity as a means of increasing food supplies for farmers and local populations. The advances of the Green Revolution led to major increases in production of staple grains, rice, maize and wheat, as new varieties identified by agricultural researchers in both developed and developing countries were adopted. However, not all poor farmers were able to benefit from these advances due to high inputs of fertiliser, the herbicides and pesticides required, inappropriate land and soil resources and local market preferences for traditional varieties over the new varieties. Additionally, many farmers depended on other crops for their livelihoods, so these developments did not benefit them. The Consultative Group for International Agriculture Research (CGIAR), whose Secretariat is hosted at the World Bank, had set up a number of international agricultural research centres (IARCs) in both developed and developing countries to tackle these issues for staple grains as well as other crops which were of importance to poor farmers. One of these institutions, the International Centre for Tropical Agriculture (in Spanish, Centro Internacional de Agricultura Tropical, CIAT), located in Cali, Colombia had a special programme on cassava, which is an important root crop for small farmers with poor soils across South America, Africa and Asia. Much of the programme's initial research was carried out in Colombia, with field-testing in other regions of Latin America and also at sites in Africa and Asia, in collaboration with national agricultural research and extension institutions. This research aimed to identify new varieties of the crop and associated agronomic practices, which could provide higher yields under the conditions typical of small farms in the regions.

## 5.3    Cassava development in Colombia

In the 1980s small farmers of the north coast of Colombia obtained 40% of their cropping income from cassava (Gottret and Ospina, 2004). The region produced 35% of the country's crop. Owing to the rapid deterioration of the cassava roots, farmers had limited marketing outlets; losses were high, marketing was risky and prices fluctuated widely. During the 1980s, Colombia followed an import-substitution economic model, where domestic production was protected by import taxes. In particular, imported cereals for animal feed industries and wheat for bread production were expensive, so opportunities for an alternative carbohydrate source such as domestically produced cassava were attractive.

### 5.3.1    New cassava varieties developed

The Government's Integrated Rural Development Programme (in Spanish, Desarrollo Rural Integrado, DRI), with strong external donor support, was present in many rural areas, providing an integrated support package to farmers including research, technical assistance, marketing, organisation and credit, among others. Farmers were encouraged to increase production of their crops. For some crop and livestock products, including cassava, this led to depressed prices, credit default and greater poverty for farmers, the opposite of what DRI had planned. In this context, CIAT evolved its strategy to identify alternative markets for cassava, as a means for farmers to maintain and increase production of cassava (Gottret and Ospina, 2004).

At CIAT, as for all the other crop-focussed IARCs, emphasis was on plant breeding, agronomy and associated productivity issues, but early socio-economic investigations identified the need to address the farmers' need for income, as well as food, and the requirement to ensure that new varieties would be acceptable to farm families as well as buyers and consumers in the market-place. Without this assurance, farmers were reluctant to adopt new varieties and were content to maintain their traditional practices. These conclusions, together with the marketing difficulties being faced by producers and DRI's search for alternatives to reduce credit defaults, led to a realignment of CIAT's research approach on cassava improvement to look at the whole production to consumption system, in order to identify key characteristics of new varieties, which would fit the system, as well as other bottlenecks or opportunities in the system, which would enhance cassava farmers' income and livelihoods. This led to the integration of R&D on post harvest handling, processing and marketing issues.

This development at CIAT was a pioneering effort to include post production or off-farm activities in agricultural research programmes aimed at improving incomes and livelihoods of poor rural families. At around the same time, the funding agency – The International Development Research Centre of Canada (IDRC), which had a Post Production Systems Programme in its Agriculture, Food and Nutrition Sciences Division since the early 1970s – was actively promoting attention to post production issues to improve food security and alleviate poverty in rural areas of developing countries. The IDRC believed that focussing research effort only on production issues would not be sufficient to bring about any significant socio-economic development for the poor in developing countries. Increasing yields could result in gluts in markets, and even lower returns for marginal farmers. Even for important major crops such as rice in Asia, significant bottlenecks in handling and storage and product quality issues had to be addressed, as new varieties with higher yields came on stream. For crops of lesser importance such as cassava, sorghum, legumes and vegetables, typically farmed by poor families in resource-poor areas, attention to post production issues to reduce food losses and generate much needed income was crucial and became a priority for IDRC's support at the time. IDRC promoted attention to income generation from small farm produce as the objective of its funding, and moved away from attention solely on production levels and food supply, thereby bringing more commercial, value adding and market factors into play in the projects it funded (Forrest et al., 1979).

### 5.3.2   Post production development

CIAT's activities in post production in the cassava sector provided an ideal opportunity for IDRC to support this important shift in agricultural research and to assist in the development of methodologies, training materials and publications based on their experience, which an international institution such as CIAT had the capacity to produce. Clearly incorporating work on product quality and market characteristics, production to consumption systems (now called value chains or supply chains), value-added, enterprise development, post harvest and processing technology as well as economics implied a number of new professional skills and resources, some of which were available at the international institution. CIAT also had the convening power to bring together other local institutions to evolve the multidisciplinary teamwork that was required. At this stage there were few other institutions with this capacity.

## 5.4    Product development strategy for cassava

### 5.4.1    Dried cassava chips for animal feeds

CIAT began its work on value-adding by focussing on the market for animal feeds in Colombia. The animal feeds industry in the late 1970s and early 1980s depended on imports of sorghum as the energy component in its formulations. CIAT economists established that if dried cassava chips could be produced in rural areas at more competitive prices and with acceptable quality, this would be a very appropriate enterprise for rural producers. There was a precedent in Asia where rural production of dried cassava chips in Thailand had grown to be a significant sector for export animal feed markets. CIAT engineers in collaboration with local university staff and students, the Colombian government's Integrated Rural Development Program (DRI) and cassava farmers' groups evaluated and adapted Thai cassava chipping equipment both at the laboratory and in the field (operated by farmers) to find the technology which gave the best throughput and most efficient drying geometry for open sun drying on concrete floors. They worked with farmers' groups in the north of Colombia who were highly dependent on cassava as their main crop as it was well adapted to the hot and dry climate and poor soils prevailing in the region. Owing to the perishability of the freshly harvested product, farmers were often limited to home use or to selling in local markets that provided very little income for them. More lucrative prices in the urban centres fluctuated widely and quality requirements were high. They did not see any value to change to new higher yielding varieties and they kept their roots in the ground until needed and when local market prices were too low in the peak season. As the production zones of these highly perishable roots were far from major urban centres, the opportunity to produce dried chips was a great incentive for farmers to increase production and adopt the higher yielding varieties being developed by plant breeders at CIAT.

To take advantage of this new market opportunity, farmers' groups were encouraged to form cooperatives or other types of associations as a means to organise themselves for testing, training and subsequently supplying, processing and marketing the cassava chips.

The chips were produced, tested by feed mills in their formulations and orders were made to farmers' organisations to produce chips under contracts. CIAT in collaboration with local institutions undertook research and provided training and technical support for cassava production, processing and drying, while DRI led the training and support to the farmers' groups for organisation and management of the business processes, loans for equipment and infrastructure.

By 1993 the number of rural processing plants grew to 101 managed by small farmer cooperatives and 37 operated by private individuals. Dried chip production was 35,000 tonnes that utilised 10% of the cassava produced in the region. Prices of fresh roots increased by 3.5% per year from 1983–90, area planted to cassava increased by 7% per year and new higher yielding varieties were grown by 80% of farmers in the region. The presence of institutions promoting cassava production and the access to drying plants each played a role in the adoption of modern varieties. Dried cassava chips became a valuable commodity in the Colombian animal feed industry. This led to reduction of poverty levels in the farming families of the region, through creating an alternative income-generating activity from selling roots, creating employment and reducing production costs through improved cassava production technology (Gottret and Ospina, 2004).

### 5.4.2    Diversification: cassava flour

As this market and value-added product focus was evolving, CIAT were conscious that diversification of markets would be an essential component of the strategy, so that farmers could reduce the risks from changes in prices or competing products in the animal feed market. In addition there was a need to reduce the risk of a glut in production as more and more farmers' groups and other enterprises entered into cassava chip production. At this time markets for consumer food products were considered. The model of producing an ingredient for the animal feed industry directly in small rural processing plants managed and operated by farmers' groups, which had been successful, was considered as a desirable goal for the food industry strategy.

The growing urbanisation of Colombia was at the same time leading to a rapid expansion of the market for breads and bakery products, and to a decline in the urban markets for fresh cassava (as a starchy staple) due to its perishability and relatively high prices when compared to other starch sources, including breads. This was leading to rapidly increasing imports of wheat and demands for foreign exchange. At the research level in Colombia and elsewhere many studies on 'composite flours' had shown that other materials including cassava starch and flour could substitute wheat and provide acceptable breads. However, there had been few experiences of implementing their production beyond the institutional level.

# 5.5    Product Development Process for cassava flour

### 5.5.1    Product development project for flour in breadmaking

CIAT, in collaboration with the then Colombian Institute for Technological Research (IIT), which had worked on composite flours for many years and was equipped with an experimental bakery, DRI and professors and students from the Food and Mechanical Engineering Department of the University of Valle

(UNIVALLE) in Cali, formulated an inter-institutional and interdisciplinary project to design, test and evaluate the feasibility of production of cassava flour by rural agro-entrepreneurs as an ingredient for flour mills to produce a composite flour with at least 10% level of cassava flour mixed with wheat flour. This composite flour would be sold to neighbourhood bakeries to produce traditional bun-type breads, known as 'pan blandito' and 'pan francés', for local consumers. These activities would also:

- evaluate different varieties of cassava for their acceptability in bread formulations
- determine the costs of production from farm to bread consumer
- analyse the economics of the wheat imports and target levels for establishing a wheat-based floor price for cassava flour and
- define the economics of various levels of cassava flour in the formulations.

The project would be a first attempt to take a total systems view of the whole production to consumption chain as a means of designing a new food commodity in Colombia. This project attracted funding from the International Development Research Centre (Canada) as its first agroindustry and product development research experience in Latin America. Figure 5.1 illustrates how the PD Process was related to the crop improvement process and how these

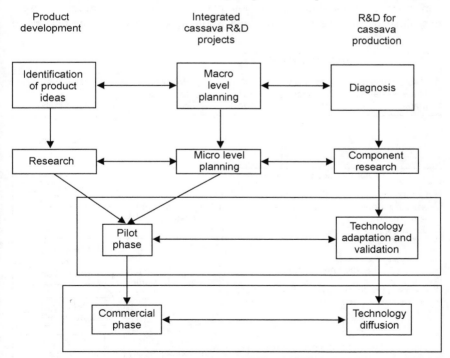

**Fig. 5.1**    Relationship between product development and integrated project methodologies developed by the CIAT cassava programme.

formed part of the integrated research and development pilot process for cassava that was evolving at CIAT at the time.

Figure 5.2 illustrates the main activities undertaken in the Product Development Process. This was a complex project to manage given the range and number of activities and the types of institutions, professionals and methodologies involved, not to mention dealing with farmers, millers, bakers and consumers in a number of different locations across a developing country.

The project was organised in three phases. Phase 1 was a 'research' phase in which experiments and studies were undertaken to generate information on the composite flour product, the process of producing and marketing the composite flour, and on cassava production costs and appropriate varieties. The technical studies and equipment development work were carried out in research facilities and the socio-economic studies were undertaken through surveys of bakers, millers, government agencies and cassava producers.

Positive results in this phase led to Phase 2 where the prototype cassava flour production process was established under field conditions with a farmer cooperative and the flour was test marketed among different potential clients. This phase also identified potential new market outlets for cassava flour.

Phase 3 was the 'commercial' phase, where improvements were made to the performance of the flour production plant and market arrangements were consolidated with selected clients with whom agreements had been reached to

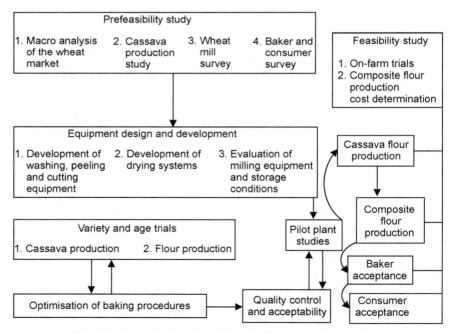

**Fig. 5.2**    Organisation of activities in the cassava flour project.

supply flour. The end result was a feasibility study on the expansion and replication of cassava flour production in Colombia's north coast region.

The principal results of each of these phases are described in more detail in the following sections.

## 5.6    Phase 1. Research on the product, the process and the market

The major results of the research undertaken on the composite flour product, the process for its production, potential markets and consumer acceptance were (Brekelbaum *et al.*, 1997):

- The sensitivity analysis of the whole system showed that feasible opportunities existed for cassava flour, priced at 20% lower than wheat imports, the prices of which were forecast to continue to grow. Blending cassava flour at the level of 15% would result in at least a 35% gross margin for farmers.
- It was feasible to plan for year round supply of cassava, taking account of varietal characteristics, low-input production technology packages developed for Colombia's north coast region by CIAT and the Colombian Institute for Agricultural Research (ICA in Spanish), the range of agroclimatic conditions in the region and the opportunity for in-ground storage of roots for up to three months.
- A small-scale processing plant was developed incorporating washing and chipping to produce a rectangular chip, inclined natural drying trays and coal-fired fixed bed dryer for the rainy season, pre-milling using a hammer mill, packing and storage. The investment was estimated at around $35,000 for a plant that processed three tonnes of fresh cassava roots to produce one tonne of cassava flour.
- Trials in commercial wheat roller mills showed that the greatest yields of flour were obtained by milling the unpeeled dry cassava chips at the wheat mill concurrently with the wheat, so that fibre could be separated and no additional milling and separating equipment was required.
- One of the four varieties of cassava tested gave superior bread quality up to the 15% substitution level in wheat flour, when tested in a commercial bakery and assessed by consumer panels. In larger-scale consumer tests in five neighbourhoods of Bogota, over 80% indicated they would buy the cassava flour breads.
- Trial production and sales in small bakeries were abandoned by bakers after two weeks as bakers complained of difficulties in handling. This occurred despite records indicating increasing sales and income from their cassava/wheat blend bread and indicated the need for training and technical support to bakers when introducing the mixed flour, enabling them to adjust their breadmaking process and to take into account the often variable quality of the wheat flour available in Colombia.
- In the course of the project, opportunities in other food markets were

identified – such as pasta, sausage, soup mix, cookies, confectionery as well as the industrial adhesives sector. These markets were deemed to be less demanding in terms of quality and also less institutionalised, and so the project evolved to consider some of these opportunities in future work.

## 5.7    Phase 2. Pilot plant establishment and test marketing of cassava flour

The second phase of the project went on to test the new cassava flour production to consumption system in a rural community in northern Colombia so as to generate reliable technical, economic and market information under the real socio-economic conditions of a cassava-growing region.

### 5.7.1    Development of the pilot processing plant

The site for the pilot processing plant was chosen from a number of possible options using a set of criteria that took into account factors such as the stability and experience of the farmer organisation, the potential for cassava root supply and the availability of essential services (water, electricity, etc.), amongst others. The Algorrobos Producers' Cooperative (COOPROALGA) in Cordoba Department, which had successfully produced dry cassava chips for the animal feed industry for a number of years, met the above criteria. The cooperative became a partner in the project and made available the land on which the plant was built and supplied the management personnel and labour required to run the plant.

The pilot cassava flour plant consisted of an area of 2,058 m$^2$ where the fresh cassava roots were received, weighed, washed, chipped and dried. The infrastructure included a warehouse where the dry cassava was milled and stored, a small office, bathrooms and a tool store. The process flow diagram is shown in Fig. 5.3. The plant was designed to dry 3 tonnes of fresh roots in 24 hours from the time of reception at the plant and operate during 10 months of the year (Figueroa, 1996).

### 5.7.2    Cassava flour production organisation and start-up

Key activities and issues that evolved during the establishment of the pilot plant were:

- site selection and construction of plant, including eventual sinking of well to supply water to the plant
- organisation of the farmers' cooperative to operate and manage plant
- evaluation of cassava production systems in combination with maize and yams to provide higher yields and incomes
- organisation of timely and economical supply of cassava across the 10-month production season

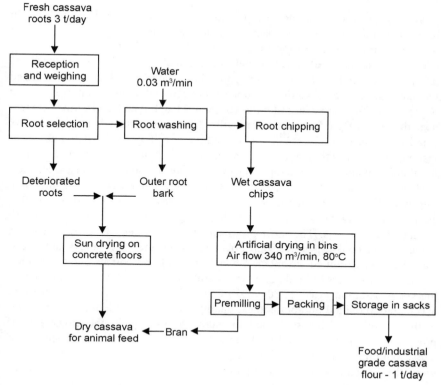

**Fig. 5.3**  Flow diagram of cassava flour process (adapted from Figueroa, 1996).

- identification of new food product and market opportunities
- adjustments to processing equipment, including more efficient drying and addition of small in-plant roller mill to produce flour for local markets
- microbiological quality assurance
- financial viability analysis of plant and key factors for viability.

A key factor, identified during this phase of the programme, was the need for a versatile and competent plant manager with the capacity to manage not only the technical operations of the plant, but also raw-material procurement, personnel motivation and management, market negotiations, sales and distribution, financial analysis and planning, etc. This was not easy to find in one person, let alone in the rural areas. This requirement initiated thinking on agroindustry management training and the need for enterprise support services in rural areas if new rural processing industries are to survive and grow.

### 5.7.3   Identification of new uses and market opportunities

As mentioned, the opportunity was taken during this phase to study a wider range of market opportunities for cassava flour to provide a range of options for producers, as well as to indicate potential constraints and pricing targets. In retrospect, this should have been done at the outset of the project before focussing on the bread composite flour market. However, this was an evolving experience in the rural development context that had had little exposure to private sector product development approaches. The market study, that included the distribution of cassava flour samples produced in the pilot plant, was carried out in Colombia's three largest cities (Bogota, Medellin and Cali), urban centres near the zone of production (Barranquilla, Cartagena, Santa Marta) and in six small towns nearby. Over 200 small, medium and large food companies were visited, provided with promotional materials and samples of the cassava flour for their evaluation. Follow-up visits were made to obtain feedback on their evaluation of the product, potential substitution level and their buying intention. The potential market volume for these companies was estimated at 15,000–30,000 t/year, with 80% of the substitution being for wheat flour in their various product formulations. The most promising target markets were for processed meats, sweet cookies, pasta, soup noodles, ice cream cones and soup mixes. Cassava flour exhibited advantages over competing ingredients due to its functional characteristics of binding, crispness enhancement and water absorption. A market penetration price of 80–90% of the wheat flour price was seen to be acceptable, provided microbiological quality could be maintained and met food quality standards. This study led to increased interest in the cassava flour product, the registration of a brand name, YUKARIBE, and the establishment of a distribution system in Medellin using a processed meat consulting company to reach the companies interested in purchase of the flour (Ostertag *et al.*, 1996).

## 5.8    Phase 3. Commercial development

The positive results of the market study in the second phase provided the impetus to initiate the third phase that was aimed at improving the economic viability of the processing plant, enhancing the microbiological quality of the product, and achieving commercial sales of the flour to demonstrate conclusively the feasibility of the project. These results would then be used as the basis for promoting similar project studies and investments elsewhere in Colombia and in other countries where interest had been manifested such as Peru, Ecuador and the Philippines. In particular, product development of cassava flour for use in industries that used adhesives was carried out, tested in companies and initial sales were made.

### 5.8.1    Plant improvement

The economic viability of the processing plant was improved by reducing drying costs through doubling heat generation and switching from the use of coke to mineral coal. These changes shortened drying times and improved flour microbiological quality. In addition, variable costs were reduced by establishing in-plant milling of the dry chips, using equipment designed and developed jointly by UNIVALLE and CIAT, which permitted the production of first grade flour for satisfying local demand for the product.

### 5.8.2   Market failure
During this phase, the Colombian government – like many others in Latin America – moved to a free market system and flour millers were able to import their own supplies of wheat; also wheat flour prices effectively decreased by 60% over two years making it difficult for cassava flour to compete.

### 5.8.3   Plant closure
The planned commercial operation of the plant and market development could not proceed. It was recommended that processing be carried out only during the local harvest season when local prices are lower, raw material supplies are high, quality is good, and nearby processed meat, bakeries and adhesives companies can be serviced.

Despite this less than satisfactory result, much was learned that has subsequently led to success in establishing rural enterprises, which bring needed income to small farmers and their local communities.

## 5.9   Problems met and lessons learnt

The specific lessons generated from this cassava flour case and which should be taken into account in developing projects with rural communities are:

- **Selection of product.** A comprehensive and systematic study on possible product opportunities and constraints should be done at the outset. The arbitrary selection of a 10% cassava/wheat flour blend for use in bread-making, based on laboratory experimentation and economic analysis, was unsuccessful because the bakers did not find the composite flour easy to use. When a broader range of products and target companies was considered, more promising markets were identified for flour, which were compatible with small-scale rural processing and marketing.
- **Selection of market.** Focus on national markets was too broad and demanding for incipient rural-based processing so smaller niche or local markets would have been more beneficial until all processing and management problems had been resolved.
- **Selection of industry partners.** Identification and working with potential industry partners who know how their food industry and product sector works would be more beneficial to ensure that product characteristics were acceptable for their needs. They could also provide practical and commercial advice on how the system could be adapted which would be more beneficial than that provided solely from technological researchers and government and non-government support organisations involved in the project.

- **Setting quality standards and quality assurance.** In the absence of product quality standards for cassava flour, or standard testing methods, the project had difficulty meeting each individual client's requirements and different laboratories often gave conflicting results on the same samples. This is important for product development, quality management, promotion and sales as well as payments to processors for raw material quality.
- **Selecting suitable sites for agrifood plants.** Experience in the project led to greater definition of criteria for selecting appropriate sites for pilot plant and ultimately rural agrifood processing plants, which added factors such as costs and proximity of fuel sources, machine repair shops and entrepreneurship capacity of the farmers' organisation to the conventional list covering raw materials, utilities and road access, etc.
- **Integration of people in projects.** Implementing rural agroindustrial projects is highly complex and demanding. The integration of innovative product development in an agricultural research programme leading to rural industry development, in a developing country context, as in this case, demanded high-level management, technical and business development skills to coordinate all the activities, disciplines, institutional actors and decision making involved. In this case many of these skills evolved over time with ongoing learning and adaptation of activities as experiences unfolded. Project staff were technically competent but lacked entrepreneurial, business and interpersonal skills. This affected the rate of project implementation and in-service training of the small-farmers' cooperative enterprise.
- **Skills and knowledge needed.** At the rural level in developing countries, entrepreneurial skills and attitudes required for business development and management are not readily found. This project showed the difficulties of trying to convert farmers into processors, marketers and businessmen. Traditional entrepreneurship courses focussed on bookkeeping and training in processing, but training for entrepreneurship was a complex issue for people who had no contact with appropriate role models. This pointed to the need for support service agencies to support rural processing groups in start-up and ongoing operations of management, marketing, negotiations, financing, etc., rather than assume, at least initially, that farmers and rural people could take on all these responsibilities. Another option would be to work with people who already had business experience or encourage their employment in the enterprise. This is a key issue that distinguishes rural product and enterprise development from larger organised corporations where all these issues can be effectively handled in-house.
- **Institutional linkages.** In a project of this nature, institutional linkages at the local level were key to supporting the integration of all the interdisciplinary research and development activities in the field. At the outset these did not exist, except for DRI staff. The national agricultural research entity ICA had no post harvest experience in cassava and it was reorganised during the life span of the project when research and extension were split. The role of the IARC to identify and bring in appropriate partners and to strengthen their

capacity was crucial and led to a greater understanding of the value of multi-institutional and multidisciplinary efforts in agrifood development.

- **Monitor political, economic and social changes.** It is important to continuously monitor political and economic issues that can impact on the sector involved and develop mitigation strategies. The project attempted to do this through its economic and sensitivity analyses on wheat and cassava prices and the various product sectors as it diversified its options. The introduction of the free trade policy at the same time as the fall in imported wheat prices had too strong an effect on the project for it to expand as planned in Colombia. However, the work on diversified product markets and niche markets did allow for seasonal processing to continue.

**Conclusion.** Much was learned on project management and the integration of experiences and results as they became available served to reorient the project to suit the prevailing circumstances. This was a valuable experience for all involved. In particular, the experience highlighted the valuable role of an international agriculture research centre in pushing the envelope to extend its research beyond the farm production sector and to act in a catalysing and coordinating role, bringing together many of the other actors in the system. In turn, the way in which these other actors perceive and influence the potential results provided important lessons for the research institutions involved.

The experience described above shares many similarities with New Product Development as undertaken by a commercial company. However, there are some fundamental differences. Perhaps the most important of these is that a commercial company's foremost aim is to optimise profit. All decisions are therefore normally taken to ensure that this is achieved, particularly with respect to the sourcing of raw materials at the most economical price, irrespective of who or where they are produced. While the profitability principal is also paramount for the success of a rural agroenterprise, the overall goal of promoting this type of activity in developing countries is as a means of improving the well being of the population in general and small farmers in particular. This difference in overall objective requires that NPD for the benefit of the rural poor, has to invest considerable resources in building the human and social capital of farmers and other rural actors so that they can effectively integrate themselves into what is now a fiercely competitive market environment.

## 5.10   Impact

### 5.10.1   The Latin American and Caribbean consortium for cassava research and development

As a result of this cassava flour project and the animal feed experience, CIAT has continued to maintain a holistic and integrated approach to its work on cassava improvement now as a member of a regional public–private sector organisation, CLAYUCA (Spanish acronym for Latin American and Caribbean Consortium to Support Cassava Research and Development), which it helped form in 1999 as a regional planning and coordination agency for cassava in the region. CLAYUCA promotes strategic alliances between cassava farmers' groups, and the private and public sector on innovation and policy in order to improve cassava's competitiveness without marginalising the small-scale producer in the process. This allows co-funding and a role in championing innovation by the private sector. A number of innovations have evolved. For example, investment from the poultry sector for feed has resulted in scaled-up artificial drying, mechanised planting and harvesting and breeding of germplasm for agroindustrial use. Joint ventures have been established where farmers provide access to their land and crop for a period in return for shares in the business, which is funded by private sector and public sector agencies.

Today in Colombia, there is increased political awareness of the opportunities afforded by cassava product development. The crop is viewed as a strategic agricultural product whose development is critical to meet the rapidly increasing demand for food, feed and starchy raw materials for the food and non-food sectors, and also for the production of bio-fuels. This heightened demand has resulted in opportunities for both medium and small-scale farmers. CLAYUCA is illustrating that by appropriate technology development and institutional interventions small-scale farmers can link themselves beneficially to these developments.

### 5.10.2   Rural agroenterprise development project
CIAT went on to establish its own rural agroenterprise development project in 1996 with a small team of specialists in Africa, Asia and Latin America, many of whom developed their skills and ideas on the cassava flour project. This team has developed several projects with a range of partners that have been the origin of a generalised methodology for institutions supporting agroenterprise innovation for the benefit of rural communities and farmer groups. More importantly, through these developments they have developed mechanisms so that many more rural people benefit from rural agroenterprises. This strategy (Ferris et al., 2006), which is underpinned by a philosophy of participation and partnership among all appropriate actors, has the following key elements:

- develop partnerships, evaluate project sites and develop joint plans
- evaluate market trends, demand and identify market opportunities
- analyse market chains and develop business plans for interventions to foster rural enterprises and improve business support services
- assess project performance and develop mechanisms for scaling-up
- share knowledge and effect institutional change towards market-oriented innovation systems
- advocate for improved market and trade policies.

The approach is designed to promote a market-chain perspective that can support the development of new productive enterprises, new services and business relationships between producer groups and other actors rather than focussing only 'on-farm'. In this way the original product development process has evolved to form part of the wider process of rural innovation, where it has become evident that market access is essential if the rural poor are to truly become agents of their own change and have the opportunity of achieving greater levels of economic progress and well being.

## 5.11   References

BREKELBAUM, T., BEST, R. and OSTERTAG, C. (1997). *Synthesis Report: Production and Marketing of Cassava Flour in Colombia; a three phase project financed by the International Development Research Centre* (Canada). CIAT, Cali, Colombia.

FERRIS, S., BEST, R., LUNDY, M., OSTERTAG, C., GOTTRET, M.V. and WANDSCHNEIDER, T. (2006). Strategy Paper; *A Participatory and Area-Based Approach to Rural Agroenterprise Development.* Centro Internacional de Agricultura Tropical (CIAT). Cali, Colombia. CIAT Publication No. 849.

FIGUEROA, F. (1996) Establishing and Operating a Cassava Flour Plant on the Atlantic Coast of Colombia. In: D. Dufour, G. M. O'Brien, R. Best (Eds). *Cassava Starch and Flour: Progress in Research and Development.* Montpellier, France: Centre de Coopération Internationale en Recherche pour le Developpement, Département des Systèmes Agroalimentaires et Ruraux. Cali, Colombia: Centro Internacional de Agricultura Tropical. CIAT Publication No. 271.

FORREST, R.S., EDWARDSON, W., VOGEL, S. and YACIUK, G. (1979). *Food Systems: an account of the postproduction systems program supported by the International Development Research Centre.* IDRC-146e. International Development Research Centre (IDRC), Ottawa, Canada.

GOTTRET, M.V. and OSPINA, B. (2004). Twenty Years of Cassava Innovation in Colombia: Scaling Up under Different Political, Economic and Social Environments. In: D. Pachico and S. Fujisaka (Eds). *Scaling Up and Out: Achieving Widespread Impact through Agricultural Research.* CIAT (Centro Internacional de Agricultura Tropical), Economics and Impact Series, 3.

OSTERTAG, C.F., ALONSO, L., BEST, R. and WHEATLEY, C.C. (1996) The Cassava Flour Project in Colombia: From Opportunity Identification to Market Development. In: D. Dufour, G. M. O'Brien, Rupert Best (Eds). *Cassava Starch and Flour: Progress in Research and Development.* Montpellier, France: Centre de Coopération Internationale en Recherche pour le Developpement, Département des Systèmes Agroalimentaires et Ruraux. Cali, Colombia: Centro Internacional de Agricultura Tropical. CIAT Publication No. 271.

# Part III

# Product Development Process

# 6

# From farm to consumer – pioneering an early nutriceutical, Stolle milk

David Woodhams, New Zealand

*This chapter illustrates the development of a tightly specified product, from the farm to the launch. Foods are biological products based on materials from the land, the sea and fermentation. The new product from these biological sources can be designed either as an ingredient to be used in the development of food products or as a consumer product. Different varieties of wheat can be bred to have qualities needed by the baker or the cereal manufacturer; different varieties of fruit and vegetables are designed to be sold directly to the consumer. Another group involves actions on the farm and also in the processing plant to give products with specific properties. Some nutriceuticals and functional foods are in this group. In this case, farm activites and processing technologies are developed together. The case study describes the early development of a nutriceutical, before even the name had been invented. It illustrates the problems in developing such a product.*

*This project involves the whole of the PD Process from the farm, through the processing to the marketing. This is a complex development as the consumer needs are recognised before the farm development, and their requirements have to be considered as processing, packaging, distribution, and marketing are built up. Also, there is a mandatory need to recognise the government regulations that the products come under and to ensure, through quality assurance programmes, that their requirements are met.*

*These are long-term, complex projects, combining the farming research with the process research, product research and market research. They have to fit in with the growing seasons, which takes extensive planning to ensure that all the necessary research is organised and completed within the months, and sometimes just weeks, of the season.*

*The chapter particularly relates to pages 111–130, 319–332 in* Food Product Development *by Earle, Earle and Anderson.*

## 6.1    Introduction

### 6.1.1    The Stolle project

In March 1988, the writer was invited to become project manager responsible for a technology transfer from the United States to New Zealand. The New Zealand Dairy Board (NZDB) had entered into a joint-venture agreement with Stolle Research and Development Corporation of Cincinnati, Ohio (Stolle R&D), owner and developer of intellectual property in the field of milk biologics. The joint venture company, known as Stolle Milk Biologics International (SMBI), was formed to develop, license, manufacture and market milk biologics products worldwide. This chapter is the story of the first two years of the project as I experienced them.

The Stolle project had three overriding objectives:

- to transfer Stolle R&D's patented technology to New Zealand
- to establish production and processing of Stolle milk in New Zealand on a commercial scale and
- to provide a reliable, high quality supply of Stolle skim milk powder in commercial quantities.

There were two major subsidiary objectives:

- to develop a seamless quality assurance (QA) framework that would make New Zealand the world's pre-eminent source for milk biologics on the basis of guaranteed quality and
- to investigate manufacture of alternative products that preserved the biological activity of the original Stolle milk.

### 6.1.2    Background to Stolle milk

It is well documented in the scientific literature that one of the important functions of milk is to transfer immune protection from a lactating mother to a nursing infant. Although particularly true of colostrum, this function of milk continues throughout lactation. The first observations that mothers transfer disease immunity to their offspring through milk were published by the German scientist (and 1908 Nobel prize-winner) Paul Ehrlich, in 1892. As a consequence of this understanding, in 1906 Emil von Behring (a 1901 Nobel prize-winner) suggested feeding human infants with immune milk from inoculated cows to protect them against tuberculosis and in 1916 the results of feeding immune milk therapeutically to adult tuberculosis patients were published. However, interest in treating human diseases orally with immune milk waned and simultaneously, blood serum from immunised animals, administered parenterally, gained widespread acceptance. The antibodies were thought to be more concentrated in serum and also, serious questions were being asked about the ability of the gastrointestinal (GI) tract to absorb antibodies intact, an issue that surfaced within our New Zealand development team more than once in the first two years of the project!

Petersen and Campbell (1955) reintroduced the idea of using immune milk from cows to control human disease and later presented an historical review (Campbell and Petersen, 1963). Different sorts of immune milk were produced, depending on which antigens were used; it is possible to immunise animals against many different antigens simultaneously and to obtain a wide diversity of antibody populations in the milk. Petersen favoured 'local' immunisation through the teat canal over both subcutaneous and intramuscular systemic vaccination. Importantly, however, he reported the absorption of milk antibodies from the GI tracts of adults of 13 different species, including humans. Petersen's human disease interests were mainly rheumatoid arthritis and allergies. At one stage he enlisted over 2000 rheumatoid arthritis sufferers, 80% of whom benefited from immune milk within three months.

In 1958 Ralph Stolle, inventor of the tear-top can and owner of a large dairy farm in Lebanon, Ohio, US, came across Petersen's work and got interested in the concept of immune milk. He first established intramuscular injections with a secret 'S100' concoction of bacteria as an alternative method of immunisation. The components and injection protocol are detailed in US Patent 4,732,757 (Stolle and Beck, 1988). He froze the resulting milk and distributed it to arthritis sufferers to ascertain that the resulting milk matched the beneficial effects that Petersen had obtained. Satisfied that his immune milk product: '... was at least the functional equivalent ...' of Petersen's (Beck and Zimmerman, 1989), he then had the milk dried at a local processing plant and established that the milk powder was equivalent to the frozen product. Over the succeeding 28-year period Stolle collected written responses from over 3500 subjects with over 31 000 months of consumption. Some 83% of the 1140 subjects who stated that they had only rheumatoid arthritis reported improvement while drinking immune milk (Beck and Zimmerman, 1989).

Both Petersen's and Stolle's trials were 'uncontrolled' in that there were no placebo controls. So Stolle contracted scientists at the Birmingham College of Medicine, University of Alabama, to design and supervise a double-blind, cross-over, placebo-controlled study. The detailed design and results of this study were published in the text of US Patent #4,732,757 granted to Stolle and Beck (1988) for the process and product. Subjects reported significantly less morning stiffness and joint pain, they took fewer aspirin tablets and received fewer shots for pain relief during the periods when they were on the immune milk than when they were on the control milk.

Although the mechanism of this benefit is not understood at the present [1989] ... [m]ore recent studies ... provide evidence that milk from hyperimmunised cows contains a natural anti-inflammatory factor [AIF] that may provide relief from the symptoms of pain and swelling due to arthritic conditions (Beck and Zimmerman, 1989).

Beck (1981) protected the anti-inflammatory principle for Stolle R&D but it took a further 18 years to characterise and patent the AIF itself (Beck and Fuhrer, 1999).

By 1987 functional foods, i.e. foods claimed to have definite medical benefits, had begun to surface. Once the patent was issued, Stolle decided that the time was ripe for immune milk to go commercial. He knew that control of the process from cow to customer would require a complicated infrastructure. Thus Stolle decided to join forces with an existing dairy production and marketing organisation that had these capabilities and selected the NZ Dairy Board to take on this responsibility.

### 6.1.3    Stolle Milk Biologics International

The features of the NZ dairy industry which made it particularly suitable for production of milk biologics were:

*   The NZDB was the world's largest exporter of dairy products.
*   The average herd size in NZ is several times larger than that in the US or Europe.
*   Most herds in NZ are managed by their owners.
*   The NZ dairy industry has a highly developed quality assurance programme.
*   NZ herds are close to factories suitable for producing skim milk powder and other possible Stolle products.
*   The NZDB could control the production of the biologics from the cow to the market.
*   NZ is among the most efficient dairy producers in the world.
*   There is an extensive supporting infrastructure of animal health and research organisations
*   There is a skilled indigenous manufacturing base.

For these reasons, the joint-venture company, SMBI, was formed, a head office was established in Cincinnati with a New Zealander as General Manager and I was contracted as Project Manager for the New Zealand operation.

## 6.2    Setting up operations in New Zealand

### 6.2.1    Project management

Initially, as Project Manager, I reported directly to the head of the Protein Products Division of NZDB. Six months later NZDB created a Milk Biologics Section within the Protein Division and appointed a full-time Business Manager, to whom I reported for the following 18 months in my revised role as Operations Manager. At operational level, each company or organisation involved appointed a Stolle Milk Coordinator who reported directly to me in addition to their normal internal corporate responsibilities; they formed a dynamic project team that was a joy and a privilege to lead.

### 6.2.2   Project location and organisation
Selection of the Waikato dairying region was comparatively straightforward.

- The Waikato valley has a dense dairying population with several manufacturing sites capable of producing a range of products.
- Hamilton, the commercial centre of the Waikato, is the location of the Livestock Improvement Corporation (LIC), a NZDB subsidiary, established to improve the genetics of dairy herds in the industry by artificial breeding. We planned to use LIC officers to manage field operations and to use technicians from LIC to carry out the vaccinations.
- In addition, the country's major research institution for pastoral farming and animal husbandry, the Ruakura Agricultural Research Centre (Ruakura), is also located at Hamilton. Critically, they offered a biochemical analytical facility and a herd of 45 sets of identical twin cows which provided an efficient research resource for investigation of the effects of hyperimmunity by eliminating genetic differences between experimental and control animals.

### 6.2.3   Selection of dairy companies
Two of the four Waikato cooperative dairy companies were chosen:

- a small company, Tatua Co-operative Dairy Company (Tatua), with a very compact collection area and a spare, refrigerated 55 000 litre milk silo, and
- a larger company, Morrinsville-Thames Valley Co-operative Dairy Company (MTV), with a manufacturing site less than 10 km from Tatua, having two small (and one large) milk powder plants.

Tatua would provide the initial milk supply, milk collection, reception, separation, pasteurisation and skim milk accumulation. MTV would provide the evaporation and drying together with bulk packing, dry product storage and routine product grading. Biological activity assays would be carried out in the Analytical Biochemistry Laboratory at Ruakura.

### 6.2.4   Selection of farms
The basis of a contract for the supply of Stolle milk was negotiated between the NZDB and Tatua which included financial protection for the selected farmers against any adverse effects of the immunisation procedures. The basis for any claim would derive from the results of concurrent trials on the identical twin herd at Ruakura. On this basis, LIC's project supervisor visited, interviewed and, with Tatua's assistance, assessed several interested milk suppliers. He identified eight farms as suitable for the establishment of hyperimmune herds. The criteria for farm selection were:

- Mutual proximity to minimise the special collection costs.
- A history of good milk quality.
- Herd testing participant in 1987/88.

- Artificial breeding in 1987/88.
- Good farm records from previous seasons.
- Keen to participate.

### 6.2.5    Visit to Stolle R&D

With these arrangements in place, in late April I visited Cincinnati, Ohio, to be briefed on hyperimmune science and technology by its developers, Stolle R&D. I also met personnel at SMBI and visited a major pharmaceutical company, considered to be an important potential client. The latter visit played an important role in the conceptual development of QA procedures and documentation for the project.

The NZ dairy industry is seasonal with milk flow peaking in October, late spring in the southern hemisphere. Cows are dried off during May and calving takes place mostly during July and August. The Stolle immunisation protocol called for all cows required for hyperimmune production in the 1988/89 season to receive sensitising injections of vaccine before calving, i.e. during June and July 1988. In the first season, we planned to immunise approximately 1900 cows and we needed to organise any new plant and identify necessary plant alterations so they could be completed during the June winter shutdown period when access to the factory is free from production pressures. May was a busy month!

## 6.3    Government requirements

### 6.3.1    Vaccine importation

Importation of the Stolle polyvalent vaccine required a permit from the Ministry of Agriculture and Fisheries (MAF). Normally, a new vaccine for use on animals would first be referred to the Animal Remedies Board (ARB), set up under the Animal Remedies Act (1967) 'to control the manufacture, importation, sale and use of animal remedies, e.g. substances used for treating and preventing animal diseases for livestock and companion animals'.

Because the purpose of the Stolle vaccine was to stimulate the production of antibodies to *human* pathogens and not for 'treating and preventing animal diseases', MAF agreed to issue the permit directly. The S100 vaccine was renamed as 'S100 Immune System Stimulant', to ensure that it did not get caught up in bureaucratic red tape as a vaccine. The loss of even a month at this stage of the project would have cost the project an entire season and we were anticipating a problem with the ARB over the intended site of the multiple injections. Stolle R&D used an intramuscular injection in the rump (*gluteus maximus* muscle). Our advice was that the ARB would put up an argument for us to inject subcutaneously in the neck area, the preferred site for all vaccinations, as it avoided possible damage to the commercially valuable rump muscle. While it was possible that a neck injection would be as effective systemically as one in the rump, we had no evidence of this and had no time to produce any. In addition, experienced veterinary technicians advised that the Stolle protocol, calling for injections repeated every two weeks through the season, would condition the animals to shy away from the technician if they were administered in the neck, unduly prolonging routine vaccination operations on the farms. The farmers, in accord with their contracts, would be able to claim for any decrease in the value of the treated animals as culls. This action caused NZDB some small difficulty with the ARB a couple of years later, when it came to light, but it served its intended purpose at the time of expediting the importation of the S100 vaccine in time for sensitising injections to be administered in June 1988 rather than June 1989. Since then, the Animal Remedies Act (1967) has been replaced by the Agricultural Compounds and Veterinary Medicines Act (1997) which defines an agricultural compound as:

> ... any substance, mixture of substances, or biological compound, used or intended for use in the direct management of plants and animals ... for the purposes of ... maintaining, promoting, or regulating plant or animal productivity and performance or reproduction.

Had this definition been current in 1988, reference to the ARB would have been essential and we would have had to live with any resulting delay.


### 6.3.2   Other approvals

Both the experimental programme at Ruakura and our intentions for field operations were submitted for approval to the Animal Ethics Advisory Committee established by the University of Waikato which was responsible for both Ruakura and LIC operations. Their approval was given without difficulty or delay and confirmed by two committee members who later attended an early vaccination session at one of the farms.

If we had intended local distribution of Stolle milk products, approval by the Health Department would have been required. This was not required for export products. Nevertheless, we ensured that the Chief Scientist at the Health Department was advised of both the background to the project and of our plans.


### 6.3.3   Disposal of excess production and by-product streams

Considerable attention was paid to the handling of any Stolle milk raw material and by-products not required specifically for biologics production. Hyperimmune farms had their milk collected first in the morning, so it could be treated separately ahead of the normal milk supply. Cream and skim milk from the hyperimmune supply were made into products where processing temperatures ensured the destruction of any residual biological activity. Whey from the casein process was disposed of by spraying on to pasture on the two company-owned farms adjacent to the Tatua factory.

## 6.4    Research and development in New Zealand

### 6.4.1    Animal vaccination research

The objectives of the research programme with the identical twin herd at Ruakura, which proceeded concurrently with the commercial production programme, were:

- To measure the effect of vaccination on milk production and composition over a full lactation.
- To determine any major effects on mastitis or general cow health.
- To observe any major effects on calving and mating performance.
- To monitor the effects of the periodic booster injections at the injection site.
- To compare neck and rump injection sites.

Of the 45 twinsets, 40 were split between an experimental group that received the immunisation treatment and a control group receiving no injections. A further five twinsets were allocated to the injection site comparison. After four sensitising injections administered weekly during the dry period, booster injections at Ruakura started on 2 August 1988, about four weeks ahead of field operations, and finished on 28 March 1989.

The twin trial revealed that:

- In almost every lactational performance parameter the experimental group marginally outperformed the control group but in no case did the mean difference between the twinsets approach statistical significance. Trends in milk antibody levels for the 10 twinsets sampled through the season are shown in Fig. 6.1. The levels were assessed using an enzyme-linked immuno-

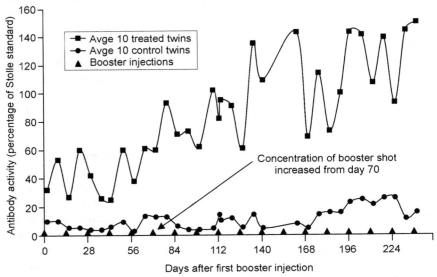

**Fig. 6.1**    Milk antibody activity on a weekly basis over a full lactation showing the gradual rise in the average Stolle titre for the milk from ten hyperimmune cows as the season progresses compared with that for the milk from their ten identical twin controls.

sorbent assay (ELISA) against a standard provided by Stolle R&D. Because the initial antibody levels reached with these 'naïve' animals were less than expected, after 10 weeks the antigen concentration was increased with improved results. This was transferred to field operations with immediate effect.

- Overall mastitis infections were almost identical at both the clinical and sub-clinical levels between the control and experimental groups. No adverse indicators of animal health were observed.
- Comparison of actual and due calving dates showed that prior administration of the vaccine had no effect on the onset of parturition. No difference in natural mating statistics was observed between twins.
- No adverse reactions were observed at the injection site other than occasional mild swellings that disappeared in a few days. Three cows were culled in February, for reasons not connected with the Stolle programme, after eight injections per side, and the six rump muscles were all dissected for histological examination; two insignificant fibrous growths were observed in one muscle.
- On average the IgG levels in the milk of the five 'neck' and 'rump' twinsets were essentially equal.

### 6.4.2    Field vaccination routines

Two different but complementary routines were developed for the two-weekly vaccination of production herds. In the first, animals leaving the farm dairy after the morning milking were diverted into a purpose-built race, provided by the project, which presented the cows' rumps conveniently for the technician. On the remaining farms, herds were held back after the morning milking close to the farm dairy and injected while in the milking position, but not during milking. This arrangement made it possible for one technician to vaccinate two herds in one morning.

### 6.4.3    Development of the quality assurance system

As noted previously, while in the US in early May, I was able to visit a major pharmaceutical company, considered to be an important potential client of SMBI. I had a lengthy and valuable discussion with the head of their Quality Assurance function, a man who had extensive experience working for the US Food and Drug Agency (FDA) prior to taking up his post at the company. I came away with a very clear picture of the need not only for superb control of the quality of products entering the pharmaceutical market but also for the capability of demonstrating that control by detailed manuals and full documentation.

As the target market for the highest returns included people with deficient immune systems – the very young, the elderly and the sick – the bacterial quality and efficacy of the product were of paramount importance. If Stolle milk were to be considered as raw material for drug manufacture, quality certification and documentation of a high standard would be essential and close attention would have to be paid to quality and hygiene at all stages, from farm to consumer. Our QA procedures were designed to meet these requirements.

The QA manual of procedures and documentation was developed and prepared under contract by MAFQual, Ruakura, a commercial arm of MAF working in the area of food safety. Five volumes, setting out procedures that were critical to the successful production of Stolle milk powder and special to that product, were prepared. They included, by reference, normal QA procedures already established in participating organisations. These volumes covered:

- Dry vaccine receipt, reconstitution, testing and storage.
- Milk production and on-farm procedures.
- Milk collection, separation, pasteurisation and skim milk storage.
- Skim milk transport, reception, powder production, packaging and storage.
- A general document linking all of the above into a comprehensive package.

For this class of product QA is a very strong selling point. Thus, establishing a gold-plated QA programme was considered to be both an operational and a strategic imperative.

Draft documents were drawn up during the course of the 1988 winter period. The procedures were evaluated critically during the course of the initial trial and amended final versions, under full document control, were issued well prior to the start of commercial production in December of that year.

### 6.4.4 Process research

The single major unknown for which we needed reliable data was the effect of temperature on the biological activity of the Stolle antibodies in commercial scale equipment. The two process areas of concern were pasteurisation and preheating:

- The pasteurisation system installed and certified at Tatua is the high temperature short time method of heating to 72 °C for 15 s. It includes a regenerative plate heat exchanger and a timed holding tube. While a higher temperature for a shorter time would probably have been less harsh on the antibody activity, it was impractical to consider any alternative because of legal requirements covering process and system certification. Thus our interest was in quantifying the expected loss in activity through a normal regenerative plate pasteuriser.
- Preheating the skim milk upstream of the first effect of the vacuum evaporator offers a little more flexibility. Milk evaporators operate under vacuum to reduce the boiling temperature and thus the opportunity for heat damage to the proteins. The temperature at which milk enters the evaporator is perhaps 10 °C higher than the boiling point so there is immediate (flash) boiling, releasing vapour which assists in the even distribution of the liquid across the tubes. The first effect boiling temperature can be manipulated, at

some cost in throughput, offering another possible way to reduce the necessary preheat temperature. Normal evaporator designs use quite small temperature differences across heating surfaces so heating times, including those in preheating equipment, can be quite long. Quantification of the damage at various time/temperature combinations was needed to guide us when setting process parameters for commercial evaporation.

Responsibility for this work was contracted to the NZ Dairy Research Institute (NZDRI), Palmerston North. They estimated a 20% loss in activity during pasteurisation and identified some rather strict limits on temperature and time combinations if we were to succeed in producing a biologically active product.

At this stage of the project we were still unable to assay in NZ for the specific Stolle antibody population by ELISA, so liquid samples were freeze-dried and, together with powder samples, sent to Stolle R&D for analysis. Stolle R&D found the freeze-dried samples difficult to handle so that we were hampered by a high level of uncertainty in the results. Limitations on the accuracy and reproducibility of antibody assay results were constant limitations through the following two years.

## 6.5    Initial production

### 6.5.1    Training of farmers

As well as the personal visit from and interview with the LIC Stolle coordinator before their selection, participating herd owners and managers, together with their wives, were invited to a special briefing on the project early in June 1988 at which we presented an overall picture of our immediate intentions and of future possibilities. At a later date we met with local farm consultants, advisors and farm veterinary officers for a similar briefing.

The importance of the participants producing milk of the highest quality was emphasised. Under their contract with the company, all farmers entering the programme were required to undergo a project-funded training course in on-farm milk quality management. The course included two farm visits by a MAF Quality Management Officer, one for a systems check and one to audit compliance. The following year, farmers who were already on the programme were evaluated on the basis of the quality of their supply in the previous season and were again visited by the MAF Quality Management Officer, both before and during the season. Each farmer received a copy of the 'Whole Milk Production Manual', Volume 3 of the 'Stolle Immune Milk Quality Management Systems'. Written cleaning procedures for each farm were individually developed and recorded by the farmer during the training period and had to be prominently displayed in each dairy shed.

For payment purposes the milk quality standards set for the participating farmers were the same as for the rest of the Tatua suppliers. However, for each quality parameter lower 'trigger' levels were set that conveyed a warning to

**Table 6.1** Milk quality standards for hyperimmune milk, 1988/89 and 1989/90 seasons

| Test | Action | | | | |
|---|---|---|---|---|---|
| | Satisfactory | Trigger | Grade penalty | Suspend HIM collection | Recommence collection |
| Standard plate count, cfu/ml | < 20,000 | > 20,000 | > 50,000 | 1 test > 200,000 or 2 tests > 100,000 (10-day period) | 3 consecutive tests < 20,000 |
| Thermodurics, cfu/ml | < 2000 | > 2000 | > 5000 | 1 test > 20,000 or 2 tests > 10,000 (10-day period) | 3 consecutive tests < 2000 |
| Inhibitory substance | Negative | Positive | Positive | Positive | 3 consecutive 'Negative' tests |
| Sediment | Finest Grade | 1st Grade | 2nd Grade | 2 × 2nd Grades (10-day period) | 3 consecutive 'Finest' grades |
| Bulk milk somatic cell count, #/ml | < 400,000 | > 400,000 | No grade penalty | > 750,000 | < 400,000 |
| Coliforms, cfu/ml | < 500 | > 500 | No grade penalty | 1 test > 10,000 or 2 tests > 5,000 (10-day period) | 3 consecutive tests < 500 |

Stolle farmers before they reached penalty levels. Also, the frequency of testing was greater. When the milk was not being processed for Stolle products, the test frequency was twice in each ten-day period, double the frequency for ordinary Tatua suppliers. When the milk was being processed for Stolle products, the test frequency was increased to once in every three-day period. However, if a trigger level had been exceeded in the previous cycle of tests, the triggered parameter was tested daily until three consecutive satisfactory tests were recorded. The main quality criteria are shown in Table 6.1. Farmers were paid for their milk at the same rate as other suppliers but we paid an additional sum per milking animal presented for vaccination to recompense farmers for their additional work and expense.

### 6.5.2  Setting up for processing

In the first season we planned to minimise capital expenditure on plant that would become surplus if the project were abandoned at an early stage. All tanks and equipment used only for Stolle milk were marked with special stickers. Tanker schedules were arranged to get Stolle milk to the factory ahead of normal milk so that it could be processed through shared plant (silo, separator, pasteuriser, etc.) first. In fact the only serious complaint we had from Stolle suppliers was that they had to have their milk ready each morning for collection by 7:00am instead of 7:30am as required for normal milk suppliers!

The daily production of the hyperimmune herds and the size of the spare refrigerated silo dictated that we could accumulate skim milk for just two days in the peak milk production season (October and November) rising to three days later in the season. The drying company, MTV, had no skim milk unloading facilities, so we planned to pump directly from milk tankers into the evaporator balance tank through flexible hoses. We planned to use MTV's powder plant #2, with a nominal capacity on skim milk powder of 2700 kg/h, because the smaller plant #1 had been unused for some time and needed some refurbishment. These circumstances combined to limit us to a maximum two-hour run on the drier, short enough to affect the consistency of properties like bulk density through the run and to affect yields adversely. Milk tankers that usually transported raw whole milk had to be cleaned to a standard that allowed us to transport and pump a pasteurised product without contaminating it. Because it was both impractical and unwise to wash and then dry the primary and secondary driers, the first 250 kg of Stolle powder would be used to remove any retained normal powder and would be diverted for use as stock food. Dry powder hoppers would be emptied and inspected before each production run and bagging lines would be dry-cleaned before and after each packing run.

Draft copies of the QA manuals were printed, distributed and used as texts for the instruction of operators and drivers. Each Stolle coordinator was required to include a critical evaluation of the relevant QA manual in trial reports.

### 6.5.3    Initial production runs

Between 30 October and 24 November 1988 we ran four trial production sequences, although the third was a repeat of the second trial, which failed to achieve our objectives. A project brief was prepared and circulated prior to each trial defining the objectives, responsibilities, procedures, sampling and analyses required. Written reports on each trial were submitted by the Stolle coordinator at each processing company to me, as Operations Manager, within seven days. One major objective of the first trial was to refine the draft 'Hyperimmune Skim Milk Powder Procedures and Specifications Manual' in light of practical experience. The definitive version of the Manual was issued on 6 November, after a detailed review of the first trial, in time for succeeding trials and operations.

Our main early processing focus was on the operating conditions in the evaporator, limiting both the temperature and the duration of preheating. During the last 30 minutes of the first run we operated with a feed temperature to the first effect 7 °C above its boiling temperature, producing biologically active Stolle milk powder with less loss of activity than expected. By the fourth trial we had reduced this 'flash' temperature differential to 4 °C and were able to produce a four-tonne batch of Stolle milk powder with a Stolle titre greater than 100%.

This trial immediately preceded a visit to NZ by Mr Ralph Stolle, Dr Lee Beck and senior SMBI personnel. Sample packs of the trial powder were made up for the visiting US party to use at the recommended dosage (90 g/d) during their visit. As a consequence of their experience with this, we were asked to improve the reconstitution properties of the powder. We achieved this during the fifth and final commercial trial on 20 December 1988 by reinstating the fines return system on the drier and produced 4.4 t of agglomerated Stolle milk powder. This product was made available for use in clinical trials and for marketing purposes.

### 6.5.4    Initial commercial production

Everyday collection of Stolle milk for commercial processing was initiated on 27 December 1988 and six batches of powder were manufactured in 17 days. A second series of seven batches was made in March/April. The final Stolle batch was followed immediately by a batch of milk from untreated cows, using the same collection, storage and processing procedures as the Stolle powder, for use as a placebo or control product during clinical trials. Because of the seasonal decline in the daily volume of milk, Stolle skim milk was accumulated for three days rather than two from 28 December and a little over 45 tonnes of export grade powder was manufactured against an order from SMBI.

## 6.6    Overall outcomes of the 1988/89 season

### 6.6.1    General

In the first season we produced 45 tonnes of biologically active Stolle skim milk powder to export standards.

The QA manuals issued on 6 November 1988 survived very well in operational use. They were comprehensively reviewed at the end of the season and rewritten to reflect our experience and to account for the numerous changes in processes and procedures introduced in 1989 to achieve full commercial production mode.

Progress of the project was hampered both by lack of capacity and by lack of precision in the measurement of the antibody strength (Stolle titre). However, this was at the cutting edge of both the science and the technology in 1988. To overcome the lack of capacity we instituted a flexible prioritisation scheme and provided low temperature frozen storage for the less important samples. These were assayed some weeks or months later; the variability of the results suggested some deterioration in storage. Nevertheless, our end of season review concluded that the activity loss through the pasteurisation step was the most significant, at around 20% of the measured titre, confirming the NZDRI pre-season prediction. On the other hand, the 72 °C for 4 s preheat conditions in the evaporator, using direct steam injection to minimise the come-up time, apparently caused little loss of activity. The loss in the drying process was minimal, as expected.

ELISA was quite a new assay technique at the time, although a degree of automation had been developed to facilitate repetitive tasks. We shared the one resource in the Analytical Biochemistry Laboratory at Ruakura with the Stolle identical twin herd trials and demand for the service exceeded its capacity to supply. Stolle R&D provided timely high level technical backup during the establishment of the assay and pressure on Ruakura was alleviated soon after November 1988 when a New Zealand science graduate, who was then working in the Stolle R&D laboratories, was appointed to work full time in the Ruakura laboratory while on the payroll of the Biologics Section, NZDB.

We also had problems throughout the development and early commercial periods with the precision of the method. If the Stolle standard curve and that of the unknown sample are parallel, as expected, the calculated concentration of the unknown sample will be the same wherever on the curve the comparison with the standard is made; a problem arises if the two lines are not parallel. Figure 6.2 shows a particularly egregious example for the milks from two individual animals. When these results are processed, the calculated concentration varies depending on the dilution selected for the comparison. The data in Table 6.2, derived from Fig. 6.2, illustrate both the inherent uncertainty of the method and also the extent of the uncertainty generated by the non-parallel phenomenon. Comparison of duplicate results at the same dilution reveals the repeatability of the test; the trend for the calculated sample concentration to increase with increasing dilution of each sample is the result of the then unexplained fact that the standard and sample regression lines are not parallel. The lack of precision in this assay and the lack of any routine assay for the AIF were the most frustrating aspects of the project.

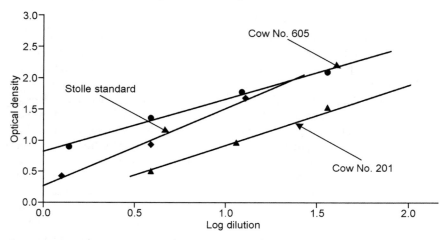

**Fig. 6.2**   ELISA curves showing the failure of the sample curves to run parallel to the standard curve. Graph of optical density against $\log_{10}$(dilution) from the ELISA of Stolle antibodies in the milk of two hyperimmune cows and in the supplied standard in the linear ranges. The unexpected fact that the regression lines for the unknown samples are not parallel to that for the Stolle standard leads to uncertainty in the results of the assay. Note that these are extreme examples, selected to illustrate the problem which was more acute for milks from individual animals than for herd milks.

**Table 6.2**   Duplicate relative activity levels of Stolle antibodies in milk from two trial animals converted from ELISA sample optical density values using a non-parallel Stolle standard curve

| Relative dilution | Cow 201 | | Cow 605 | |
|---|---|---|---|---|
| | Duplicate 1 | Duplicate 2 | Duplicate 1 | Duplicate 2 |
| 1/1 | 18% | 17% | – | – |
| 1/3 | 29% | 34% | 76% | 80% |
| 1/9 | 31% | 42% | 133% | 160% |
| 1/27 | 41% | 51% | 227% | 231% |
| 1/81 | – | – | 246% | 277% |
| Sample average | 29% | 36% | 171% | 187% |
| Reported average | 33% | | 179% | |
| Standard deviation | ±11.8% | | ±77.4% | |

### 6.6.2   Alternative products

We also completed our plan to investigate alternative products that preserved the biological activity of the original Stolle milk. During the first season we produced:

- 2400 kg of high protein Stolle milk powder by ultrafiltering skim milk. Production required hygienic tanker transport between three processing sites

instead of two, so we did not plan for early commercial scale operation. This concept had two markets in view:
- therapeutic use where the strength of the dose needed to be higher and
- prophylactic use in adult populations known to be intolerant of lactose.
- 80 kg Stolle whey protein concentrate, made by ultrafiltration of rennet casein whey for the same two markets. The mismatch between the small volume of Stolle whey and the large throughput of the commercial ultrafiltration plant made control of both the ultrafiltration and the drying processes quite precarious. The product manufactured was useful technically but no steps were taken to devise a commercial scale process for early production.

During the second (1989/90) season:

- Work was done developing a process for the manufacture of Stolle milk powder with most of the lactose hydrolysed to glucose and galactose. This development was designed to meet a difficulty with initial lactose intolerance being encountered in the Taiwanese market. A special spray drier was needed to manage the thermoplastic and hygroscopic properties of the product. Formal storage trials showed that, when stored at higher temperatures, this product very quickly exhibited Maillard browning. We did not plan for commercial production of this product.

## 6.7    Review of the 1988/89 season and planning for the 1989/90 season

### 6.7.1    Review of Stolle milk production on-farm
At the end of the 1988/89 season, two of the participant farmers took their herds out of the district. As part of setting up for the new season we planned to increase the number of hyperimmune cows in milk from 1940 to about 2400, bringing peak Stolle milk production to around 40,000 l/d. Thus five new farmers, with a combined herd of 840 cows, were selected and inducted into the project. Including replacement stock on the continuing hyperimmune farms, essentially half of the new season's animals were hyperimmunologically naïve and required pre-calving sensitising injections.

It was noticeable that the Stolle titre improved as the 1988/89 season progressed and the hyperimmune state of the animals increased. We were thus interested in the new season to observe the comparison between second season herds and naïve herds. Second year herds produced higher titres earlier and maintained their advantage throughout the season.

During the 1988/89 season, routine vaccination of the herds was trouble-free. One morning late in the season I escorted the Chairman of the NZDB, Jim (now Sir James) Graham, to visit one of the participating farms during morning milking and vaccination. After watching the senior LIC technician inoculate the first few lines of animals, he turned to me and said, 'Dave, next time I need an injection in the backside, I am going to get [this man] to do it!' After some 30 to

40,000 injections through the season, execution of each injection was perfect and not one cow flinched.

Milk quality was mostly very good throughout the first season although there were occasions when the thermoduric bacteria count or the somatic cell count exceeded trigger levels. No changes were made to the farm milk quality programme.

The farmers themselves seemed pleased with their results. One, whose family had been drinking unpasteurised Stolle milk throughout the season, told me that none of his family had had a cold or influenza through the 1989 winter. This gave him much confidence in the eventual outcome of the project.

### 6.7.2    Review of operations at Tatua Co-op Dairy Company

Problems identified at Tatua and their solutions for the new season were:

• With no separate Stolle milk reception silo in 1988/89, congestion during the morning was minimised but not eliminated by early collection of Stolle milk so it could be processed first to release the reception silo for normal milk. The need to have their milk ready for collection by 7:00am was a continuing and legitimate source of irritation for Stolle suppliers.

   During the 1989 winter shutdown, a new refrigerated 140 000 litre Stolle skim milk silo was installed, freeing the existing 55 000 litre silo to act as a dedicated Stolle reception silo, with its own tanker-unloading pump. This brought the Stolle farmers back on to the same schedule as other farmers, with a 7:30am deadline, and eliminated constraints on factory operations at both Tatua and MTV. At 140 000 litres, skim milk from the larger herd could be accumulated for three days at peak production and for four days later in the season. Thus we would be able to operate the MTV drier for longer periods with consequent gains in yield and product uniformity.

• Stolle skim milk leaving the pasteuriser was about 18 °C. The refrigeration available in the 55 000 litre silo was inadequate for chilling milk to below 7 °C for transfer to MTV. This was unsatisfactory both for hygiene and for processing, as a lower temperature would facilitate operation of the evaporator.

   The chilled water production capacity at Tatua was increased to permit the chilling of skim milk to less than 7 °C directly off the pasteuriser.

### 6.7.3    Review of operations at Morrinsville-Thames Valley Co-op Dairy Company

Problems identified at MTV and their solutions for the new season were:

• The use of the transport tankers as skim milk silos caused an unavoidable loss of yield because they could not be fully drained without causing an airlock in the transfer pump. It was also a high risk strategy hygienically.

   The system was upgraded for the new season by the installation of a dedicated Stolle skim milk unloading facility, giving direct access from the tankers to the existing skim milk silos. Costs, yields and convenience were all affected positively by this installation.

- In anticipation of an increased demand for agglomerated whole milk powder, MTV had upgraded their powder plant #2 by installing a system to return fine powder recovered from the cyclones and fluid beds to the region above the atomising disc in the spray drier. While fines return above the disk was a positive development for the improved production of agglomerated powders, the demand for whole milk powder and the associated problem of avoiding cross-contamination between whole and skim milk powders, meant that this drier was unlikely to be available for Stolle milk powder.

    The Stolle project was left with a forced choice of using powder plant #1 with its less desirable fines return point below the atomising disc. This plant was therefore prepared for Stolle service, including the necessary re-piping of the evaporator preheat system.

- The pneumatic conveying system for both driers was not suited to the transfer of agglomerated powders, leading to some loss of reconstitution properties.

    MTV installed a larger fan with speed control and upgraded the conveying system to minimise powder attrition during transport to the packing hall.

- For commercial production in 1988/89 we could fit only 20 kg of powder into the multiwall bags bought for the product. Once some agglomeration was introduced to the process, the bulk density had decreased, of course.

    Bag design was reassessed and for all of the 1989/90 season the powder was packed in 25 kg bags.

### 6.7.4    Financial approval and process plan

In order to meet our various deadlines during the winter, we needed both approval to proceed and assured access to funding before 31 May 1989, some months before the planned product launch and thus well before demand for the product could be assessed. Fortunately SMBI agreed to underwrite the necessary capital costs to provide a routine production capability in the 1989/90 season and the work went ahead as planned.

Normal skim milk powder production capacities for powder plants #1 and #2 are 1800 and 2700 kg/h respectively. However, for production of agglomerated Stolle milk powder, these rates are reduced to about 1450 and 2150 kg/h respectively. Some benefit was expected from the use of the smaller plant as, combined with the greater skim milk storage volume at Tatua, the length of a drying run would increase from around 2 hours to about 7 hours; we expected improved yields and a greater uniformity of powder properties. Consistency of powder properties through a production run was important because any blending of the powder before consumer-packing would cause further breakdown of the agglomerates with a consequent loss of desired reconstitution properties. The improved yields we expected would result from the smaller proportion that would be lost during start up and shut down procedures.

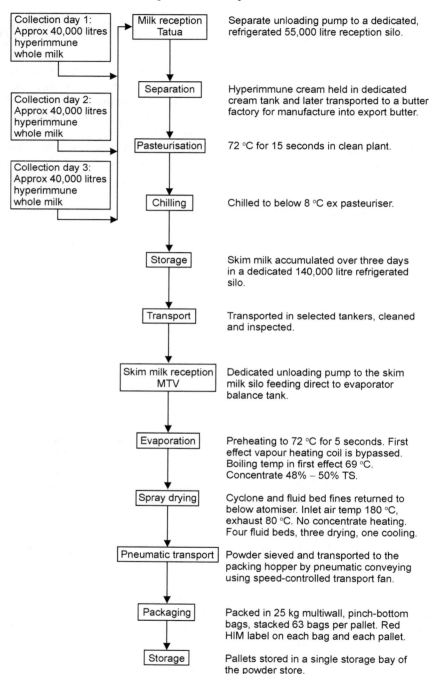

**Fig. 6.3**   Flowchart for the manufacture of hyperimmune skim milk powder at the start of the 1989/90 dairy season.

Figure 6.3 is a flow chart showing the processes involved in the manufacture of Stolle milk powder during the 1989/90 dairy season.

## 6.8    Commercial production 1989/90

### 6.8.1    Farm production and milk quality

Production in the new season was scheduled to begin on 11 October 1989 but late installation and commissioning of the new skim milk accumulation silo at Tatua delayed the first production run until 28 November, when the concrete foundations reached full strength. However, once started, routine production continued without a break until 12 March 1990. After 18 December we accumulated four days' production for each run.

We had 11 hyperimmune herds under contract, totalling 2394 cows. The peak daily Stolle milk collection during the processing period was 33 170 litres on 9 December. Considerable attention was paid to milk quality during the season, as discussed in Section 6.6.1 above. Two farms in particular had more quality problems than the others, apparently arising in part from an inadequate culling programme for cows with long histories of mastitis. However, the overall response to the on-farm milk quality programme was not as enthusiastic as had been hoped. All milk supplies exceeded trigger values from time to time and no supply was entirely free of second grades, entailing a financial penalty. The whole thrust of the quality management programme was to supply good quality milk consistently by taking appropriate preventative action before getting a 'grade'. In my judgement we were not meeting the quality objectives of the programme, so towards the end of the season I visited each participating herd owner to discuss their results. It was apparent that there was a body of resistance to some aspects of the milk quality regime, particularly the opening up of the plate milk cooler monthly, and to the frequency of testing of their milk. Clearly, a change of attitude was needed which called for a change of approach by NZDB.

Consequently, on my recommendation, a substantial monetary quality premium for consistently high milk quality was included in the schedule attached to the supply contract negotiated with Tatua for the 1990/91 season. With this in place, I made a further individual visit to each of the Stolle suppliers to acquaint them with how this premium would be applied and, more importantly, the reasons for NZDB's insistence on a high quality raw material. While the manufacture of Stolle milk powder is reasonably forgiving of the occasional lapse in milk quality, the manufacture of a whey-based product is much less so because there is more opportunity for microbial growth during the process. Also, products destined for use as pharmaceuticals, the highest returning market, would come under stringent scrutiny from bodies such as the USFDA because the intended consumers are likely to be ill and have compromised immune systems. It was to prepare for this sort of future development that the NZDB placed such emphasis on milk quality.

### 6.8.2    Milk powder quality

In the course of 28 drying runs during the operational period we produced 234 tonnes of Stolle milk powder, 177 tonnes of which was suitable for export. The main reason for product being downgraded was the presence of scorched particles in the powder (45 t), with high moisture and high thermophilic spore counts accounting for the remainder (12 t).

The scorched particle problem was persistent and, in the end, unresolved. The first five runs produced 37 t, with all 'A' pads for scorched particles, indicating minimum and tolerable colour change. Of the next eight runs there was only one that was completely clear of 'B' pads, which indicate a greater degree of browning than is acceptable, in spite of heroic efforts by a dedicated MTV team. Attention was concentrated mostly on the fines return system but the air disperser and the drier chamber itself came in for thorough cleaning and inspection without solving the problem. Thus, on 11 January 1990, we returned production to powder plant #2, which was no longer in demand for whole milk powder. Further work on plant #1, using normal skim milk, convinced us that for future production the plant should be rebuilt with an all-new fines return system.

Sampling throughout the production season again showed, on average, that an activity loss of around 20% had occurred during pasteurisation. Differences in activity between the pasteurised skim milk and the powder produced were not statistically significant. However, ELISA results for the finished product varied over a much greater range than would be expected from any variations in processing conditions. Given that all of the pasteurised skim milk was drawn from a single, well-mixed silo, it is apparent that the precision of the assay results was still a problem.

### 6.8.3    Milk powder yields

Yields were measured and assessed at both Tatua and MTV. Because of the short run on the separator each day, around one hour, the yields of fluid skim milk were around 4% lower than would be expected from a normal commercial operation.

Payment to MTV was made on the basis of a site-specific cost model, negotiated with the company before production started. While much information and many costs were well known, some uncertain input factors had to be estimated. Of these, one of the most critical was the yield of skim milk powder, estimated as $91.8 \, kg/m^3$ skim milk received. Careful measurements revealed quite early in the season that we were falling a little short of the estimate because of an under-estimate in the model of the quantity of fine powder being vented from the packing bin to the fines recovery unit. The production-weighted mean yield was $91.2 \, kg/m^3$, 0.6% below the model estimate and the payment model was adjusted retrospectively.

## 6.9   Product launch in Taiwan

Nutriceutical products presented something of a problem for marketers in the late 1980s; they were so new they were not catered for by existing legislation covering either foods or medicines. Indeed, the word 'nutriceutical' was not coined until 1989, about the time of the Stolle launch in August of that year. At the time we used the term 'functional food'. Taiwan was one of very few countries where existing laws governing the importation of food products were sufficiently flexible to permit entry of Stolle milk powder. Thus it was in Taiwan that the product was first launched commercially as 'Ultra Lac Functional Milk' by Société Commerciale Eurotaiwanaise (Eurotai). Launch activities included a seminar for local pharmacists, a scientific symposium for the local medical and scientific community and a press conference attended by about 25 journalists. Questions and answers at the latter function lasted 1½ hours. The product was targeted at the over-35 age group and was sold through pharmacies and hospitals. Initially, the product was packaged in sealed cans containing 454 g (1 lb) of powder, equivalent to five days' recommended consumption of 90 g/d. The first sale to consumers was to be three months' supply, or 18 cans containing 8.1 kg of powder. Some time later the product presentation changed to sachets containing 90 g of powder and later again to 45 g sachets. The marketing strategy also changed to a pyramid selling model, relying on word of mouth advertising, which was particularly successful.

Some initial problems were encountered with lactose intolerance, as would be expected in a predominantly Chinese population. However, by starting with half a dose, the original impetus for packing in 45 g sachets, it was found that the regular consumer developed a bacterial population in their gastrointestinal tract that was able to secrete lactase enzyme, thus compensating for the loss of that facility in Chinese adults.

## 6.10   More recent history

In 1998, two years after the death of Ralph Stolle in January 1996, SMBIntl. was sold and became SMBInc., an integral part of Spencer Trask Holdings, a long-established US venture capital and private equity firm with the resources to continue the research begun by Ralph Stolle almost 50 years ago. (Spencer Trask, a 19th-century New York financier, originally financed Thomas Edison's invention of the light bulb in 1879.)

In 2001 NZDB merged with major NZ dairy companies to form the Fonterra Cooperative Group. Tatua remained outside the merger but the increased volume of Stolle milk needed to meet market requirements had earlier exceeded their ability to supply with comfort in view of other production commitments. Thus production has now moved to Fonterra which continues to produce and manufacture Stolle milk products in the Waikato region.

In the 2006/07 season the milk from about 30 000 hyperimmune cows in 103 Fonterra herds was processed into three products, Stolle milk powder, Stolle milk protein concentrate and Stolle whey protein concentrate.

Stolle milk powder is still produced at the original MTV site. The season's production will be about 400 tonnes and will be exported to eight countries in Asia, North America and Europe. Consumers purchase the products in a variety of delivery forms that include single-serve sachets, pills, capsules, chewable tablets and beverages. Health claims for the products include: immune support, muscle recovery, joint health and lowering of blood cholesterol levels.

## 6.11    Problems met and lessons learnt

The Stolle project was, by any measure, a major success. In less than two years, starting from scratch, we were producing Stolle milk powder routinely on a commercial basis. We could process up to 55 m$^3$/d of Stolle raw milk and had produced over 275 t of Stolle milk powder with measured and acceptable biological activity. However, some of the problems encountered and lessons learnt call for a little more comment.

*Motivating farmers for quality production*
Our intention was to encourage the farmers to see themselves not only as a specially chosen group, which indeed they were, but also as an elite group who had to meet high quality standards to justify their selection. However, while two farms had particular quality problems during the 1989/90 season, all farms at some time fell short of the QA ideal. For example, only one farm in eleven completely avoided any 'grade' penalty for thermodurics during the 1989/90 season. This did not measure up to the standard achieved in 1988/89.

In a factory, managers can ensure their workers' consistent attention to quality by encouragement, by daily oversight and through their hierarchical authority and responsibility. With farmers, there is no daily oversight and no hierarchical authority. Pride and encouragement, backed up by financial penalties for failure, were apparently insufficient to ensure the level of commitment we sought. Thus, for the 1990/91 season, we offered a financial incentive, in the form of a monthly 'quality premium' paid for consistent quality from December through March, the productive months for Stolle milk.

*The scale of process equipment*
Matching the limited throughput demands of a product development project to the scale of commercial equipment is often an issue. We based the scale of our initial operation on achieving a minimum two hour run for the #2 spray drier and evaporator, a normal operating run being around 20 hours. Anything much less than two hours could have been misleading from both the process and the product viewpoints. We had to reach steady-state operation and stay there long enough to permit reliable conclusions to be drawn regarding the loss of immune

activity due to processing. We also had to be sure that any product that was made and used for market development truly represented the physical and biological properties of the eventual commercial product. In planning, to get 5400 kg skim milk powder per two-hour batch we needed about 60 000 litres of skim milk, or about 69 000 litres whole milk in three days. Getting 23 000 litres whole milk per day, from cows producing an average 12 l/d each over the projected production period, required a minimum herd of 1900 cows. Had the only available spray-drying plant been larger, we would have had to contract more hyperimmune herds.

### Developing quality standards

Developing quality standards suitable for a pharmaceutical-base product needed a highly professional approach. By 1988 quality documentation in dairy factories had become reasonably sophisticated as they pursued ISO certification. Even so, in these comparatively early years of QA, the expertise needed for the higher drug-base level was not widely available. MAFQual, the commercial arm of MAF, was aspiring to develop excellence in the process, starting from a substantial auditing base. As it was likely that MAFQual personnel would be auditing compliance with Stolle QA protocols, the possibility of misunder-standings between the writers and the auditors was, of course, considerably reduced. MAFQual personnel performed with distinction and the resulting QA documentation met our requirements precisely.

### Analytical methods

The analytical methods at our disposal at the time were either imprecise (the Stolle assay for antibody activity) or not available to a normal laboratory (assay for AIF). Another time, and given the opportunity, I would put more resources into this area because knowing the biological activity with some precision is so important both to satisfactory processing and to providing a product with consistent properties, batch to batch.

### Scaling for higher production

During 1989/90 the team was asked to put forward plans for scaling the Stolle production herd up to 10 000 hyperimmune cows. Based on hyperimmune herd production records, and assuming the continued availability of tanker transport and silo capacity at MTV, 10 000 cows could produce about 1600 tonnes of Stolle milk powder between 1 October and 15 March, more over a longer season. At peak production, the expected daily whole milk volume was 163 000 litres, producing 140 000 litres of skim milk, which neatly fitted the available Stolle skim milk silo at Tatua, filling it in one day, but requiring the use of or new installation of a larger whole milk silo. However, even at full production, the operating hours of the #2 plant would never exceed six hours per day unless further skim milk accumulation space were made available. We concluded that scaling up by a factor of four was feasible with minimal capital expenditure in the factories unless a new whole milk silo were needed. As field operations could be scaled-up similarly, all we needed was the word in sufficient time to turn plans into reality.

The time needed to achieve such a scale-up, or indeed production of Stolle milk powder at any lesser scale, was also assessed in terms of the NZ dairy season. Provided the Operations Manager was informed of the quantity of Stolle milk powder required during the following June-to-May season before the end of the preceding February, we would be ready and able to give sensitising shots to naïve cows in June and to hyperimmunise up to 10 000 cows from July, within the scope of timely revisions of the QA manuals. Given adequate silo capacity, production of 1600 t or more could be made for export with the facilities put in place for the 1989/90 season.

## 6.12    References

BECK, L.R. (1981) 'Method of treating inflammation using bovine milk'. US Patent 4 284 623; filed 9 November 1979, issued 18 August 1981.

BECK, L.R. and FUHRER, J.P. (1999) 'Anti-inflammatory factor, method of isolation, and use'. US Patent 5 980 953; filed 22 July 1997, issued 9 November 1999.

BECK, L.R. and ZIMMERMAN, V.A. (1989) *Stolle Immune Milk*, Cincinnati, OH, Stolle Milk Biologics International.

CAMPBELL, B. and PETERSEN, W.E. (1963) 'Immune milk: a historical survey', *Dairy Science Abstracts* 25: 345.

PETERSEN, W.E. and CAMPBELL, B. (1955) 'Use of protective principles in milk and colostrum in prevention of disease in man and animals', *Lancet* 75: 494.

STOLLE, R.J. and BECK, L.R. (1988) 'Prevention and treatment of rheumatoid arthritis'. US Patent 4 732 757; filed 9 December 1983, issued 22 March 1988.

# 7

# From concept to international company – development of a new pet food product

Brian Wilkinson and John Palamountain, New Zealand

*This chapter describes the creation of product concept and company, through raw materials and processing to market: the total Product Development Process (PD Process) and organisation. Is the PD Process different for a small company starting with two people, and a large national or a multinational company? The overall structure is similar. But the knowledge, the activities and the risks taken are very different. The small company has limited knowledge and needs to seek it or ignore the need. It cannot undertake all the activities and have to select what they do – ignoring consumer research, using equipment that they have or can contract, curtailing storage tests, launching without sufficient market research, and in particular not predicting competitive action.*

*The first thing is to understand the new product they propose to develop and the consumers' or industrial customers' needs in the food product. These need to be translated to properties that can be measured by chemical, microbiological, physical and sensory testing procedures. Often the small company does not have a laboratory or has only a small laboratory, so they need to identify a consulting laboratory or a university laboratory. Also there is a need to recognise the price range dictated by the customers, by competitors or by the supermarket. The price determines amongst other things the cost range of the raw materials.*

*Raw materials are often a problem: identifying the sources, the properties of the raw materials and their relation to the final product, the price range, the quantity needed in the product and the quantity available, any barriers to use such as import regulations, the suitability of the product for the process. Small manufacturers often cannot buy directly because of the small quantities needed, and therefore have to use agents. They need to be aware of the security of their raw material sources.*

*Processing can also present problems because of lack of knowledge and equipment. It is easy to ignore a processing variable and even a processing operation but resultant variation in the product quality can make it unacceptable to the consumer.*

*In developing the commercialisation of the product, there are four areas to consider – marketing, product qualities, physical distribution and the production plant. Storage and transport testing is important to determine how the product will retain its qualities during handling and selling.*

*In launching of the product, timing has to be considered – the timing of the market entry, the speed to market and the emergence of competition. The production and marketing need to be monitored so that changes can be made quickly to ensure the desired outcomes.*

*The chapter particularly relates to pages 96–130, 287–299 in* Food Product Development *by Earle, Earle and Anderson.*

## 7.1    Introduction

In September 2001 John and Ruth Palamountain started their small pet food supplement company, Vita Power New Zealand Ltd, to produce a dietary supplement for working dogs. This product was to supply the working dogs with all their daily vitamin, essential mineral and calorie requirements.

In his previous role, John Palamountain had worked for Heinz Wattie's New Zealand Ltd and was responsible for marketing their dog food to the farming community. In his numerous meetings with farmers, he was told that existing commercial pet foods, and in particular dog biscuits, could not supply the dogs with enough energy whilst they were mustering sheep and cattle. Most dogs had to be rested every other day, as they could not sustain their heavy working commitments. The palatability of most commercial pet foods was poor and as a consequence many dogs refused to consume their recommended daily ration allowance, which for an average sized dog is 12 biscuits. As a consequence the dogs were getting insufficient nutrients to sustain their considerable energy expenditure. Consultations with a leading pet nutritionist, Professor Grant Guilford, at Massey University, Palmerston North, New Zealand, suggested that the problem was twofold: inadequate energy to sustain the dogs' work load and lack of vitamins, and in particular B vitamins. As a consequence of his advice, Heinz Wattie's tried to increase the fat content of their dog biscuits but this proved futile. The dog biscuits with greater than 15% fat crumbled and rapidly spoiled due to mould development.

The aim of the new company was to produce and market a nutritional supplement for working dogs in New Zealand that would meet Guilford's nutritional guidelines, particularly for animals fed exclusively dog biscuits.

In this chapter, the product development specialist activities – formulation, processing, consumers, product testing and marketing – are described separately, but of course they were intertwined throughout the project.

## 7.2    Background

John Palamountain, a friend for whom I, Brian Wilkinson, had undertaken numerous research projects over a 25-year period, approached me after he and his wife had left their jobs at Heinz Wattie's with the idea of developing a supplement for the New Zealand working dog.

In his travels around the country, he had seen one product that was manufactured by a company in Christchurch that could possibly solve some of the issues mentioned above. This product was a sprinkle and included protein, a small amount of fat and essential minerals and vitamins. This product was aimed at domestic pets. He wanted me to develop a high fat version of this product that could be sprinkled on the dog biscuits of working dogs. A prototype high fat version of this product was developed by spray drying a high melting point (43 °C) tallow fraction which was then mixed with the correct amount of a pre-mix of vitamins and essentials minerals together with meat and bone meal and garlic oil to act as the pet attractants. A medium-sized working dog was supposed to be fed 25 grams of this product with its dog biscuits. The product was trialled on two large sheep stations each with at least 30 working dogs. Unfortunately, the dogs didn't take to the product with only 30% of the dogs showing any appreciable interest in the product and the shepherds didn't really show any interest either.

The product concept was changed after this trial to an emulsion-style product that would be poured over a commercial pet food. A recommended dose of the supplement would supply all the daily vitamin and essential mineral requirements and would include enough fat to satisfy the dogs' extra calorific needs. The product would be based largely around a vegetable oil and would contain pet attractants that would prompt the dogs to eat their recommended commercial pet food ration. The oil from the emulsion together with the extra biscuits would ensure that the working dogs ate enough nutrients to sustain them through high intensity working days. The dog owner would simply have to squeeze the prescribed dose on the animals' biscuits or other food.

This chapter then deals with the problems that were encountered in formulating, manufacturing and finally marketing this product.

## 7.3    Product concept

The product was to be a water-in-oil emulsion and 30 ml of this product was to supply an average-sized working dog with its daily vitamin, essential minerals and lipid requirements. The essential minerals were to be micro-encapsulated to prevent oxidation of the oils. The product was to be packaged in 1 litre high-density polyethylene containers. It was to be stable at ambient temperatures (15–40 °C). When presented to a panel of 30 working dogs the product had to be acceptable to over 85% of the dogs.

The product concept was predicated on the following assumptions:

- The working dogs were currently getting insufficient lipids in their diet to cope with the energy demands that were being made on them, particularly during mustering.
- Their current diets were not supplying adequate amounts of the essential vitamins and minerals either because the dogs would not eat enough of the commercial foods fed to them, or in the case of the vitamins, the heat processes needed to commercially sterilise these products were destroying the heat labile vitamins. The essential minerals were included as some farmers were only feeding their dogs with raw meat and potentially these dogs could be missing out on some essential nutrients, particularly selenium and cobalt – New Zealand soils are deficient in these particular minerals.
- Many of the essential minerals are recognised lipid pro-oxidants and needed to be isolated from the lipid fraction by encapsulation. From the viewpoint of lipid oxidation, the essential minerals should have been left out of the formulation because they are powerful pro-oxidants and could as a consequence limit the product's shelf life.

## 7.4    Product formulation and process development

This section of the chapter describes a number of problems that were encountered firstly with the ingredient selection, secondly with the emulsification/encapsulation of some of the minor ingredients in the formulation, and finally with the development of the process.

### 7.4.1    Ingredient selection

The main ingredients were oil, water, powdered yeast extract, powdered meat extract, an acid regulator, a preservative, an antioxidant system, an essential minerals mix, an essential vitamin mix, and pet attractants.

*Fats and oils*
Analyses of dog biscuits showed that these products had high saturated fat contents, correspondingly low unsaturated fat contents and an exceedingly poor omega 3 to 6 ratio of 1:12. Moreover, the fats failed to supply all the daily essential fatty acids of a dog. The fatty acid profiles of a number of vegetable and seed oils were evaluated and for reasons of stability, canola oil was selected for the intended product. However, a small amount of salmon oil was also included to provide the needed EPA and DHA fatty acids as these were known to have a number of nutritional benefits to dogs. Sourcing the fish oil proved difficult as most salmon oils were highly oxidised. A Norwegian company was finally selected as the preferred supplier.

*Oil/water ratio*
The product was designed to be a water-in-oil emulsion. The product would contain only enough water to dissolve all the water-soluble ingredients, as the

main purpose of the product was to supply the working dogs with enough lipids to satisfy their daily calorie deficit.

*Acidifier selection*

To minimise the likelihood of microbial growth, it was important to ensure that there were a number of 'hurdles' in the product. The main hurdle was of course the antimicrobial ingredients and the second the pH. A final pH of 4.8–5.0 was selected for the product as this pH would minimise the growth of most pathogens and food spoilage micro-organisms, but not too low as to cause palatability or fat hydrolysis problems. Low pH accelerates fat hydrolysis and subsequent autoxidation of the released free fatty acids. Sodium acid pyrophosphate was selected as the acidifier for the product.

*Palatability enhancer selection*

Most of the palatability enhancers were selected on the basis of previous unpublished research that had been carried out at Massey University over many years. The primary enhancers selected for the proposed product were hydrolysed meat extract, yeast extract powder, salmon oil, and garlic oil plus a number of individual amino acids.

*Antimicrobial compound selection*

A range of compounds was considered for the product: These included sodium metabisulphite, potassium sorbate, sodium diacetate, and a range of organic acids and their salts. Sodium metabisulphite was excluded because it is known to reduce vitamin B12 levels in pet food. The organic acids and sodium diacetate were excluded because they were not particularly effective against *Clostridium botulinum*, a serious hazard for this particular product because it contained hydrolysed yeast, meat proteins, peptides and free amino acids, many of which were imported from countries where clostridial food poisoning is a major issue. Potassium sorbate was selected for the product.

*Antioxidant selection*

The antioxidant system for the product was finally selected after a large shelf life trial on the product. In this trial a range of primary and secondary antioxidants, both singly and in combination, were evaluated. The most effective antioxidant system for this product, from a shelf life perspective, was a combination of equal parts of propyl gallate and *tert*-butylhydroquinone (BHQ). However, the study showed how important consumer testing, in this case pet testing, is in the development of any product. When a panel of dogs and cats were fed fresh emulsion with and without the antioxidants acceptability dropped from 95% to 35% for dogs and from 78% for cats to zero. Clearly the animals did not like the taste of the product with these two antioxidants present. In further panels comparing antioxidants (see Section 7.8.2), alpha tocopherol was finally shown to be the most effective compromise antioxidant – palatability was extremely high, though shelf life was diminished compared to the original antioxidant system.

*Final ingredient selection*
The final formulation ingredients selected were: canola oil, salmon oil, sodium acid pyrophosphate, hydrolysed meat extract, yeast extract powder, garlic oil, individual amino acids, potassium sorbate, alpha tocopherol.


### 7.4.2   Emulsification

The aim was to develop a stable water-in-oil emulsion; the apparent viscosity of the emulsion had to be high enough to keep the encapsulated minerals in suspension, whilst also being pourable at ambient temperatures as low as 0 °C. At the start of the project there was insufficient knowledge of emulsion technology and encapsulation technology in the team to rapidly advance the development of the new product and so two of the foremost experts in the respective technologies in New Zealand were consulted.

A 12-factor Plackett and Burman (see Box and Wilson, 1951) screening factorial design was set up on the advice of the emulsification expert to look at various emulsifiers, homogeniser settings and temperature conditions, thinking that the Plackett and Burman screening design would identify the important factors responsible for the production of the emulsion. This design could then be followed with a full factorial design to identify the optimum concentrations, settings and temperatures for the manufacture of the emulsion. The screening design proved to be useless – no matter what emulsifier or combination of emulsifiers, homogeniser settings or temperature conditions were used no stable emulsion resulted. This was one of the few times that the Plackett and Burman screening experimental approach had been found inadequate in almost 30 years of practical food product development.

It took almost a year to finally arrive at an emulsification system and process that would produce a stable emulsion that had the desired viscosity properties to suspend the micro-encapsulated minerals. All the emulsifier suppliers in New Zealand were contacted and asked to provide samples and these were systematically evaluated singly and in combination, along with a range of homogeniser settings and temperature settings in an attempt to find the combination of emulsifiers and process settings, but to no avail. At the end of about eight months the company was just about ready to abandon the project. So it was back to the literature, exploring a range of food systems where different emulsion systems and processes were used to stabilise other food emulsion systems. Appropriate emulsifier systems were ordered that had worked in these other food systems and these together with the appropriate equipment and temperature conditions were systematically evaluated. Finally one set of process conditions and emulsifier combinations was found that gave a stable emulsion. When the first stable emulsion came off the process line, there was a celebratory dance around the pilot plant. It took a further three months of experimentation in the pilot plant using factorial designs to optimise the process and emulsifier concentrations before the client and I were finally happy with the product.

### 7.4.3   Encapsulation

The product concept required that a 30 ml dose of the product would supply all the daily essential mineral requirements of an average-sized dog. Many of these essential minerals are potent pro-oxidants and so had to be encapsulated in some way to ensure that they did not come into contact with the polyunsaturated fatty acids in the canola and salmon oils, especially at the oil/water interfaces. We wanted to develop a micro-encapsulating system that would stop any leaching of the essential minerals into either the oil or water phases. Any leaching of the minerals would lead to rapid development of rancidity in the emulsion. This phase of the work was not commenced until we had actually produced a stable emulsion. The micro capsules had to be small so that they would not settle out of the emulsion.

The task of developing a micro-encapsulated, essential mineral complex was given to Dr Jim Jones, Massey University, an acknowledged expert in the area. A large number of gum/protein and gum/mono and disaccharide and even gum/protein/mono-and disaccharide systems were spray dried at various inlet and outlet spray drier temperatures to find an encapsulating system that would prevent the leaching of the minerals from the micro-capsules. Initially a number of Plackett and Burmann screening designs were used to identify some of the more promising gum/protein/carbohydrate combinations and spray dryer parameters that could be used in subsequent experimental designs. Once the more promising admixtures and the operating ranges for the process variables such as inlet and outlet temperature, feed flow rate, disc nozzle speed and air flow rate were established, an eight variable half factorial experimental design was used to optimise the parameters and ingredient combinations.

Each of the products produced in the large experimental design was evaluated in a small storage trial. The micro-encapsulated material was incorporated into the emulsion prior to its formation and stored in aluminium/foil laminated pouches at ambient temperature (approximately 20 °C). Samples were taken on a daily basis for analysis to follow the concentration build-up of cations in the water phase of the emulsion and also the development of rancidity. None of the micro-encapsulation systems prevented leaching of the metal ions and so the concept of incorporating the essential minerals in the emulsion was finally abandoned as the emulsion from some micro-encapsulation formulations showed significant rancidity even after two weeks of storage at ambient temperature. Minerals were therefore not included in the final mix.

## 7.5   Storage tests

Over the course of the development of the product a number of storage trials had to be conducted. The first of these was to establish the most appropriate antioxidant system for the emulsion product. The second was to identify the most appropriate mineral encapsulation system for the product and the third was to assess the effectiveness of various packaging systems. The temperatures at which the various products were evaluated at were largely determined by the marketing objectives for the product. These had established that the product should:

- be microbially stable at ambient temperatures,
- last for at least six months and possibly longer at ambient temperatures and
- withstand temperatures as high as 50 °C (temperatures inside unrefrigerated containers when vessels were crossing the equator) and as low as −5 °C.

A number of packaging formats were also investigated. Initially the product was packed in 1 litre opaque, high-density polyethylene containers with aluminium foil laminated impulse seal and high density polyethylene screw cap. The initial storage trial was set up to investigate the stability of the emulsion using these opaque polyethylene containers. Storage trials were commenced using pre-production material, but the results of the trial were not available to the company until after the company had been marketing product for at least two months. This proved costly because the packaging had to be changed, as it could not ensure a six-month shelf life. The trial showed that the product had to be stored in an aluminium foil laminated gusseted pouch. The pourer was sealed by an aluminium foil laminated film by impulse sealer and then with a high-density polyethylene screw cap. It also showed that where possible the product should be held at chilled temperatures (0–4 °C), particularly during overseas transport, to minimise the rate of oxidation. The trial also forced a design change to the plant – the oil storage silos were changed so that they could accommodate a blanket of carbon dioxide at the top of the tank.

About this time the antioxidant system in the product had to be changed because test cats at the Heinz Wattie's Feline Unit at Massey University refused to consume the product (see Section 7.8.2). The TBA and TBQ antioxidants were replaced with natural alpha tocopherol and so a second storage trial was instigated. This trial showed that product stored at 20 °C or less lasted for over six months and the 0 °C sample lasted for over a year.

Ideally the storage trial should have been completed before the plant design phase was completed and different packaging materials should also have been tested as well. Unfortunately, the company needed a cash flow as quickly as possible to pay expenses, and needed, as a consequence, to make decisions without the availability of better data. In hindsight, if the experimental evidence had been available before the critical commercial decisions had to made, then different decisions would have resulted, but John Palamountain is quite happy with the way he approached the whole development given the financial imperatives of the project for the company.

## 7.6   Equipment selection and production development

As soon as we had developed a successful process to produce the supplement in the pilot plant, a start was made on designing and building the full-scale plant. We did not follow a classic optimisation approach for the pilot process before proceeding to the commercial stage. John Palamountain was convinced that the process would work on the full-scale plant and so stopped any further development work on the pilot plant and immediately started sizing the full-scale plant to meet projected forecasts for the next two years for just the rural dog market. Once this was done he sought out a factory site for the plant which happened to coincide with the development of a small industrial park near the centre of Wanganui, New Zealand. Four small, just completed, 1200 m$^2$ warehouses were available for lease at the moment we completed our first pilot plant run. Three weeks after we had completed the most successful run, a three-year lease was signed for one of the completed corrugated iron shells with concrete floors, offices and store rooms above the offices.

The regional dairy companies were visited in search of two 200-litre stainless steel jacketed vessels with mixers, and a smaller 100-litre jacketed tank with lid which were subsequently purchased. A piston filler was obtained from a local second-hand equipment dealer and a local engineer was appointed to install the equipment, do the piping and control systems plus design and build any other equipment for the plant. The plant was installed and the factory was ready for commissioning runs within 14 weeks of the first successful pilot plant run.

The commissioning/process optimisation trials were commenced two days after the final piece of equipment was installed. Minimum batch size (50 litres) for each trial run was determined by the configuration of the 200-litre stainless steel tanks. The first trial run was based on the process specifications established in the pilot plant and the product from this run was superior in consistency and appearance to any product produced in the pilot plant. John Palamountain had made a calculated gamble when he had decided to suspend any optimisation work on the pilot plant and instead go straight to the full-scale plant. I had strongly voiced my concerns about the approach he was adopting, but was thankfully proved wrong in this case.

However, in spite of the success of the first run, we systematically looked at each process step and optimised each parameter. We had to establish the optimum stirrer rate, initial oil temperature, addition temperatures for the various additives, the optimum cooling rate to form the emulsion and the sequence of addition of the various ingredients, filling temperature and post-packaging cooling rate to establish the optimum fat crystal size in the emulsion. Product from each run was evaluated for emulsion stability and consistency. John Palamountain carried out most of the commissioning work. In the fourth week after the commencement of the commissioning trials, production volumes were increased to batches of 160 litres. Small amendments to the process specifications were made at this stage. Product from these trials was used in the working dog, greyhound and equine market tests and in the first storage trials.

Within six weeks of the commencement of the commissioning trials, product rolled off the production line for sale to the rural market. The company remained in the initial premises for 18 months, but was forced to move to much larger premises as sales in the various niche markets took off and some overseas markets started to develop as well. A decision to move to new premises was taken after 14 months of commercial production. A major shareholder was brought into the company to assist with payment for the new facility and to finance the overseas market development.

The company took over an existing factory, stripped it bare, installed more purpose-designed equipment including two 1000-litre oil storage tanks within four months of the decision being made to move to new premises. Once the plant had been installed a week was set aside for commissioning trials. Production was stopped at the old plant at the end of the week and commenced at the new plant without any loss of production. The company is still producing from this second, larger plant.

## 7.7    Changing consumers

Initially this product was aimed specifically at the New Zealand working dogs, though the owners would have liked to also target domestic dogs as well. However, because of the size of the company, and its limited financial resources, it could not afford to pay for TV advertising and so could not market its products through the two major supermarket chains to domestic dog owners. The company has expanded by developing other niche markets and the development of overseas markets has been to the same niche markets that were originally identified in New Zealand. In the main it has adopted a non-penetrative marketing strategy for its products to date, domestically and in its overseas marketing. The company initially started with working dogs and when sales to this market were significantly less than forecasted, they moved to the equine market, and then the greyhound market and soon after the racing pigeon market. They are now targeting cattle breeding centres, which sell semen for insemination of dairy cows, and the large dairy cattle market.

### 7.7.1    Working dogs

John Palamountain had many years of producing and marketing dog biscuits for this market and as a consequence had built up contacts at the highest level amongst the retail companies that service the farming community. He was on a first name basis with many of the leading sheep and beef farmers in New Zealand and also the people involved in dog trials. In summary, he had a complete understanding of the market, he knew who the movers and shakers were in the market, what their concerns were and used this information to market his product. His wife, too, is a very good marketer and she handled the intricate marketing details such as media plans, packaging design, promotional material, in-store promotions, advertising, and did much of the day-to-day selling that is necessary to get a product launched.

It was thought that working dogs on New Zealand farms would be the main consumers for the product. These dogs are mainly on sheep and cattle farms, mustering the animals on often hilly country, so they are very hard-working dogs.

### 7.7.2 Greyhounds

Greyhound trainers and breeders all have their own idiosyncratic training and feeding regimes and are most reluctant to change either unless there is very sound evidence to demonstrate otherwise. They are extremely sceptical of commercial diets, preferring to feed their animals with 'natural' diets formulated to their own specific recipes. It was decided to take on this sceptical bastion with product made during the equipment and process optimisation phase.

### 7.7.3 Domestic dogs and cats

When sales to farmers did not take off as predicted, the two owners of the company decided to investigate the potential of the domestic cat and dog market for their product. No formulation changes were made to the dog product, other than recommending that domestic dog owners feed 5 ml of product/day as opposed to 30 ml because of the obesity problem amongst domestic pets. Vitamin levels were adjusted to ensure that the product provided 50% of their daily requirements. In the case of the cat market, a multivitamin mixture specific to cats replaced the dog multivitamin mixture and the recommended daily usage rates were obviously changed to suit the domestic cat.

### 7.7.4 Race horses

During the commissioning of the plant, an electrician was making some adjustments to the controllers whilst we were carrying out a production trial. This person owned a number of unsuccessful racehorses. After previous trials he had watched us dump product because it was not up to specification and during these trials he had heard us talk about the encouraging results that were coming in from the large sheep stations and kennel owners. He was particularly taken by the reports about increased stamina and coat health and so asked the owner whether he could take 10 litres of the product we were going to dump to try on his horses that had never won a race. The owner immediately told him to go ahead and take as much as he wanted. He was not sure of the outcome, but was always willing to see what effects it would have on different animals. An appropriate dose for the horses was provided to the electrician and he had his trainer feed the two horses for a period of months. One of the horses was a particularly vicious stallion and the other a mare. Within three weeks, the horse trainer reported that the stallion was a good deal calmer and both horses were working far harder and appeared to have more stamina. At their next starts both horses were placed and both won in their next three starts.

As soon as these results came in, John Palamountain realised he might have a winner. He contacted a number of horse trainers around Wanganui and leading horse jump and eventing riders around the country to get them to trial his product. Only one horse trainer was prepared to try the product. However, all of leading horse jumpers and eventers contacted gladly offered to trial the product as the appearance of the animal is very important in these sports. Any improved stamina would be a bonus. All the horse jumpers and eventers were voluble in their praise for the product at the end of the trial. One of the country's preeminent retired eventers was so pleased with the product, he immediately offered to act as an agent and sell the product to the horse fraternity in New Zealand.

A new product was formulated for the horses. A specific horse vitamin supplement was added and the meat extract was removed and replaced with creatine. Production volumes at the plant quadrupled as a consequence of sales to the horse market. Within 15 months the client had to move to new premises with larger production capabilities in order to meet demand for the product.

### 7.7.5   Racing pigeons

When John Palamountain was in the UK market in November 2005 to attend a large London dog show to promote the dog and cat products, he was approached by a man who wanted to know everything he could about the products. During the conversation, John mentioned the anecdotal results the company were getting with greyhounds and race horses. The man, a racing pigeon enthusiast, took a number of 250 ml pouches home with him to try on his pigeons. Some eight weeks later, John received a phone call from the pigeon racer who was in raptures about the product. His pigeons had performed well in a number of European pigeon racing events. The upshot of this chance meeting was the development of a product specifically for racing pigeons. It comes in 4 mm diameter soft gelatine capsules, each of which contains the recommended daily dose for a pigeon.

### 7.7.6   Other animal species

Once products had been developed and specifically marketed to each of the niche markets above, John decided to target the deer velvet industry in New Zealand in early 2005, with the aim of boosting deer velvet production. The freeze-dried ground velvet is used in Chinese medicine and now arthritis remedies. Approximately two years ago in a veterinarian controlled trial the product produced significantly faster growth and more velvet compared to the control.

Later in 2005, a major artificial beef insemination company was approached and they ran a trial. This company constantly rejects bulls with excellent genetics because they either do not produce enough semen or produce semen with numerous defects. In a controlled study involving 12 bulls, the supplement dramatically improved both sperm counts and sperm quality to the point that the company was prepared to collect semen from the bulls on the supplement. Trials are currently underway to assess the feasibility of supplying dairy cows with the supplement to boost milk production.

## 7.8    Product testing

In an ideal world, the product should have been tested in a scientific manner during the development cycle, with an independent laboratory conducting the tests. Minimum acceptability criteria would have been established and the product would not have proceeded to the next stage of development until it had met the acceptability criteria. However, such an approach was not feasible when this product was developed. The company's owners had little cash and had to allocate a large portion of their funds to the construction of a new plant and pay for all the associated costs of getting a product to market. They could not afford to adopt the classic approach in the testing of a product through its developmental cycle. Instead they used market testing extensively with a view to proving the product concept and also developing the respective niche markets.

### 7.8.1    Market testing

*Working dogs*

In the early phases of the development of the new product when we were concerned about acceptability of the product to dogs, we used the Palamountain's Great Dane, Beau, as the expert panel for the product. Initially he was fed just one formulation per test day, as we struggled to come up with the right flavour combinations for the dog, and a process that would consistently produce the desired product. On days when multiple samples of product were produced, he was presented with small amounts of each product on a saucer and the order in which he approached each of the saucers was recorded and future formulations were based on the first two samples. This approach was continued until we felt that we had an acceptable product that needed to be evaluated by a larger panel of dogs. Beau became one of 30 dogs that acted as the taste panel for the latter phases of the product development.

Once we had developed a successful emulsion in the pilot plant, a number of variants, differing only in flavour combinations, were produced in the pilot plant and these were evaluated by a kennel owner in Wanganui. Thirty of his dogs were selected for the panel and each was fed three different product formulations, and the order and amount of emulsion consumed by each dog was recorded. The test ran for two weeks and the order of presentation was randomised each day. In hindsight, a veterinarian should have been involved in the tests as well to record the condition of the animals at the start and end of the trial, as some discernible changes to coat health and stiffness in the arthritic dogs was readily apparent. Photographic evidence of the appearance of the dogs before the commencement of the trial and after was collected. The most acceptable formulation was then produced commercially. A product was only released for further testing if it reached an approval rating of ≥85% by the panels.

John Palamountain then approached lead users (large sheep farm owners and key dog triallists) and convinced them to trial the product. They were supplied with enough free samples to last up to three months, i.e. time enough for the owners to notice a difference in their dogs. The only proviso for the free product was that they had to write an endorsement for the product if it proved to increase dog stamina and coat appearance in particular. If the product resulted in any other benefits, then those comments, too, were also welcomed. In most cases the owners were more than willing to endorse the product as a consequence of the changes that had been observed. The endorsements were then used in the company's advertising brochures and promotional material.

*Other species*
Products produced for horses, racing pigeons, deer, and cattle, have not gone through the same rigorous preference testing. The same basic approach was used of getting lead users in each niche market to test the product before marketing the product to each of the niches. The same strategy was followed in each of the overseas markets as well, hoping that word of mouth would lead to rapid uptake of the product in each of the markets. However, this approach has not led to the hoped for outcomes and as a consequence the company has been forced to get much of their testing carried out at independent scientific laboratories both in New Zealand and overseas.

### 7.8.2   Scientific trials
In 2004, the company's owners were finally convinced that they should start to use independent scientific tests to substantiate any claims that they wished to make for their products. To be fair to them they could not afford them during the first three years of the company's existence.

The first scientific trial that was carried out was a cat acceptability trial. In 2002 the company had launched a domestic cat product, but sales were rather lacklustre. The cats would trial the product, but soon stop eating the supplement. We wanted to establish why, and so an acceptability trial was conducted at Massey University's Heinz Wattie's Feline Unit which is principally set up to evaluate feeds for Heinz Wattie's, a major pet food manufacturer in New Zealand. A panel of 50 cats was used to evaluate six product variants. The trial showed that the cats did not like the antioxidants BHA and BHQ and one other ingredient in the formulation but particularly liked the product with alpha tocopherol. Once the formulation changes were made to the product, sales increased dramatically. This trial changed the owners' perspective about the value of scientific trials.

The next trial was relatively simple and could be achieved at negligible cost to the company. Anecdotal evidence had suggested that the supplement improved digestibility and stool size and so a Latin square designed trial was set up using Basset hounds to evaluate the effects of the supplement would have on the digestibility of a range of manufacturers' dog foods. This trial was successful and indicated that the supplement did in fact significantly improve digestibility and stool volume. A summary of the report was produced in brochure form and sent to New Zealand veterinarians and major distributors of pet food in both written and electronic form.

In 2005, a major trial was then undertaken to prove one of the major claims that the company wanted to make about its product: namely that it boosted energy reserves and stamina. Twenty Basset hounds were hired from a local hunt club for the trial. Ethical permission for the trial was sought and obtained and financial assistance for the trial was obtained from the New Zealand government. The hounds were required to run on a specially designed treadmill. Load and speed were varied over a period of two hours according to the experimental design. Blood parameters were measured before and after the trial and heart rate, blood pressure, $VO_2$ (the volume of oxygen utilised per unit of time) and $VO_{2max}$ (maximum volume of oxygen that the animal can utilise/unit time) and oxygen consumption were measured at prescribed times throughout the trial. The dogs were also divided into four different feeding regimes, two of which included a positive and negative control as well as the supplement. The trial confirmed all the anecdotal evidence, namely that the dogs' stamina was improved by the supplement and that the animals on the supplement were able to handle repeated days of exercise, whereas animals not on the supplement could only cope with being worked every second day. A summary of the evidence was produced in brochure form for distribution to the trade and veterinarians. However, the veterinarians would still not stock the product.

The Basset hound endurance trial threw up a totally unexpected result, which triggered the next major scientific study for the company. During the endurance trial, the blood results continuously showed that the anti-infection side of the immune system was being enhanced in those animals that were on the supplement and that this immune improvement continued to increase over the six-week period that the dogs were participating in the study. A major trial on both cats and dogs was undertaken to see whether the results could be repeated on both species. The results confirmed the results from the initial trial and the information has been published in brochure form.

Whilst much of the research has been undertaken in New Zealand, the company has commissioned research in other countries as well. In 2005, a leading Kentucky equine specialist was asked to conduct a study on the amelioration of ulcers in horses by the supplement. These results have been published. Other studies are currently being undertaken in Germany to see whether the supplement can cure a particular bronchial infection to which European horses are prone during the winter months.

In summary the owners of the company have moved from a position of reluctant users of independent research to very keen users. However, they are also very keen observers of comments being made by users of their products and practically all the independent studies have been initiated by anecdotal evidence.

## 7.9   Market development

When the company went into production in 2002, it had a staff of five – the husband and wife and three part-time process operators. John was responsible for production, recruitment of agents and distributors and a sizeable portion of the personal selling of the product. Ruth was office manager, the marketing manager and sales manager. The two were working exceedingly long hours. Staff numbers have gradually increased in line with revenue increases and the company can afford to employ a few more specialists. However, the two principals still undertake all the overseas market development, i.e. visits to agents, major retailers, animal shows, trade shows, race meetings, etc.

### 7.9.1   Working dog market

The company started by developing the farm working dog market in New Zealand after they had received all the endorsements from the lead users and a number from other people who had supplied unsolicited product testimonials. Appropriate promotional material was developed to advertise the product together with stands displaying their products. Meetings were arranged with the senior managers of the rural service companies in New Zealand and these were persuaded to stock the dog supplement. Launch dates for each service company were then arranged and key outlets for each service company were also established. The company then followed a typical roll-out launch of its product. Each major outlet in New Zealand was visited and the staff were provided with some product training. They also attended the major working dog trials where they used personal selling to encourage farmers to try the product on their working dogs.

All registered veterinarians in the country were approached in the hope that they would endorse the product and thus promote the product to both rural and domestic dog owners. However, without exception, the veterinarians refused to stock the product because they believed that the product was 'snake oil' and could not possibly do what the company's brochures were claiming.

The views of the veterinarians did not change until they saw the results from the independent scientific tests and not until they participated in a major market survey study carried out by the marketing department at Massey University. The lead market researcher managed to convince each veterinarian to test his or her own dogs and cats on the appropriate supplement. The veterinarians were given product to feed their animals for six weeks. At the end of this time they were asked to comment on the health status of their animals compared to their status at the start of the trial. Over 60% of the veterinarians participated in the trial and many became enthusiastic users of the product. The market survey also asked a range of other questions which enabled the company to better target their markets, improve packaging design and promotional material.

### 7.9.2    Other species

A similar strategy was adopted in the development of the other niche markets. Product was trialled by lead users and others. Endorsements were sought from the lead users and others were provided by anyone in the particular niche who had used the product on their animals. This material was used in the production of promotional material. The company used word of mouth as the main strategy to boost product sales. Whilst this worked for some niche markets, it was an abject failure in the greyhound and race horse industries. Trainers in the respective niches were most reluctant to tell other race horse owners or grey-hound owners what their winning recipes were. They wanted a competitive edge and so word of mouth was never a successful marketing tool in these industries. As a consequence the company was forced to target a much larger number of the major trainers in each industry to get them to trial the product before there was a noticeable increase in sales to these niche markets.

Stands were also set up at race meetings, greyhound meetings, domestic dog and cat shows and equine events. Free samples of products were handed out together with the appropriate brochures with the aim of getting as many people as possible to trial the product on their animals.

### 7.9.3    Overseas marketing

In mid 2003, the company started targeting the New South Wales working dog market as the company's owners felt more at home in this market. They knew how it operated, what buttons to push to get results. But this market did not really develop as expected for the same reasons as in New Zealand. The company quickly changed strategy and started to target the equine market, and domestic cat and dog markets later that year. They then targeted Victoria in April of 2004 and Queensland in August 2004. These three Australian markets have been very slow to develop for the following reasons:

- difficulties in finding an agent/distributor
- lack of time to devote to the market because of commitments to the New Zealand market
- trying to open too many niches at the one time without the resources to cope with the demands of each niche
- unwillingness of horse and greyhound trainers to pass on their secrets for success
- lack of financial resources to fund either radio or television advertising
- the actions of the major pet food companies telling retailers that people do not need to feed their pets the supplement as these large pet food manu-facturers already produce fully balanced, nutritious foods.

The company next targeted the United Kingdom market in late 2004 before either the New Zealand or Australian markets had fully developed, and in the same year the company initiated market development in the Western USA, followed by other EEC countries, southern Europe, South Korea and finally China. South Korea has been very successful, as the company was able to employ a very astute veterinarian entrepreneur as their agent. This individual owned a large number of veterinary practices and has a very effective national distribution company that ensured shelf space in most outlets. This company also had the resources to effectively market the company's products in South Korea, something that has not been possible in any of its other markets.

The selection of suitable agents in overseas markets has proved to be the Achilles heel of many New Zealand export companies and the client's company was no exception. Before an agent was selected in a market, John Palamountain would firstly visit the country and talk to leading retailers and veterinarians in the capital as well as in some minor towns. He would find out who the main agents/distributors were, their strengths and weaknesses, their coverage of the national market, pricing strategies and margins. He would select the agent/distributor who best met the company's needs and expectations and would then approach this agent/distributor with a view to handling his company's products. Usually a large, professionally run company was selected, but he was invariably knocked back because these companies would baulk at handling product for which claims being made were not substantiated by independent scientific research. These companies were not prepared to accept unsolicited responses from happy customers as proof of claims. As a consequence, the company was then forced to approach smaller agents/distributors that did not necessarily cover the whole country. Inevitably the company ended up with one company handling their dog and cat products and a separate company handling their equine products.

In a number of markets, the agents/distributors were changed within six months because of non-performance. In the United Kingdom the company has had four different agents over three years. One of the United Kingdom agents was selected after visiting the client at a major dog show in London. This person promised the world, but as it turned out had little knowledge of the pet food business, little ambition and drive and couldn't get the necessary market traction. He was replaced within two months as sales started to falter. The difficulty that most agents face is that the client company's products make up a very small proportion of their sales and so they find it difficult to warrant the time and expenditure on marketing that he expects. The products never get the attention and resources needed to push an innovative product.

## 7.10    Patenting, trademarking and licensing

As soon as a stable emulsion was produced in the pilot plant, a local patent attorney was approached to carry out a thorough investigation of all patents in the area and, if there was no similar patent, to apply for a provisional New Zealand patent. In retrospect, the patent search should have been carried out at the start of the project, but it did not seem important at the time, since the market was going to be small and just in New Zealand, namely the working dog market. However, as the target market for the product rapidly expanded within months of starting production, the wisdom of undertaking a patent search as part of the desk research for the product became ever more obvious. It is something that should be done for all new products before the company has invested significant funds in developing the new product. The company, once it realised that there were no comparable products in any overseas country, started to file patent applications in 100 countries, starting with Australia, US and Canada, the EU countries, four South American countries and South Africa and then progressively applied for patents in other countries as and when finances became available to pay for the patent applications. At present the patent costs exceed $NZ600,000. The European patent has been finalised: European Patent 1 446 025 B1.

During the first visit to the patent attorney, he advised against patenting the process as he felt that the process could not be readily copied. His advice was to trademark the brand, but keep the process confidential. The attorney was also concerned about the financial ability of the client to fight any patent breaches should they arise. However, the company decided to continue with the patenting exercise as it was recognising that the product had significant export potential and they wanted to develop some key international markets as soon as possible after launching the product in New Zealand. Patenting the product could work favourably when approaching key agents and retail chains to distribute and sell the products.

The attorney's advice was followed on registering the company and trademarking the brand and product name before patenting the product and process.

In late 2005 the company was forced to consider licensing their technology to an American company when sales to that market started to challenge their production capabilities. The owners of the company approached a number of prospective companies in early 2006 with the idea of having them manufacture their products in the United States. This was whittled down to one company within six weeks. Though this selected company was extremely knowledgeable about emulsion products, they had no previous experience in the pet food industry and have taken longer than expected to start production. The company has had problems sourcing raw materials, particularly the type of yeast and meat extracts that are currently produced in New Zealand and also the emulsifiers. Consequently, they have had to go through an extensive testing and development programme to ensure that their product is very similar to the parent company's products. There were also some distribution issues that set back market development. Sales were expected to be close to $180 million in 2007, but all the problems that the company has encountered in getting product to market will probably mean sales may only be a quarter of their expected level.

## 7.11   Problems met and lessons learnt

The project raised a number of issues that need to be considered in the development of any new product.

- The project highlighted the need to adopt a flexible approach to the evolution of the product concept. The product concept must of necessity change in the light of consumer research, technical abilities, production capabilities and financial resources of the company. The evolution should be managed to ensure that that the concept meets the new product strategy of the company. Failure to allow this organic evolution could result in increased new product failures.

- The failure to get traction with veterinarians showed how important independent scientific studies were to the company to enable them to overcome resistance in the veterinarian market by demonstrating to this group of professionals that the supplement did actually achieve what had been claimed for the product. It is vital that any company wishing to develop functional foods must carry out these trials, preferably before launching their products, if they want to gain credibility in the market place. Professionals demand proof of concept before they will stock an item. Testimonials from happy users seemingly have little impact on professional people such as veterinarians and health workers. Sales to professionals and large distributors in the United States and Europe suggest that the reputation of the research establishment carrying out the independent research may also have a bearing on sales of a functional food to veterinarians and major pet food stores. The company found that the reputation of Massey University, a small university in the South Pacific, had little impact on the target groups in the United States and Europe. As a consequence the company is initiating a number of research studies in the United States to counteract these perception problems.

  It is very difficult for a small company to afford such expensive trials, but failure to undertake them could lead to the company dealing with peripheral players in the market and at worst eventual demise of the company. Small companies need to explore every government funding opportunity to obtain the funds to carry out these independent trials demonstrating the efficacy of their product. They may even have to consider selling some equity in the company just to conduct the trials. Failure to do so often means the company only has access to peripheral market players – major retailers and distributors will not handle the product until they see proof of concept. This will result in very poor sales growth, as demonstrated by the company in New Zealand and many of their other markets. Consumers and professionals are quite sceptical of any claims made for a product unless there are these independent scientific tests to back up their claims; even then the consumers may not necessarily believe the independent tests if the research has been paid for by the manufacturer of the new product.

- The study also showed the importance of maintaining a flexible approach to the target market and the need to look for new market niches when sales to

the anticipated target market are below forecast. In the case of the company, sales to the owners of working dogs never met forecasts for a number of reasons including:
- cost of the product
- unwillingness of the owners to believe that their animals were not currently getting all their nutrients from reputable commercial pet food manufacturers
- a degree of scepticism about the claims being made for the product and finally
- the fact that dog welfare was not particularly high on the farmers' priority list.

When the product was targeted at other niches where the owners were far more concerned about the health status of their animals, sales were high in spite of the product's cost.

- The study also highlighted the need to carry out pre-development consumer research amongst the target market. The company had identified a target market and an opportunity to satisfy some seeming deficiencies in the diets of working dogs. However, they failed to carry out the necessary market research to find out whether farmers would buy the product, what format they expected of the product and how much they were prepared to spend on such a product. Had this research been carried out, it is conceivable that the product would never have been targeted at the owners of working dogs. Such information needs to be sought out before the initiation of any new product project and should also be obtained regularly throughout the development of the product and even after the launch of the product. This information obviously comes at a cost, but the cost of not getting it can often be product failure.
- The study also highlighted the importance to small-scale manufacturers, who have limited technical expertise within the company, of exploiting professional expertise outside of the company to quickly surmount commercial difficulties. These people come at a cost, but most are only too willing to assist a fledgling company to succeed.
- The issue of patenting and its costs was also highlighted in this project. As pointed out in the text, the company has already paid out over $NZ600,000 in filing patents in most countries. Should a large company decide to challenge these patents, then the costs of defending the patent could conceivably break the company. Patenting is clearly a significant decision that has to be made by small companies and the company must have a clear idea of what they want to achieve from the patents. It may be more prudent in cases where a technology is difficult to copy to not patent the product/process and rely on franchising or license agreements to generate extra revenues from the invention. If, on the other hand, the company wishes to sell the company, then patenting may be a suitable strategy for the company as it will have its intellectual property to sell to the purchasing company.

- The current project highlighted the difficulties that small companies have in export marketing. The difficulties are exacerbated by the fact that company personnel do not have an intimate understanding of the export market, i.e. they do not have the personal contacts in the regulatory authorities, retailers, suppliers, that they have in their own local market These have to be established and thus takes time. They often do not know the distribution network for their product(s) intimately and who matters in the distribution system for a specific niche market. Finally, their production capabilities are too small to supply the larger export markets and so they adopt a not-penetrative marketing strategy, which may not be the most effective strategy for their product(s). It clearly takes time to gather all the information and so gain an in-depth appreciation of the intended export market. In the outlined case study, finding good agents and distributors in each of their export markets has proved to be a monumental nightmare for the company and is likely to continue for the foreseeable future.
- The case study showed that systematic experimentation failed to produce a satisfactory emulsion for the supplement. However, systematic experimentation can only work when the ingredients/process variables have been defined correctly and when the limits for each ingredient/process variable have also been correctly established. In the case of this study, the selected emulsifiers and process conditions for producing a satisfactory emulsion were inappropriate for the product system (i.e., compatibility with other ingredients) and only when the emulsifier system and process variable limits were changed did a satisfactory emulsion result. Technologists must know something about the system they are working on and the likely effects of each variable and ingredient on the resulting product before carrying out systematic experimentation.

## 7.12    References and bibliography

BOX, G.E.P. and WILSON, K.B. (1951). On the experimental attainment of optimum conditions. *Roy. Stat,. Soc. Ser. B,* **13**, 1.

EARLE, M. D., EARLE, R. and ANDERSON A. (2001). *Food Product Development*. Cambridge: Woodhead Publishing Limited.

FENNEMA, O.R. (1996) *Food Chemistry* (3rd edn). New York: Marcel Dekker.

KOTLER, P. and ARMSTRONG, G. (1999). *Principles of Marketing* (8th edn). Englewood Cliffs, NJ: Prentice Hall International.

# 8

# From kitchen to market – first came the oat cake, now the oat bake, a traditional product for today's consumers

Liz (Ashworth) Bowie, Scotland

*Invention is based on observation of what is happening and is not limited to scientific knowledge. Inventors rely on their accumulated practical knowledge and their own intuition. Invention requires some conceptual or imaginative creativity. It often occurs in 'kitchens', seldom in large industrial research and development laboratories. Invention, despite all the resource and financial difficulties, brings innovation into the food market.*

*But the innovation has to go through the PD Process to change it from the invention to the innovative product. Risks may be taken and steps (or stages) left out, but the main stages of product concept, product design, process development, packaging design, commercialisation, launch and evaluation of launch have still to be carried out to a greater or lesser extent. Success of the product depends on this.*

*This can prove difficult to the lone inventor, who most likely does not have the resources and the finances for the costs, nor manufacturing equipment nor any strength in the marketplace. They can either start on a small scale, find a venture capitalist, form an alliance with a food company or sell the idea to a food company.*

*The new product idea can come from a food often eaten: an ethnic food or an old traditional food; a technology outside or in another part of the food industry: extruding or computer industrial design; a consumer need: simple food or safety.*

*The chapter particularly relates to pages 95–130, 156, 223–236 in* Food Product Development *by Earle, Earle and Anderson.*

## 8.1    Introduction

I had developed a new oat snack in my kitchen, which was attractive to family, friends and small boys. But how was I to develop this into a product for the marketplace? I was a food writer with few assets, so there was little finance and few resources. Also I lived in Elgin, in the north-east of Scotland, well away from the main industrial areas to the south where there might be spare manufacturing space to rent and some expertise to help me on my way.

## 8.2    Initial idea

The initial idea came through publishing *Teach the Bairns to Bake* and a personal appearance at the Edinburgh Book Festival.

The famous TV Chef Jamie Oliver has helped to turn the spotlight onto what children eat and their lack of cooking skills. However, there are many less well known mortals who have been banging the same drum for many years. In 1996, Scottish Children's Press published two traditional Scottish cookbooks written by me aimed at encouraging children to cook and eat the wholesome foods, which have sustained our nation over the centuries. *Teach the Bairns to Cook* and *Teach the Bairns to Bake* met with wonderful book reviews such as 'F. Marian McNeill Made Easy'; 'Hats off to the Little Chef's Guide!' and 'Must have books for Kids'. To explain – F. Marian McNeill wrote the most famous book on Traditional Scottish Cooking and my book was described as a simpler version for young cooks or chefs, recommended to encourage youngsters and their Mums to cook together. The two 'Bairns' books hit the top ten Christmas favourites that year. As a result, I was invited to give children's workshops at the Edinburgh Book Festival in 1997. Keeping youngsters interested for an hour is no mean task, so I hit on the idea of making loads of Scottish 'goodies' for the children to taste during my stint. In among the spread were small round oatcakes made to an old recipe and cut out using a miniature round cutter given to me by a local baker when I was a child. The children loved the oatcakes, which disappeared in double quick time.

Afterwards, many adults came to ask me where they could buy the oatcakes and over the days and weeks following the Festival I was contacted several times with similar requests. My son Alan (then 5 years old) suggested that I try to make the oatcakes thinner like real potato crisps and to give them flavours. Over the next two years I was kept busy baking the oatcakes (christened 'Zips' by the children who reckoned they zipped in, zipped them down and then zipped out to play again). They took all morning to make, but disappeared in about 30 minutes at the end of school time when the kitchen filled with hungry little boys who demolished the 'goodies' with relish.

## 8.3 Protection of the idea

But how was I to protect the idea and ensure that no one stole it before I put it onto the market. Buying a patent was out of the question, so taking advice from various business friends, I did two things to protect the product concept.

- I wrote a letter to myself detailing the concept and recipe and had it delivered registered mail. It has remained unopened and is thus proof that the initial idea was mine should anyone else try to steal the idea before it hit the market place.
- I borrowed £100 from my mother and had a commercial lawyer in Edinburgh draw up a basic confidentiality agreement to be signed by any potential business partner/manufacturer before any details of the new product were revealed, to protect my interests. Without this safeguard, unscrupulous food companies could have taken the new product for their own – a lone person with little finance has a slim chance of proving otherwise. In the end I was to be very lucky indeed when Simmers of Edinburgh agreed to take on the project; over the years of development a very good relationship has been established which has led to a much wider working partnership than was first envisaged.

## 8.4 Home kitchen development

Baking has always been a great pleasure to me so it was with alacrity that I began the first trials on what we had by then christened 'Zips'. The plan was to make a basic recipe and then to try out flavours, savoury and sweet, till I had a product I felt was possibly good enough to show to someone. Taking the basic oatcake recipe (stone ground oatmeal, pinch of salt mixed with water to a stiff pliable dough), I experimented with different oatmeals.

Using my trusty mini-baker's cutter, which was made and presented to me by a local baker at the young age of 6 years old, it took many a long morning mixing with a wooden spoon, hand rolling with a wooden rolling pin as thinly as possible, cutting and baking in a domestic gas oven. I learned a lot about home kitchen development.

- Make sure that you have a reliable timer with you; set at all times – oats burn easily. I mean take it everywhere with you – to the 'loo' if necessary.
- It takes a lot of rolling, cutting and baking to produce a relatively small quantity which – if liked – disappears very quickly indeed. I later learned that the indication of a potential winning food product is the speed at which it disappears down the tasters' throats and if they ask for more.
- Ensure a reliable supply of your raw ingredients – oatmeal in particular varies widely, there is no such thing as one oatmeal. It varies depending on where it was grown, the weather when it was growing and harvested, the method by which it was cleaned, the temperature to which it was heated during the kilning and the degree to which it was milled. Medium ground oatmeal from one mill may be the equivalent of fine oatmeal from another.

- Establish a standard recipe, method and baking regime and be able to reproduce the same product time and time again.
- Then start with flavours. I discovered that the very nature of oats causes them to be rather temperamental. Some flavours work – others do not. I have had many a disappointment with a flavour idea that turned out to be a major flop! Oatmeal absorbs and kills some tastes and is not affected by others – a case of trial and error!
- It is also very important to have a wide cross section of family and friends who will taste and try and be completely honest in their opinion of the product. I am very fortunate in that respect and extremely grateful to my long suffering taste panel.

## 8.5    Market research

Eventually I decided that it was time to approach the local enterprise company MBSE (*Moray, Badenoch and Strathspey Enterprise*), now re-named 'Highlands and Islands Enterprise' (HIE for short). Their Food and Drink Executive was so impressed that they commissioned a market survey by an Edinburgh company to carry out a preliminary feasibility study. 'Niche Marketing' delivered a glowing and enthusiastic report which firstly gave an in-depth description of the product and then identified the various areas of the market in which it could be placed. The report follows:

*Product concept*
The product is ideally suited to be sold solo or along with a complementary product such as a dip. The product is very versatile. It can be eaten on its own or as an accompaniment to a savoury pate or cream cheese. The basic nature of the product means that it can be oriented towards a number of different markets, all depending on the slant you wish to give it. For example, this could be a healthy eating slant or a children's snack or an adult's appetiser.

The status of the product was described – packaging, manufacture, finance (or lack of it); pricing at that time dependent on too many variables to make an accurate calculation:

- the market into which the product is sold
- the agreement with the joint venture partner
- the distribution arrangements
- the pack size and weight
- the price comparisons with competitive products.

*Customers*
The versatility of the product means that it has potential in several markets:

- Specialist – hospital catering – therapeutic non-allergic/diabetic ranges.
- Youngsters – retail multiples – (Tescos, Asda, etc.) and convenience stores – (Alldays, Costcutter).

- Adults – as youngsters.
- Travel/Leisure – airlines, ferries, trains, motorway service stations, hotel/restaurant/café/wine bar.
- Catering.

The geographical spread of the market quite literally could be all over the world, realistically, however, it should be looked at in a number of stages, that is, Scotland, UK and then export overseas.

*Promotion*
Essentially a brand new product which, if launched into the snacks market in direct competition to crisps and other similar snack foods, would need to find considerable sums of money to compete. Liz Bowie, through her writing has developed media contacts and should use these to the full at the appropriate time.

*Sales*
The success of the product will largely be down to the sales drive of those entrusted with this responsibility. The requirement would be to locate a company able to take on board the sales task and add it to the bakery production side or to have an independent company/s handle this.

*Competitors*
Competitors can be classed as any company offering a rival product which could be sold in any of the markets detailed in the foregoing. Overall the list is too numerous to mention.

*Summary*
The retail market place is on the constant look-out for healthy options – we are rapidly becoming a nation who eats (snacks) on the run and until recently snack products have been aimed at satisfying immediate needs of hunger and convenience. Now the emphasis has moved to health as the top priority.

Children have, for the crisp companies, been the target of successful campaigns to increase crisp sales, but these are now being viewed with some distaste by some ABC1 (middle class) parents who would prefer their children to try alternatives.

*Conclusions*
1. The product concept is first class and meets the needs of so many different markets and customer groups, by offering such versatility.
2. The breadth and diversity of the opportunity creates a real problem, in that all opportunities cannot be capitalised on at one go, therefore the operation of introducing the product to the market must, by necessity be phased.
3. Liz Bowie has no personal funding to develop, launch and sustain the product, therefore it is imperative that a co-sponsor(s) be found without delay.

4. The domestic kitchen trials have demonstrated that the product can be a success, but this trial must be taken to higher production levels with a suitable bakery manufacturer.
5. Once the product can be proved in a limited manufacturing environment, the next stage is to establish buyer reaction.
6. The success of the final product is largely down to the effectiveness of the sales operation to secure and obtain repeat purchases for the product(s). This input must either come from the bakery partner or from another source.
7. Liz Bowie would appear to have first class publicity connections within the media and these should be used to the full at the right time.
8. The snack market is one which has many big players with a great deal of cash, this presents a *major* obstacle for newcomers who will have to contend with their opposition.
9. Given the primary contribution which Liz Bowie has made to this project in that the product is her idea, this must be protected through legal means. Recipes cannot be copyrighted or patented, but, to avoid being 'ripped off' at the early stages there needs to be a non-disclosure document drawn up and thereafter a legal agreement between the joint venture partners to agree and spell out their various requirements.

*Contributions*

The product has an excellent opportunity of 'making it' in several markets; however, without substantial amounts of investment in the early stages, it is very unlikely to make in-roads into the major consumer children's snack market, as a consequence other niche markets should be targeted in the first instance.

This report was written in 1998 before the absolute focus had begun on the food we eat and the diet that our children are following. The year 2006 has been the best and most opportune time for the product to emerge onto the market with the focus on oats as a super-food. The growth in sales of oat meals, oat flakes and related oat products has rocketed and is still rising rapidly.

## 8.6    Initial factory trials

A local Shortbread manufacturer and friend kindly tried some experiments on one of the biscuit lines using a small moulding roller fitted to their shortbread rotary moulding machine. The product successfully transferred from the roller onto the conveyor and baked well in the travelling oven without undue disturbance by the fans. The final product emerged crisp and ready to be packed and eaten.

The initial trials were done to establish a plain basic dough using the base ingredients of oatmeal, salt and water which then could be adapted by method and recipe and flavoured as required. The manufacturer having kindly helped with preliminary trials decided that the product was not in their remit and declined to pursue the project. They did, however, introduce me to John Smith (a well-known north-east master baker) who specialised in the baking and manufacture of oat cakes and other oat products in his family bakery at New Pitsligo.


## 8.7   Further kitchen work

Having proved that the product was indeed capable of large-scale production, I continued kitchen trials for the following reasons:

- To identify an oatmeal which would provide consistent texture, flavour and colour and which did not vary too much with harvesting and milling conditions.
- To find an oatmeal which is lipase inactive. Lipase is a natural enzyme present in fresh oatmeal, which if left active dramatically affects the shelf life and flavour of the meal. Heat treating during the milling process kills this, improving the flavour and extending the keeping qualities of the meal. Lipase also affects the fats used in the baking process – if still active, will react with any vegetable fat, producing a nasty sour soapy taste and rendering the product inedible.
- Research into nature identical and natural flavourings (sweet and savoury) which would give good flavour and texture. Interestingly not all flavours respond to being mixed with oatmeal which seems to mask and alter several flavours and it really is trial and error as far as this is concerned. Many flavours which were tried did not work.


## 8.8   Co-operation with Aberdeen company

John Smith – a well known authority on baking in general and oat cakes in particular – was approached, with a view to forming a working partnership with him to produce the oat product. He had just taken over the old family business in New Pitsligo, Aberdeenshire and I worked with him over the next two years – developing several bakery products and learning a huge amount about baking and the food industry in general which was to stand me in very good stead when after two years, Scottish Food and Drink's Alan Stevenson persuaded me to become self-employed as a food product development consultant. John Smith was not in a position to develop the oat crisp; however, I learned a huge amount during the time that I worked in the bakery.

In the early months of 2002, John Smith and I attended a Scottish Food and Drink 'one to one' surgery with Julian Mellentin, health-enhancing foods consultant to Scottish Food and Drink. He was very impressed with the whole concept of the product and helped in the initial approach to Simmers of Edinburgh in November 2001. They decided to go ahead with the development of the product.

## 8.9    The commercial partner: Simmers of Edinburgh

Simmers of Edinburgh was formed in 1996 by a management buy out from United Biscuits. The company purchased the two brand names, Simmers Biscuits and Nairns Oat Cakes. In 2001, sales had reached £6.4 million and they employed 100 people. Rebranding of the Nairns range with the distinctive 'N' standing for the Natural goodness of oats, extending the range of oat cakes and producing a new oat biscuit had prepared the way to interest Simmers in my idea.

Why were Simmers interested?

- Excellent fit with the Nairn's range.
- 'Out of home' snacking opportunity.
- An opportunity with existing trade customers.
- Opportunities in the convenience and food service markets.
- Liz's factory trial proved that production was a possibility. This was a major key to their positive decision to develop the product further.

## 8.10    Initial product and packaging development

It has been a slow process to develop a product with a good texture and taste while ensuring a low sodium outcome with all the health benefits of oats. To begin with Corinne Dalaudier (product development manager, Simmers of Edinburgh) worked on the product in the test kitchen at Queen Margaret's College in Edinburgh, while I tested out ideas in my home.

### 8.10.1    Texture

To achieve a low fat product, I had omitted all added fat or oil; however, on a larger scale the manufactured result proved to be lacking in crunch and inclined to be hard and dry in the mouth. To give more shortness, Corinne added some high quality sunflower oil; however, this resulted in a product which, although very nice to eat, was too much like a very thin cocktail oat cake. How to get a snap into the texture? Back to the drawing board.

I recalled working with finely milled barley flours, used as a partial fat replacement in the manufacture of low fat cakes and over the next few months Corinne and I worked on an amended dough with this concept in mind. Market and consumer tests still revealed that the texture needed more work – what next? Six consumer research groups held in 2003 revealed that the market was looking for: 'an enjoyable, wholesome and sustaining snack for discerning healthy eaters'.

With this remit in mind product development then focused on the following:

- No added fat vs. less dry eat – too dry an eat is not popular.
- How to achieve crispness?
- How to form a very thin dough piece?

- How to optimise the bake? Particularly difficult considering the size of each small biscuit/crisp.
- What flavours? Again the question of what flavours 'go' well in an oat medium.
- How to pack them? Snack packs – bags, tubes or mini-packs? What does the customer prefer? Eye catching and consumer friendly.
- How to achieve six months shelf life? We wanted a natural product, so the secret had to be in the recipe, the baking and the packing.

Reading an old edition of *British Baker*, I read an article about a starch used in the meat industry. Would it work in the crisp product? Only one way to find out – and it did! The motto to be learned from this is, 'Never be afraid to try out "off the wall" ideas'. Especially with new ingredients, you just don't know what you may discover – and if it is a disaster at least you tried it out!

Corinne still had to carry out a lot of baking in the Queen Margaret College test kitchen and then squeezed into the tight factory schedule as trials at odd hours and weekends. Gradually things were taking shape.

### 8.10.2    Flavours
Oatmeal is a funny thing to work with when it comes to flavour – many of the tastes that you think are bound to be tasty don't seem to work. The very nature of the grain means that its composition seems to blunt the sharpness of some herbs and spices and the result is a sort of 'cardboard' uninteresting bland eat. Not what we wanted at all.

So far there are three flavours:

- lemon and cracked black pepper
- cheese
- tomato garlic and basil.

With ongoing research into others as time permits.

### 8.10.3    Oatmeal source and milling
The oats are grown in Scotland with conservation in mind and milled traditionally. They are heat-treated to inactivate the lipase enzyme which is naturally present in oats and can affect the shelf life of any baked product if not so treated. Hogarth's Oatmeal Mill situated in the picturesque Border town of Kelso supplies the oatmeal, which goes into the making of the product.

How are the oats processed?

1. The oats are thoroughly cleaned, de-stoned and then steam-treated and kiln-dried to over 100 °C which gives the oatmeal its distinctive flavour and also kills the enzyme lipase.
2. Cooled, shelled, further de-stoned and colour sorted. These shelled, kilned oats are now referred to as 'Groats'.

3. The Groat is chopped into 3 or 4 pieces to make pinhead oatmeal and then further processed into coarse, medium, fine or oat flour or steamed, rolled and flaked.

### 8.10.4   Design of protective packaging

What does the consumer identify with? What is handy, easy to use, eye catching, convenient, while giving maximum protection to the product so that it delivers all that is implied on the outside when the pack is open? After long deliberations, it was decided to go for a foil bag similar to that used in the crisp market.

## 8.11   Manufacture

Every journey begins with that first step and so it was that the very first trial was carried out using a piece of equipment based on the small tester previously made by the shortbread manufacturer. A plain dough mix was made and then tried on the existing production line and that very first attempt proved the possibilities of the product.

### 8.11.1   Equipment design and building

Initially it was planned to manufacture the product by traditional rotary moulding. However, after many trials using a test roller, it was decided that a thinner, crisper product could be achieved using an adaptation of that method. The diameter, shape and depth of the cutter required to achieve the final thin crisp result was designed 'in-house' by Gavin Love, the technical director but it became very clear that the existing equipment was not adequate to produce the volumes required. Over the previous 12 to 18 months, the 'oat' has gained celebrity diet status as one of the 'super-foods' resulting in a massive increase in demand for oat products. To meet the surge in business Simmers have re-designed and extended their premises including equipment to make the new range of products. By 2007, Simmers of Edinburgh's sales had risen to £13 million, now employing 120 people.

### 8.11.2   Production line design and testing

The new factory is now up and producing – bearing in mind that new, large-scale manufacturing equipment takes a lot of fine tuning and gathered experience from actually working the lines, to come into its own. This has taken a lot of work and dedication from the staff.

### 8.11.3   Production trials

In an already busy production environment, it is very difficult to programme in trials, especially of a completely new product. The product is made by a method that has evolved through a lot of hard work and constant trials to make a product, which will match the consumer demands. This has involved working strange hours and weekends to fit in with the busy factory schedule. To get there it takes time, dedication and hard work.

### 8.11.4   Final production and quality assurance

The final production and quality assurance only arrived through a lot of testing not only on the texture, taste and appearance of the product but also the shelf life and overall presentation. It takes time and patience – don't be happy with second best – you need a picture in your mind and taste buds of what your customer wants – the one that you know is the winner – don't stop till you are there.

### 8.11.5   Costing

Be competitive. A great product that costs a fortune will not make it. Study the market and make sure that you are in there with the section leaders at an attractive level.

The product has been priced to meet well with the crisp and snack competitors in the market place that have been pitched to feature in the snack biscuit sector. This gave a target direct product cost that would give the margin required to cover fixed over heads and profit requirement. The recipes were developed by Corinne within these cost guidelines.

## 8.12   Shelf life testing

This is something that cannot be hurried.

- The recipe itself – the ingredients, method of mixing and processing and the final bake all play a part. One of the most important is the way the product is baked.
- The packaging – what will give the optimum product quality at the end of the required shelf life.

Eventually Corinne's painstaking work paid off. A combination of the bake and packaging proved to be the key.

## 8.13   Marketing development

This has taken quite a long time – firstly to decide which sector is the most suitable for the placing of the product. Is it a crisp, a biscuit, a snack? Where would the most sales be generated by the buying public identifying the new product with the well-known brand of Nairns? The N of Nairns has already become established as the symbol for food that is primarily made with natural, nutritious oats. Oats and oat products have proved to be one of the main growth areas in the food market and are still growing.

### 8.13.1    Finding a name for the product

This took some time of brain storming to decide. In my work with local primary schools, I encouraged the children to come up with ideas for names and slogans for the new product. The general public was invited to taste panels and to offer their ideas on the matter. In the beginning I had called the biscuit 'Zips' but it was decided that this may send the wrong message to the consumer and that a more sensible approach would be to use a name suggesting that the product was related to Oat Cakes. Thus the product was named the Nairn's 'Oat Bake'.

### 8.13.2    Final packaging design

Tayburn, an Edinburgh-based packaging company, created the packaging design, which draws from the recently developed Oat Cake redesign, and applied this for the first time to a foil bag with colour variations selected to bring the individual flavours to life. As a bagged, crisp-style pack, this was the company's biggest departure from core products, and involved new packaging technology. The new Oat Bake has instant brand recognition, complements the existing range and reinforces Nairn's proposition of providing naturally nutritious foods.

### 8.13.3    Pricing

Six qualitative research groups were conducted across the UK to gauge response to initial ideas. It became clear that the product had to meet a similar price point to those of the competitors in the market place.

### 8.13.4    Market identification and distribution

The research into packaging design, preferred brand name and reaction to a number of proposed flavours were tested (Dragon Research, March 2004). This was augmented by trial and taste tests at a number of consumer events. Clear guidance was provided on the preferred flavours, the packaging and where the consumer would expect to select the product in the market place.

With so many products vying for shelf space in the snack market, the design for the Oat Bakes needed to be as relevant and eye-catching as the Oat Cakes and Oat Biscuits already established in the Nairn range with consistent branding and positioning.

Just like a Crisp – except tastier and oh, made from oats!

### 8.13.5    Test market in health food chain

Initially the Nairn's Oat Bakes were launched on a test market in a leading health food store. Sales were analysed, combined with continuing trial and taste tests at a number of consumer events and also within schools across the UK. Simmers of Edinburgh were confident enough in the results to fully launch the product.

## 8.14    Launch of product

It was decided to present the product to the trade in February, 2007. Firstly adding to the range already sold by existing customers and then moving into food service outlets through the services of an agent who specialises in this sector.

## 8.15    Evaluation of launch

Sales have started well with this range being quickly taken up by some of the multiple grocers. The products are now listed in two large British supermarket chains, Morrisons and Sainsburys. Reactions by buyers at trade shows has also been overwhelmingly positive. The Nairns Oat Bakes has also won the Scottish Food and Drink Excellence Award for the Best New Retail and Food Service Product 2007 in the Healthy Eating Category.

## 8.16    Problems met and lessons learnt

The initial concept – the idea – I was so sure that it was a winner but how to convince others? – You need patience, tenacity, a packet of tissues and a good sense of humour. Along the journey you will learn more than you ever dreamed, meet many people and make wonderful friends – your life will never be the same again. It is like an artist taking a blank canvas and wondering what to do to make it the way he sees it in his mind's eye – but making something that someone will choose to eat is even more, it involves all of you and all of them – eye, taste, smell, health but most of all it has to come from your HEART.

For, if it comes from your HEART and you really believe in what you are doing, the problems along the way – and believe me there will be many – will be but stepping stones.

## 8.17    Conclusions

August 2007 is my 10th Birthday – it has taken 10 years since the publishing of *Teach The Bairns to Bake* in 1996, to see this dream come true. I pursued this dream because I believed in it. Once I had proved to myself, in 1998, that manufacturing the oat crisp product on a large scale was feasible, I could talk about it and show others with more knowledge that I than it could be a good one!

More than that I really wanted to make a difference, if only a small one, to the way that our children, in fact all of us eat – we are what we eat at the end of the day. We are the great snack munching generation and it is now obvious that snacking on high fat foods loaded with preservatives and additives is not a good thing; it affects our health, behaviour and life expectancy. Nairn's Oat Bakes are only a small contribution but I hope and pray that they will be like the dripping tap that wore away the stone and be part of the a change in the way we eat our food. Nairn's Oat Bakes share the same formula as all Nairn's Oat Cakes and Nairn's Biscuits made with only whole grain oats with no artificial ingredients – just what I dreamt of – a product that is wholesome and really and truly 'good for you'.

Ten years seems a long time to eventually realise a dream and take it to market but I have it on good authority that it can, and often does, take that long to 'get it right'. Through the 'journey' my life has been enriched by all the people who have helped me on the way – it has been a privilege to work with them. My thanks to John Smith the baker from New Pitsligo who shared his valuable experience and knowledge. To Julian Mellentin who championed the cause.

I am most appreciative of the support I received from Alan Stevenson and his team at Scottish Food and Drink and very grateful in particular to Simmers of Edinburgh and their wonderful team who believed in my dream and made it come true.

Through my little Oat Cake idea ten years ago I now have such an interesting life working in NPD, helping children to have fun enjoying healthy food and have been honoured to be made a Member of the Guild of Food Writers of Great Britain.

# 9

# From basic research to marketable product – success and failure of instant baked potatoes

Buncha Ooraikul, Canada

*In this chapter, possibilities for solving a major industrial problem were explored systematically, and the design process followed to produce an acceptable product; but ended in failure because of local managerial and commercial circumstances. In PD Process diagrams there are stages with their activities, outcomes and decisions, and usually they are described as occurring inside a company. But often the activities are outside the company, especially in medium- and small-sized companies who do not have the people, resources and knowledge to organise the activities. The activity can be specific, such as a consumer survey or design of equipment, but it can be major such as the development or the launching of the product. For a smaller company, it can be the total product development project.*

*The problem is to unite the outcomes of the outside activity with the company management, personnel, strategies and internal activities. Even with a specific activity such as a consumer test of a product, there is often grumbling that there is not the correct information to select the product or redesign the product. But where the major part of the development is taking place outside the company, the final product and process may be beyond the company's abilities and resources; there is no way it can be transferred, and the whole project is a failure.*

*To ensure success in innovation transfer, the company must develop the innovation and the product strategies and, with the group who have responsibility for the product development project, develop an outline for the PD Process. There needs to be a clear definition of where decisions are to be made, what the decisions are and who makes them. Timing of the project, resources needed and available, financial constraints, company constraints, need all to be identified and understood by the company and the development group. In the end, the responsibility for the PD project rests with the company, who are using the knowledge and skills of the other group to carry out the activities.*

*During the development of the innovation, company personnel need to be involved and, especially towards the end, they need to be working with the development group so that there is no break in communication, and the technology can be transferred to the company without major difficulties.*

*Very often the PD groups are university departments and government research centres – because they have knowledge, equipment and often are not so expensive. But, no matter where they are situated, it is the cooperation with the company that matters in innovation transfer. There can be a brilliant innovation but the company cannot develop it further; or the innovation can be a 'fizzer' – too early for the market to accept, too impractical, or too costly. Innovation transfer has difficulties that need to be recognised.*

*The chapter particularly relates to pages 46–64, 111–123 in* Food Product Development *by Earle, Earle and Anderson.*

## 9.1    French fry industry and waste problems

The frozen French fry industry is a large industry in North America, supplying the product not only to the fast food outlets which are sprouting up at a very rapid rate all over the world, but also to a sizeable retail market. French fries are produced mainly by large and medium-size companies using a well-developed, highly automated processing line, which includes a series of washers, peelers, slicers, blanchers, fryers, freezers and packagers (Somsen *et al.*, 2004a). In the process a large amount of waste materials is produced, e.g. wash water, culls, peels, potato pieces (nubbins) from the slicers and inspection tables, and used oil. Somsen *et al.* (2004b) reported production yield of French fry to be as low as 30–45%, indicating a large amount of solid waste is being generated by the processing line. Therefore, it is not surprising that waste treatment costs can cut quite deeply into the profit margin of this highly competitive industry. Most companies remove solids from the waste stream before sending the effluent either to municipal or their own waste treatment facilities (Gelinas and Barrette, 2007). The solid waste is either hauled to landfills or sold cheaply as animal feed.

## 9.2    Product development challenge

A medium-size French fry manufacturer in Edmonton, Alberta, Canada was faced with a similar waste problem from its production. The company dumps nubbins and culls together with other solid waste and contracts a trucking company to haul it to the municipal landfill. The then president of the company was quite forward looking and had regularly supported university research and development efforts. Nubbins and culls formed a significant portion of his solid waste, and he had been for some time looking for ways to recover and use them to make value-added products, which would not only reduce his waste load but might also help improve the company's profit margin. The author was asked to consider the challenge and come up with ideas for new value-added products with good market potential.

## 9.3   Product conception

Nubbins and culls are pieces of peeled and washed potatoes of varying sizes, which could readily be used for making food products. Several products came to mind that could be made from this recovered waste. These included shoestrings, potato patties, mashed potatoes, crisp-fried or extruded snacks, and even potato flour. The common problem with these product ideas was that they all had already existed in the market, most did not have a high profit margin, and the company would have strong competition from the existing manufacturers. A more novel product idea must be found for this valuable waste.

At one meeting, the president wondered out loud if he could cook the potato waste and put it in some form of container and market it as precooked potatoes in retail or fast food outlets, or even vending machines, he might be able to make some money from it. The idea was feasible, but what form of 'cooked potatoes' and what kind of container needed some further thoughts.

More visits to supermarkets were made to carefully observe products made with potatoes in order to inspire new ideas. Friends and strangers alike were buttonholed for their opinions on the kinds of potato products they liked. Two kinds of potato products had been making quite a stir among the consumers in North America during that time, i.e. potato wedges and refilled potato shells. Potato wedges are made from pre-fried pieces of thick potato shells from baked whole potatoes with the centre part of the flesh scooped out. Refilled potato shells are made from baked potatoes cut in half with the flesh substantially scooped out and refilled with mashed potatoes, and may be garnished with butter or cheese, bacon bits, and spring onion or parsley. Well-known potato manufacturers, e.g. Simplot and McCain's, among several others, were making these products. The process was tedious, time-consuming and labour-intensive. For potato wedges: it involved baking the potatoes, cutting them to desired size, scooping out the flesh, frying, freezing and packaging, and for refilled potato shell: baking the potatoes, cutting them in half, scooping out the flesh, filling the half shells with mashed potatoes, with or without garnish, freezing, and packaging. These products were available on retail shelves and in restaurants, commanding good prices.

The refilled potato shell idea was quite attractive and interesting. The nubbins and culls, after removing defects, could be cooked and mashed, and may be garnished, quite easily. However, an edible shell to replace the real baked potato shell had to be developed. After several informal brainstorming sessions we concluded that our 'artificial' shell should possess the following characteristics:

- Be totally edible, safe and nutritious.
- Be organoleptically acceptable to the consumer.
- Be strong enough to hold the mashed potatoes.
- Be able to withstand freeze-thaw cycles.
- Be microwaveable.
- Be easy to produce and adaptable to mass production.
- Be cost competitive.

This posed a considerable challenge. The author had long experience in potato research, having developed and patented a process for making potato granules and worked on improvement of French fry and potato chip quality. However, development of a container in the form of potato shell with all the above properties was a novel idea in which he had no prior experience. Nevertheless, the company's president believed in the power of research and development and was prepared to support the project.

An R&D plan was developed outlining the steps through which artificial potato shell would be constructed, potato waste would be cooked and mashed and filled into the shell, and what the finished product should look like. With seed money from the company, a Canadian federal research-funding agency called 'Industrial Research Assistant Program (IRAP)' was approached for a matching grant. The application was successful and we were on our way to assemble a small research team to carry out the R&D.

## 9.4    Development of prototype product

### 9.4.1    Assembling a research team

A very important part of a successful product development is able research personnel with suitable experience and temperament for the intended job. Initially, thinking that the R&D would be relatively simple, a research assistant, a recent MSc Food Science graduate, was recruited to work directly under the author's supervision. However, several months of trial and error later a lot more intricate work was encountered than anticipated. As technical problems became more complicated, it was concluded that a research assistant with limited research experience who had to wait for directives from the project director most of the time would not be sufficient to take the project forward. A more senior researcher who could be devoted full-time to the project was needed. Meanwhile, a postdoctoral fellow who had recently graduated with the thesis on edible films was looking for a job. A colleague introduced him to us, and after an interview he was hired. The match was perfect, and the new recruit proved to be worth all the additional financial outlay.

### 9.4.2   Development of the potato skin

There were three aspects to be investigated: raw materials, film casting and shell forming.

*Searching for right materials for edible film*

The first and foremost characteristic of the artificial potato shell was that it had to be wholly edible, and secondly it could be formed into the shape of a potato shell. It should also be microwaveable or be able to withstand reheating in the oven. The initial step of the development was to find a suitable material that would form a relatively thin sheet, which could later be compressed into a shell. The material had to be at least partially soluble to form a solution or suspension from which a sheet could be cast. Several colloidal materials were tried, including pectin, gelatine, gums, starch, carboxymethylcellulose, etc. Most of them could form acceptable films or sheets. However, when they were dried and rewetted, either in the mouth or in contact with water, they would not hold their shape and would literally melt away. Therefore, they could not be expected to hold mashed potatoes, which contained about 80% moisture.

What we needed was a material that could form a film that would withstand moisture without breaking down, and would be chewable without leaving solid gristles or gumminess after chewing. It could also be modified to resemble the texture, colouring and appearance of real potato skin. In short, we wanted a potato skin without potato skin.

Our initial search for the right material was expanded, but without acceptable result. Fortunately, our newly hired Postdoc was able to steer us in the right direction. After being sufficiently briefed on the aim and objectives of our R&D, he introduced us to the materials, which he had been working on, i.e. alginate and carrageenan. Both gums are water soluble and can be cast into films and dried into sheets. They still dissolve when they come in contact with water, but the films can be made water insoluble by polymerising them with a suitable polyvalent ion, e.g. $Ca^{2+}$. The polymerised alginate film is stronger, more flexible and more resistant to water while polymerised carrageenan film is rather short and less water repellent. The alginate film is also tougher to chew than the carrageenan film. A combination of the two materials at different proportions was investigated to obtain a film with acceptable chewing property. The result was a film that was less chewy, but was somewhat less water resistant as well. This seemed very promising.

However, the initial alginate–carrageenan film was just a thin opaque sheet with no character. We wanted something that had the colour, texture and general appearance of potato skin. Thus, individual artistic imagination and resourcefulness was relied upon to come up with ideas to transform a plane white sheet into a brown film, preferably with speckles similar to those found on the skin of Netted Gem or Russet Burbank potatoes.

The colour could be easily fixed with food-grade colourings to the nearest shade, but the speckles or 'net' appearance required some thinking. Dietary fibre just came in vogue about that time, and several products were available, e.g.

oatmeal, oat bran, wheat bran, Fibrex (dietary fibre from sugar beet), etc. After considering their colour, appearance, particle size, sensory characteristics and cost, wheat bran of medium-fine particle size was chosen.

Thus, several 'potato skin' formulations were developed with various combinations of the ingredients, i.e. sodium alginate, carrageenan, brown colouring (we finally chose caramel #10 for colouring to stay natural), and medium-fine wheat bran. With the addition of wheat bran, and probably caramel as well, to the two main hydrocolloids, the desirable appearance seemed to compromise the film strength somewhat. To compensate for the slight loss of strength, we added sodium caseinate to the formula. Thus, an edible film that looked very much like real potato skin was achieved. With the addition of sodium caseinate, we could still claim 'all natural' ingredients as well as high nutritional value for our shell due to the combination of high quality protein (sodium caseinate), high dietary fibre (wheat bran for insoluble fibre, sodium alginate and carrageenan for soluble fibre), and no fat.

*Film casting*
How to cast a film with uniform thickness and consistency was a problem. A gel of sodium alginate–carrageenan mix, together with other ingredients, was produced by slowly dissolving the mix in warm water and was then poured into a flat tray, on which it spread till it reached the confinement. The thickness of the film was, therefore, determined by the size of the tray and the amount of the gel poured into it. This process was rather slow and tedious, and the film thickness was not easily and consistently controlled. A new casting method was devised using a thin-layer chromatography plate spreader, modified by making the guide-plate on the spreader adjustable to control the thickness of the gel (Fig. 9.1). This proved highly satisfactory.

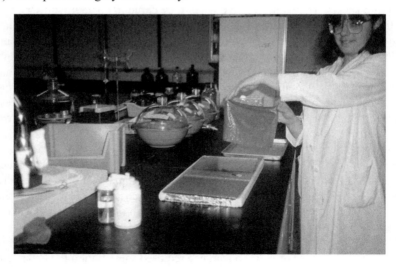

**Fig. 9.1**   Edible film cast on a TLC plate and polymerised with CaCl$_2$ spray.

*Shell forming*

We first toyed with the idea of producing a ready-to-eat whole potato. This could conceivably be accomplished by extruding mashed potatoes into a balloon-style edible shell, or by coating preformed tuber-shape mashed potatoes with the edible gel and polymerising it with $CaCl_2$. However, under close scrutiny, though the idea was considered technically feasible, to develop equipment to produce such product would be too complex and costly. Therefore, it was eventually decided that the final product should be simpler to produce and should take the shape of a half potato tuber, similar to the stuffed baked potatoes already available in the market. What we needed was an edible shell that had the shape and general appearance as close as possible to the half baked-potato shell into which mashed potatoes could be filled.

Obviously, a mould was needed to transform the flat edible film into a half potato shell. A machine fabricator who had had experience making baking moulds for local bakeries was consulted. He was asked to sign a confidentiality agreement so that we could divulge the detail of our product idea, show him our initial development and tell him what we wanted him to do for us. This was an important safeguard to protect our intellectual property and the chance for eventual patent application.

The fabricator proposed an aluminium mould similar to a waffle iron, electrically heated on both sides. The size, shape and thickness of the shell were carefully decided upon. The shell should be a regular oblong shape, approximating a medium-size tuber, capable of holding about 100 g mashed potatoes. The thickness of the shell should be no more than 1 mm so that the shell is strong enough to hold the mashed potatoes without much distortion, yet thin enough to resemble potato skin on chewing.

After a few adjustments on heating mechanisms and clearance between the male and female parts of the mould, a hand-operated edible potato shell press was born, capable of making four shells at a time. The mould was preheated to about 150 °C and a relatively moist edible sheet was placed on the female mould and the male mould was brought down and held tightly pressed for about 15–20 seconds then released. The time and temperature of the 'baking' were adjusted so that the shells would not be too dry or they would be too fragile. The thermoplastically formed shells were then trimmed of the excess film around their edges, ready for filling with mashed potatoes.

## 9.5    Market assessment of prototype product

The idea for the product seemed fine, and our team had proved that it was technically feasible. But would it sell? Would the consumers accept 'artificial' potato skin? What should the final product look and taste like? What should be its acceptable price?

These were among the main questions we discussed with our supporters, i.e. the company president and his senior staff. We could not answer these questions ourselves, and we were not equipped to conduct a consumer or market survey. The company wanted professional opinions on the product before it would agree to support us to proceed further. It had access to a consulting firm, which specialised in consumer study. The firm was, therefore, hired to conduct the study and produce a report to answer the foregoing questions.

The consulting firm set up a 'focus group' of four persons consisting of a restaurant chef, a delicatessen owner, a fast food operator, and a trained consumer researcher who served as the group leader. They were asked to evaluate the product from their individual perspective, give their opinion on market feasibility, and make recommendations for product modifications or improvement. We were asked to produce our 'ideal' product and serve it to them in the manner we thought it should be served. They would review and discuss the product in private. We were called in only to answer some questions or clarify some points. There were two such sessions. The focus group made their report to the consulting firm. The consulting firm conducted an additional consumer survey with consumers picked randomly among the shoppers in a supermarket and produced two reports to the company, with a copy to our team.

The first report was the result of the focus group's evaluation of the product. As expected, their views varied based on their respective business; but on the whole they provided very useful comments and suggestions. The chef didn't think the product would be suitable for up-market restaurants, where customers expect genuine baked potatoes, but might be a good and relatively cheap substitute for regular restaurants or institutional catering. He commended us for the looks and chewability of the shell but wanted it to be somewhat thinner for a little less work on chewing. He also stressed on the taste, suggesting adding butter and/or cheese to the mashed potatoes and garnishing it with parsley or chives and bacon bits or similar materials. The delicatessen owner and fast food operator wanted to know if the product could withstand freeze/thaw cycles and whether it could be microwaved as well as reheated in regular ovens. They considered these as important product features in their businesses. In their opinions, the product should be quite suitable for retail and fast-food outlets if the taste and price were competitive. They suggested that considerable promotion and consumer education would be needed for a novel product such as this.

The second report dealt with the results of the interviews with consumers. The consumers, picked at random, were shown our product and asked their opinion on such a product, whether they were willing to buy it, how much they would be willing to pay for it, and how often they consumed such products. Some personal information was also collected. Briefly, most consumers were unfamiliar with such a product and some were quite sceptical about the 'artificialness' of the potato skin, wondering if it was really edible and whether there could be any health hazards. Data analysis showed that about 33% of the consumers interviewed were quite intrigued by the product and would definitely buy it if it were available in the market, since they were already quite fond of baked potatoes. Another 33% said they would try it and see what it was like before committing themselves, while the rest were either non-committal or definitely against such a product.

The company and our team were certainly not elated by the results but were quite encouraged by the recommendations in the reports. After all, 66% potential buyers could form quite a sizeable market, and the product had room for improvement, which could attract more consumers in the future. We were, therefore, given a green light to proceed further with the development.

## 9.6   Fine-tuning the product

Making the edible shell thinner as recommended by the focus group was not a major problem, though thinner shells came at the expense of product strength since after filling with mashed potatoes the product would be softer and more delicate to handle. Therefore, there was a limit to how thin we could make the shell without sacrificing the product firmness, especially after reheating.

Garnishing of mashed potatoes, however, offered a much greater challenge. The following were factors to consider:

- taste and flavour of the product
- firmness of the mashed potato after mixing with butter and/or cheese
- product colour.

These factors were all interrelated, as addition of cheese or butter not only affected the taste and flavour of mashed potato but also its firmness and colour. Taste and flavour were the most important factors influencing the acceptability of the product. Some people might object to the taste and flavour of butter and/or cheese, while some people might prefer certain types of cheese and not others. Some might not want parsley or chives on their potatoes, and some might prefer just plain potatoes.

How much, and how cheese or butter should be added for best taste and flavour without unduly softening the mashed potatoes, were other issues to consider. Mixing butter or cheese, which contain fats, into mashed potatoes would soften the product and make the mash smoother. Some people would prefer their 'baked' potatoes to be mealy, but not too smooth and, therefore, may object to this. Some people just put butter or cheese over their hot baked potatoes and let it melt into the product, thus not significantly affecting the product firmness.

We decided to forego bacon bits and parsley or chive as garnishes in the initial development, since there were sensory and technical reasons to keep the product as simple and as plain as possible. Of course in industrial production, any garnishing could be easily incorporated into the lines. Also, we felt that the consumers could add butter to the product themselves after reheating, if they so desired, but cheese should be mixed with mashed potatoes during mashing and before filling into the shell. Therefore, our next step was to determine the type and amount of cheese to be added to mashed potatoes.

Informal polls among friends and colleagues revealed that cheddar cheese was probably most compatible with potato products. We experimented with mild, medium and strong cheddar and contracted the Sensory Division of the government's Food Processing Development Centre (FPDC) to conduct a consumer test for us, using plain mashed potatoes as control. Some 30 consumers were recruited from the Centre's personnel and factory workers in the area to serve on the panel. Results indicated strong preference for 5% medium cheddar among those who liked cheese in their mashed potatoes. However, when asked to choose between plain and cheese-incorporated products the majority picked the plain product with comments that they preferred to garnish the product themselves. The consumer test also confirmed the focus group and consumer survey reports that about two-thirds of the consumers would buy our product.

At this point we felt that we had gone as far as we could in fine-tuning the product for consumer acceptability. The next step was to design a complete processing flowchart to show all the steps involved in manufacturing the 'instant baked potato' product. This is essential, especially for a novel line of product, if an industrial-scale production is anticipated. Indications from our industrial supporter were for us to proceed with the development.

## 9.7    Process development

There were two separate processing lines involved in manufacturing our product; one was for the production of potato shells and the other for the production of mashed potatoes. The two lines met where the filling of mashed potatoes into the shells took place.

The shell production line consisted of mixing dry ingredients in warm water into a paste, casting the paste on a flat plate, partially drying the thin layer of the paste, polymerising the partially dried paste with $CaCl_2$ by spraying it with or dipping it in $CaCl_2$ solution, peeling the sheet from the plate, and placing the sheet in the heated mould to form the shell. Excess material might be trimmed off the edge of the shell, if necessary.

The mashed potato line consisted of cooking potato slices (assuming receiving potatoes in a ready-to-cook form), and mashing the potatoes (cheese may be added to the potatoes while mashing). The mashed potatoes were then filled into the shell and the finished product frozen before packaging in a suitable form of package and kept frozen. The processing flowchart is shown in Fig. 9.2.

## 9.8    Pilot plant development

### 9.8.1    Funding for pilot plant development
It is interesting to recall the mood of both the research team and the supporting company at that particular time of development. Deep down, neither side expected the project to reach that stage. It had been over a year since the project

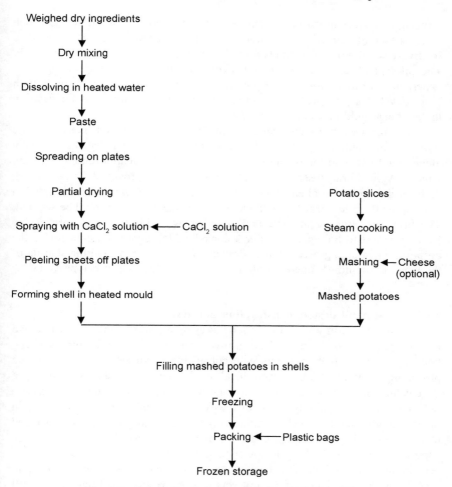

**Fig. 9.2**   Processing flowchart of instant baked potatoes.

started and the research team had to overcome many technical obstacles, especially in the development of the shell. We were not altogether so confident that we could accomplish the task that the company's president had challenged us to pursue. The author was quite sure that the president's original thought was to support the project as an intellectual exercise and did not anticipate any real success. Though he had been regularly apprised of our progress, when he was finally presented with the finished product he was still quite taken aback. The company's executives were quite amazed at how real our product looked as potato, and even complimented us about its texture and taste. We had presented them with a real product, the future of which had now to be decided by the company, whether or not its development should be pursued to the end.

We informed the company that if the decision were to continue with the project, the next step was pilot plant development, and that would mean a much more substantial financial investment. We had till that time worked with a relatively small amount of funds provided by the company, matched by IRAP. The pilot plant development would involve contracting an engineering firm, experienced in food processing equipment design and fabrication, to work with us to develop a pilot-scale processing facility. An estimated cost for such an undertaking could be as high as half a million dollars.

The company's president made it quite clear that he would like to see the development continued to the end, but he was equally clear that the company alone could not fund the whole project. A provincial research-funding agency, Alberta Agricultural Research Institute (AARI), was approached for assistance. At the same time, a local engineering firm was asked if it would be interested in developing the pilot plant for us if and when adequate funding could be secured. The firm was quite enthusiastic and, together, we submitted a comprehensive R&D proposal to the agency. After a considerable negotiation concerning the proportion of funding between the company and the agency, and the sharing of the benefits should commercialisation succeed, the funding was approved.

### 9.8.2    Pilot plant design, construction and testing

The engineering firm was recommended to us by a government agency that had had equipment made by them. The firm was run and operated by two brothers, both highly competent engineers with considerable experience in constructing processing equipment for several local food processing establishments. We visited the company several times to take a look at their operations to assure ourselves of their capability and to discuss with the two brothers what we had in mind. The idea and the level of engineering challenge they had to face intrigued them. In fact, they were quite eager to accept the project as another learning experience and were willing to lower their fees, which would otherwise have been quite high had the project been the usual equipment design and fabrication.

After a confidentiality agreement was signed, we disclosed our idea in full detail and demonstrated to them the processing methods step by step. The research team proposed at first that we should break the whole processing operation into sections, i.e. the making of the paste, the production of the sheets, the forming of the shells, the cooking and mashing of potatoes, the filling of the shells, the freezing of the product, and the packaging of the frozen product. In doing so we thought we could make use of most of the equipment already available in the market to perform the jobs in each processing section, thus avoiding the need to design and construct new equipment and, hopefully, shorten the project time.

But the two engineers had a different idea. They instead proposed combining the various processing sections as we had originally visualised into one con-tinuous operation to accommodate the making and the filling of the shells in one compact pilot plant unit. To test the feasibility of their idea they experimented with various parts of the shell making operation to see whether or not some could be combined and make the main processing line truly continuous. The idea was an exciting one, but the research team had some reservations as to its technical feasibility. The most difficult part as we saw it was how to combine the sheet making, the shell forming and the edge trimming into one continuous operation. There appeared to be an irreconcilable difference between the visions of the mechanical engineers and those of the food technologists/scientists as to what was possible in food processing operations.

To prove us wrong, the two engineers set out to experiment on their idea of making a half imitation potato shell in just one operation. They looked closely at the method of paste making and at various physical properties of the paste, i.e. viscosity, spreadability, adhesivity, polymerising ability, and drying characteristics. The paste viscosity (or consistency) is very important, since it controls its transport characteristics, the amount it can adhere to the mould surface and, hence, the eventual film thickness, and the evenness it can spread over the mould. The viscosity also affects the drying characteristic of the paste. With what the engineers had in mind, the adhesivity and spreadability of the paste were also very important in determining whether it would adhere to a certain type of surface and how easily it could spread over the surface. Of course, all these properties are affected by the paste temperature. To complicate the matter further, the paste had a considerable amount of solid particles, i.e. wheat bran, suspended in it, and this would precipitate if not kept afloat by some mechanical means.

The paste had consistency similar to honey or thick syrup, i.e. about 3,500 centipoise (cP), thus it did not flow easily and a special pump was required to move it around, especially with suspended particles. This could be improved by keeping the paste heated at a certain temperature and the solid particles suspended by a mechanical stirrer.

The first thing the engineers needed to know was what kind of surface the paste would adhere to. A simple experiment was conducted by dipping various types of materials, e.g. glass, stainless steel, plastic, etc., into the paste at a certain temperature to see if it would adhere to their surfaces. If it did, how thick was the layer of the paste and whether it could be spread evenly over the surface. Results showed that the paste would adhere to most surfaces including metals and plastics.

What the engineers had in mind for the design of the equipment, to make a complete shell in one continuous operation, was nothing short of brilliance. They would make a 'male' mould, the shape and size of an average half potato shell, dip it into the paste, pull it out and rotate it around to spread the paste evenly, partially dry it, dip into calcium solution to polymerise it, dry it in a stream of hot air till it could be removed from the mould, eject it into a 'female' mould, and fill it with mashed potatoes.

To test the practicality of the idea they made a male mould from a Teflon block. They first considered an aluminium mould but decided against it for a few reasons, i.e. its low heat capacity, machinability, and high cost as compared to Teflon. Being a good conductor, an aluminium mould would heat up very quickly and, with high expansion quotient, might create mechanical problems on a continuously moving belt that they had envisaged, not to mention a possible adherence problem between the paste and the hot and shiny aluminium surface. A Teflon block, on the other hand, with a much higher heat capacity and lower expansion quotient would be easier to handle, and it could be machined more easily as compared to aluminium. The Teflon mould could hold the paste on its surface, the thickness of which was dependent on the viscosity and the temperature of the paste. By constantly flipping the mould from side to side the paste could be dispersed quite evenly over its oblong surface.

The sequence of drying and polymerisation of the paste on the mould surface was also crucial. If the freshly spread paste was polymerised immediately with $CaCl_2$, the paste layer would shrink and pull away from the mould. Therefore, it was necessary to partially dry the paste layer first before polymerisation. This way the layer would retain its size and shape and would continue to adhere to the mould surface after the polymerisation. The mould had to be constantly flipped around to keep the paste, which was still fluid, evenly spread on the mould surface till it was dry enough for polymerisation without excessive shrinkage. The polymerisation process also needed to be rationalised. If the whole process was to be enclosed in a hot air chamber, as visualised, then spraying $CaCl_2$ solution on to the paste was out of the question since much of the solution would be vaporised before it reached the mould surface. Besides, it would produce the level of humidity in the chamber that would render the drying process impossible. Therefore, dipping the mould with partially dried paste on its surface into $CaCl_2$ solution appeared to be the only viable alternative. Dipping would substantially reduce the consumption of the solution as well, as compared to spraying. The concentration of $CaCl_2$ and the dipping time were important factors affecting the polymerisation.

It was quite a memorable sight to witness the engineers using a mock-up processing line, with the paste in a saucepan kept heated on a hotplate, a Teflon mould attached to a ruler to control the depth when dipped into the paste, a hairdryer to provide a stream of hot air to dry the paste on the mould surface while the mould was manually flipped back and forth, and a beaker of $CaCl_2$ solution into which the mould with partially dried layer of paste was dipped for a pre-determined time. From this unlikely beginning, the engineers drew up a blueprint of our pilot plant. They estimated the fabrication time to be six months, but it actually took eight months before the pilot plant was completed and ready for testing.

The finished $8 \times 4 \times 6$ ft height pilot plant consisted of 25 rows of two male moulds carved out of Teflon blocks and screwed on to rectangular aluminium bars attached to a carriage chain on each side via metal shafts and gears. The moulds were to pick up the paste from the paste tray, to be partially dried before being polymerised in the $CaCl_2$ solution tray. The paste and the $CaCl_2$ solution were held outside the chamber and continuously pumped onto the trays. The shells on the male moulds were dried as they moved along inside the chamber before being pulled into the female moulds, filled with mashed potatoes and

**Fig. 9.3**  Interior of the pilot plant showing rows of moulds inside the chamber and mashed potato filler at the front end of the unit.

ejected onto a moving belt to be frozen and packaged. A computer program controlled the movement of the moulds, the paste and $CaCl_2$ solution trays, the female moulds, the mashed potato filling section, and the air temperature inside the chamber. The chamber was enclosed in stainless steel panels (Fig. 9.3). Human intervention was needed only to prepare the paste, the $CaCl_2$ solution and the mashed potatoes, and to load the product into the freezer and subsequently package them in plastic bags. If everything functioned properly, it would take 18 minutes for one cycle of the shell making and filling operation.

## 9.9    Production trial

After the fabrication of the pilot plant was completed and several dry runs were carried out to ascertain its mechanical worthiness, the unit was moved to the Food Processing Development Centre for production trials. The Centre had the necessary utilities for the pilot plant, i.e. three-phase 220 V outlets, 90 psi air supply, food-grade steam, cooking kettles, a Zip Freezer, a packaging machine, and cooling and freezing storage. The engineers worked closely with our research team and the company's personnel to set up and test-run the pilot plant. There was a high level of anxiety on everybody, not only for the research team and the company's workers to learn to operate the new equipment and to understand its idiosyncrasies, but also for the engineers who were not altogether confident that their totally new machine would work as it was designed to do.

As is usually the case with most newly developed equipment, nothing went smoothly in the first few attempts. Most problematic in our case was the movement of the moulds, which could be out of step after an hour or so of operation, due possibly to the expansion or contraction of the various parts of the pilot plant, or the slippage of the belt, or irregular air pressure, etc. For example, sometimes the moulds did not rotate properly after dipping into the paste, making the shells thicker on one side than the other. Or, the male and female moulds did not synchronise properly and the shells were not drawn off the male moulds, or the shells were not properly dried and stuck too firmly to the mould surface to be removed by suction. When this last problem occurred, the moulds with already one layer of shells continued on their way through another cycle, thus forming thicker shells of two or more layers before they could be removed. Occasionally, the mashed potato-filling unit misbehaved and the shells were not properly filled, or not filled at all. Sometimes, the main mould moving mechanisms slipped out of line and the whole pilot plant had to be shut down for repairs.

Nevertheless, most kinks were ironed out in time and we had a long series of successful test runs. The product, with and without cheese, looked and tasted excellent. Indeed, it was a steep learning curve for both the research team and the engineers throughout the test runs. For the research team, it was the excitement of seeing the two years of hard work on bench scale metamorphosed into a pilot plant that churned out the product continuously that made it all worthwhile. For the two brothers, the 'instant baked potato' pilot plant confirmed their engineering capability and strengthened their innovative spirit (Fig. 9.4). They were convinced that they had learned a lot more about how such a plant worked and reassured us that the second-generation plant would be much more reliable.

**Fig. 9.4**  'Instant baked potatoes' being produced by the pilot plant.

## 9.10    Consumer test

Unavoidably, the product produced by a pilot plant on mass scale would be quite different from those manually produced on a bench scale. Therefore, it was essential that another consumer test be carried out to determine whether the results previously obtained by our focus group, the consumer survey and the FPDC's Sensory Division were still valid. We once more contracted the Sensory Division to conduct the test for us. About 40 persons were recruited from the pool of factory workers in the area to serve on the consumer panel. Their report reiterated the fact that the majority of the consumers liked the looks and the taste of the product. However, they complained that some of the samples were too soft and the shell could not hold the shape of the product after reheating in a microwave oven. Also, there seemed to be some inconsistency in the thickness and, sometimes, colour of the shells.

These problems were attributed largely to the difficulties experienced with the operation of the pilot plant, especially during the early trials. By and large, the problems were overcome after the pilot plant had been adjusted to run more smoothly. As for the shape of the product being distorted after being heated in the microwave oven, plastic trays similar to egg cartons, were developed to hold the product in place during reheating and appeared to help solve the problem to some degree.

## 9.11    Intellectual property development

Considerable publicity was generated from this project through local newspapers and TV since we had completed our bench-scale development. The research team and the supporting company felt that our project had commercial value and, therefore, the intellectual property (IP) needed to be protected. The university's Industry Liaisons Office (ILO) was contacted and after being convinced that the IP we had developed deserved protection, they assigned a patent lawyer to work with us to prepare a patent application.

The patent application process was another valuable experience to us. A patent lawyer hired by the ILO spent several days with us to learn about what we did in our research project and whether it was new knowledge that would withstand the scrutiny of patent officers. As with most other patents generated by the university, it was decided that we should first apply for a US patent, and if the application was successful a Canadian protection should follow. It was reasoned that the biggest market for our type of product in North America was in the US. We should, therefore, try to get protection in that market first. The Canadian market was smaller, but we needed protection, nevertheless, if we wanted to set up production in Canada.

The lawyer first wanted to find out who were the inventors. This was determined through the role each participant played in the project. For example, who came up with the idea; who developed the research plans; who gave directions to the research activities; who contributed valuable suggestions to the research, etc. It was interesting to note that the lawyer did not think that laboratory assistants, who just followed their supervisor's directions in conducting the experiments without any intellectual contribution, should be included in the list of inventors.

The 'invention' was then thoroughly reviewed to determine whether what we had 'discovered' or done was obvious to the practitioners in the field, or whether there had been any 'prior arts' similar to ours. Literature search, especially patent search, was conducted thoroughly. What should be patented was another very important question, which may eventually determine the success of the application. There appeared to be three aspects of the research outcome that we could apply for patent, i.e. the product, the process, and the equipment. We certainly had developed a product, which we thought unique, and as far as we knew, nothing like it had been done before. We had developed a processing method, which was, again, unique to this product. However, we began the patenting process long before the pilot plant was developed, therefore, we had no equipment that we could patent at that time. That left the product and the processing method, and the lawyer suggested that we should combine both in one application and make the claims as broad as possible to prevent possible future encroachment.

Thus, the title of our patent was 'Synthetic food product' (Ooraikul and Aboagye, 1986, 1987), and the application, prepared by the lawyer, covered the method of making food products, which involved filling 'synthetic' but edible wrappers, or casings, or containers with food products. The food products (the filling) could be potatoes, rice, meat, vegetables, fruits, or combinations thereof. In the application, the list of ingredients for the shell was given, together with the method of making it, but in the claims a range of the weight or proportion of the ingredients or temperature or time of heating, etc., was given to prevent patent breaking by slight modifications. Some possible alternatives for the ingredients were also given. For the US patent, it took just over a year for our application to be issued, and a year later the Canadian patent was issued. The patent was assigned to the university, thus university's rules governing intellectual properties (IP) developed in university laboratories applied with regards to the sharing of benefits between the university and the inventors, and how the IP should be commercialised. In our case it was quite clear; the exclusive right to the IP was given to the company that supported our project.

## 9.12    Commercial trial

### 9.12.1    Certainty of success went awry

Our product development project was quite unique in that, unlike most other PD projects conducted by either universities or corporations, ours was a combination of both. That is, a private company requested and supported us to conduct the research and development in university laboratories; therefore, we expected that the company would want to commercialise the product and/or the technology themselves. We were led to believe by the company that this would be so. The engineering firm that developed the pilot plant had already made a plan to build a commercial plant that would produce up to 10 000 units of the product per hour, using even better design and newer technology. They were ready to develop a blueprint for such a plant and even envisaged the need for setting up factories in the US.

The company asked us to estimate the production cost per unit of our product, fully packaged, and compare it with the retail price of the refilled baked potatoes available in the market at the time. Our best estimate for the cost of our product was C$0.25/unit, and an average retail price of commercial products was C$0.89/unit. The company calculated that they could make a good profit if our product was retailed at C$0.65/unit. Excitement was in the air, and we dreamt of our share of the profit.

Several months after the pilot plant study was complete we learned that the company's president had announced his retirement. The company had also accepted a takeover bid from an English firm. We were stunned. We had lost our champion whom we believed had enough faith in the product and would try his best to take it all the way to the market. Nevertheless, we remained somewhat hopeful that the new owner would consider our product and technology viable enough to proceed with its commercialisation.

That was not to be. The president informed us that our technology was not part of the takeover package. He was of the opinion that the new owner would have a lot on his plate to deal with after the takeover and would not want to worry about any new venture and, therefore, he decided not to include our pride and joy in the package. Since he had always been supportive of the university he would sign his right to the technology back to the university, with the hope that the ILO would be able to license it out to some companies or individuals. We were deflated. Our years of hard work, dogged determination to overcome many obstacles, and the joy of technological success were obliterated with this news.

We knew the limitations of our capability, and commercialisation of technology was certainly not among the list of our expertise. Giving the technology back to the university could only mean one thing, in our experience, and that was we had to learn to be salesmen. The university was not in the business of marketing our technology for us, not without outside assistance. Therefore, if we did not want to see our technology remain on the shelf for the rest of its patent life, we had to try to market it ourselves or help the university to do it.

### 9.12.2  Attempts at technology marketing

Substantial publicity followed our project, from bench scale to pilot scale. A number of people had tasted our product and often enquired when they would see it in the market. However, since the sale of our supporter's company not much happened with regards to the technology, and the interest seemed to have waned. The pilot plant was still left at the FPDC, and they were anxious to know its future, as they needed to clear the space for somebody else.

We encouraged the ILO to contact some companies or individuals, both in Canada and the US, whom they thought might be interested in the technology. We would be happy to show them the product and the pilot plant. This they did, which resulted in several inquiries from a number of small potato-manufacturing companies in the US and some entrepreneurs in Canada. A few came to visit us to review the product and discussed licensing terms with the ILO. One small US company brought their own refilled baked potato products to compare with ours and was quite impressed with the novelty of our product and technology. He was intrigued by the ability to mass-produce the product at a much cheaper cost. He promised to discuss our technology with his executives and let us know whether they were interested in licensing it. Nothing was heard from him again. A similar scenario repeated itself a few times and as years passed we became increasingly discouraged and resigned to the fate of another good idea remaining on the shelf because of an unfortunate turns of events.

### 9.12.3    How to choose wrong partners

Several years later a small group of entrepreneurs who had just acquired a 'shell' public company was looking around for new products or technologies that they could commercialise. The idea was that they acquired these technologies, tried to promote their commercial potential in the stock market and raised funds from the public or private investors for the commercialisation of these technologies and/or products. They searched university patents and found that our technology was ripe for the picking. They approached us and we put them in touch with the ILO. They agreed on a relatively small fee to 'review' our technology. In the agreement, we had to produce enough products with our pilot plant for them to conduct a marketing trial. We, again, set up the pilot plant at FPDC with the help of its original developer. This time we also designed a packaging label, hired a few workers, had plastic bags and cartons of appropriate sizes made, purchased ingredients, and signed a space leasing agreement with FPDC. Over a period of two months we produced some 30,000 pieces of the product, both with and without cheese. During that time the company's officers came and visited us and took pictures of the pilot plant churning out the product. Apparently, this was for publicity and promotion of their stocks. A freezer trailer was rented to store the products, which were neatly packaged and boxed. All was ready for the company to carry out marketing trials.

The company made some attempts to contact consulting firms specialising in market trials for new products. One of them who had some experience in the Calgary area gave a quote of C$50,000 for the job. We were told that the company was negotiating the fees with them. Several months later, the company informed us that they could not afford the fees and were in the process of negotiating with another group of entrepreneurs from an Eastern Canadian Province to help them explore the market potential for the product and the technology in their Province. This Eastern group visited us to evaluate the product and the pilot plant. They eventually signed an agreement with the university and the company to move all the product and the pilot plant to their Province. We were quite apprehensive since we knew that they would need the help of the engineers who developed the pilot plant to set it up and get it operational on their premises, wherever that might be. Nevertheless, it seemed to be the only way out at the time.

Five months later we learned that this Eastern group of people had decided to abandon the project, without ever setting the pilot plant up or attempting to market test the product. This group of entrepreneurs led everybody to believe that they were serious in their effort to commercialise the product and the technology. Apparently, they expected financial assistance from their Provincial government for the undertaking and when this did not materialise they simply left the 30 000 pieces of product to rot, cut their losses and informed us that we could take our pilot plant back. In their agreement with the ILO, there was no clause to require them to pay for the return shipment of the pilot plant, or for any negligence or failure on their part. In the end, the university had to pay the shipping cost and the pilot plant has been mothballed in the university's warehouse ever since.

## 9.13    Problems met and lessons learnt

This case study is a classic example of the entrepreneurial naivety on the part of university researchers. We are happy and content to use our scientific and technological ability to explore new frontiers and solve problems, especially when we are well supported. However, when it comes to the commercialisation or marketing of our knowledge or discoveries we should be well advised to leave it to the experts. It is best to get the industry involved right from the beginning of the project if we feel that our R&D outcome may have commercial value. Even so, there is no guarantee of success. Perhaps, research scientists should take some business courses to understand the private sector better and minimise their rate of failure when they have to venture into the outside world.

Our instant baked potato project appeared to have every chance of success. Indeed, we were very successful in finding a way to solve a solid waste problem for the French fry industry by developing a value-added product from the waste material. We had brought the project from product idea conception right through to the development of a technologically advanced pilot plant which could be quite easily scaled up for commercial production. Unfortunately, we did not anticipate the turn of events, which could be quite common in the industry, to turn the fortune of our project upside down. And when we had to cope with the commercialisation problems ourselves we were totally ill-equipped to deal with them. Food industry is the biggest and probably the most important industry in the world, but its profit margin is low and not as attractive as compared to most other industries. This could be one of the reasons for our failure to commercialise our technology and product, which, though novel and exciting, is still a lowly food product for which most people refuse to pay too much money.

## 9.14    References

EARLE M, EARLE R and ANDERSON A (2001), *Food Product Development*, Cambridge, Woodhead.

GELINAS P and BARRETTE J (2007), 'Protein enrichment of potato processing waste through yeast fermentation', *Bior Tech*, 98, 1138–1143.

OORAIKUL B and ABOAGYE NY (1986), *Synthetic food product*, U.S. Patent No. 4,582,710.

OORAIKUL B and ABOAGYE NY (1987), *Synthetic food product*, Canadian Patent No. 1,219,165.

SOMSEN D, CAPELLE A and TRAMPER J (2004a), 'Manufacturing of par-fried French-fries Part 2: Modeling yield efficiency of peeling', *J Food Eng*, 61, 199–207.

SOMSEN D, CAPELLE A and TRAMPER J (2004b), 'Manufacturing of par-fried French-fries Part 3: A blueprint to predict the maximum production yield', *J Food Eng*, 61, 209–219.

# Part IV

# Technological development

# 10

# Radical process development – cutting techniques for an aerated sugar confectionery

Rex Perreau, UK

*This chapter covers technological aspects of the knowledge base for product development, in particular process change. Process change is an integral part of New Product Development (NPD). Most new products are first produced using batch processes. Although the labour costs of such processes are usually high, they are cheap to establish. Process refinement is justified only when the product is established in the market and volume has grown. A very significant change comes when the process is developed into a continuous one.*

*The new process technology has to be explored so there is a need for initial study and experimentation into alternative methods. This may require the construction of pilot plants on which to undertake the necessary testing and proving, before the final production plant is built and commissioned. It can take a great deal of time, money and resources. A further stage in the life of a product can then be reached where the market has grown to the extent that the first generation of continuous plants are no longer able to economically meet the market demand. A radical reappraisal of the process becomes necessary in the search for economies of scale, reduction of waste, and use of modern computerization.*

*For the purposes of process development, the total process is divided into standard steps called unit operations, which are individual processes such as heat transfer, drying, evaporation, size reduction and separation. Each unit operation is taken separately, related to appropriate theory and experience and to the circumstances of the problem. Available alternatives are explored; the selected one is built up on paper and if necessary modelled to meet the requirements. Systematic experimentation is used to find the optimum process-ing conditions. The unit operations are then strung together to build up the total process. The product material has to move into and out of each operational unit, so flow patterns, material control and storage have to be considered. Material balances are measured as components enter and leave.*

*In this chapter, where a radical reappraisal of a well established continuous process was undertaken, the key unit operation researched was size reduction and in particular, cutting. A straightforward multiplication, to meet the required volumes, would have presented great problems in dealing with the waste generated. So a novel cutting technique was tried, found a possibility, and then various particular difficulties explored thoroughly and overcome. The optimum cutting method was then combined with the other unit operations such as evaporation (sugar boiling), mixing (magma foaming), and heat transfer (magma cooling) to form the production line.*

*The chapter particularly relates to pages 150–153, 165–191 in* Food Product Development *by Earle, Earle and Anderson.*

## 10.1    Introduction

In the manufacture of confectionery, one of the most difficult unit operations is size reduction or cutting. Some materials can be very sticky, some fragile or brittle, and some collapse under the weight of a knife. This is particularly true where a firm material has been layered on top of a softer one, or where firmer inclusions such as raisins or nuts have been included in a softer base such as light nougat, sometimes referred to as nougatine. This chapter is about the development of cutting techniques for an aerated sugar glass. When this project was initiated, it was recognised that the existing methods of cutting produced high levels of waste and therefore contributed significantly to the costs of production and the size of plant necessary to produce the tonnage of product required. Process development was essential if the generation of waste was to be minimised or eliminated, and the costs of production reduced. This aerated sugar glass typically has a density of 0.4. Its moisture content is approximately 4% and at this level, it is very hygroscopic. It is also extremely brittle at room tem-perature. It tends to fracture if subjected to temperatures at or below 0 °C. The confection is Crunchie manufactured by Cadbury Schweppes. The author had the opportunity to see the earliest methods of manufacture in production. He took part in the tuning of the first of the continuous plants, and was part of the team responsible for the design and development of the latest one with particular responsibility for proving the capability of liquid jet cutting.

## 10.2    The history

The product described here was initially developed in about 1935.

### 10.2.1 The first commercial process

The earliest method of making this product consisted of boiling sugar and confectioners glucose syrup to about 150 °C in a copper pan over a gas flame. Sodium bicarbonate slurry was then beaten into the hot syrup by hand with a wire whisk and the foaming mass poured onto a cooling table. Here it was made into a flat slab by placing a weight on top. When cool, the sugar foam was transferred to a cutting table. Tungsten wires were strung between two holders, one at each end of the table. They were mounted above the product at a pitch which corresponded with the desired thickness of the bar to be produced and were heated by passing a low voltage current through them. The frame holding the wires was then lowered onto the product to be cut and the current turned on. The hot wires melted the sugar glass and fell slowly through the product much in the way that a cheese wire works. Bars were then separated by hand before enrobing with chocolate.

*Problems*
- This was a very labour intensive batch process.
- Plant capacity was very limited.
- Waste levels were high.
- Broken and damaged bars could not be recycled back into the product.

In spite of these handicaps, this process was used for the next 30 years in some parts of the world.

### 10.2.2 The first continuous process

This was developed in the early 1950s. Sugar and confectioners glucose syrup was continuously cooked in a microfilm cooker or coil cooker. In a microfilm cooker the syrup is continuously fed into a steam jacketed barrel where it is wiped onto the inner surface by a series of rotating blades. A coil cooker consists of a steam chest containing a simple coil through which the syrup passes as it is being concentrated. The cooked syrup was then passed through a mixing pot where the bicarbonate slurry was added using a variable speed peristaltic pump or small piston pump with an adjustable stroke so that the rate could be varied. This was necessary as the amount of slurry required varied depending on the particular characteristics of the sugar and glucose being used at the time. The hot magma was then poured onto a plastic coated canvas belt and passed through a cooling tunnel. Reciprocating side frames were used to control the edges of the magma slab until it was set. On some plants two belts were used, the product being transferred from one to the other part way down the cooling tunnel.

The slab once cooled to about blood heat (the magma becomes very brittle if it is too cold) is cut across using a travelling band saw mounted across the travel of the magma slab. The cut slabs pass through a re-warming cabinet to make them more malleable and more able to withstand physical shock. They are then pushed through a gang saw. Here the saws are mounted at the desired distance apart to give the correct size of bar. The first saws in use were of the sort that would be found in any carpentry shop. However, there was another type developed and used on some plants. This consisted of two gang saws. One mounted below the cutting table and the other mounted above and were slightly offset. The saw blades in these gang saws were made of thin metal with simple triangular teeth. Each blade consisted of segment of a circle. The segment was less than 180 degrees so that when two blades were mounted opposite each other to maintain balance, there was a significant gap between them. The saw revolved at very high speed and balance was crucial. Bars produced by the gang saws were then either placed by hand onto a belt feeding an enrober or they were fed to a series of guides that turned and spread them before feeding them to the enrober.

*Problems*
- Saws produced large quantities of sawdust. Early processes used filter bag mechanisms to capture it. The dust, however, was very hygroscopic and high temperatures were necessary, in areas where dust was being handled, to keep the relative humidity down. Later plants developed scrubbers where the dust was captured with water or dilute syrup sprays.
- High levels of breakage were suffered, particularly if the magma was too cold at the time of cutting.
- The height of the magma slab was difficult to control. The foam, when hot, was two to three times higher than when it was cool. There was a significant time delay between the sugar foam being made and the final height being reached. This time delay made it difficult to ensure that the height after cooling was always correct. Some designs of saw were fairly intolerant of variations of magma height.
- The side guides used to restrain the edges of the magma in the cooling tunnel had a tendency to foul it. Magma would stick to them and cause the sheet to deform and tear.

### 10.2.3    The challenge
To design and develop a single plant that would be capable of producing 10 000 tonnes of product per annum and replace three existing plants operating in one factory. The development programme looked at all major aspects of the process. Here we will concentrate on the problem of cutting this material at the rates required while minimising waste.

### 10.2.4    The organisation
*The factory*
The site where this plant was to be installed had been making this product since it had been marketed back in the 1930s. They therefore had a deep under-standing of the product and the current process. They also had significant process engineering, plant design, and computer control resources. A process

engineer in this factory who reported to the factory manager developed the basic design and concept for this plant. He led the team charged with the development, building and commissioning of this multi-million-pound plant. The concept was a 21st century one. It was to be a fully computer controlled plant capable of automatic start up and shut down triggered by a time clock. Parts likely to fail were to be duplicated to improve reliability.

### The research development department
This was based 90 miles away from the client factory and managed by a technical director on the company's board. There was an extensive pilot plant facility staffed with fitters, mechanical, electrical and process engineers, chemists and technologists.

### Research facilities
The company also maintained a separate facility for more fundamental research. It was staffed with scientists and equipped with an electron microscope, mass spectrometer and other equipment necessary for investigative research.

### External consultancy
An external consultancy was employed to provide an overview and source of independent advice to the factory manager. This organisation specialised in the design and sourcing of specialised equipment to solve difficult problems. They were particularly good at producing and developing innovative ideas.

Meetings were held at key stages in the development of this plant. Relevant people from the factory, research and development and the external consultancy attended.

## 10.3    Methods explored for cutting of sugar glass foam

Existing production techniques for cutting foam produced typically 20 to 25% of waste as dust and a rather more variable percentage of broken bars and selvage, which had to be recycled or disposed of in some way. Therefore the major, but not only objective for the development of the next generation of production plants was to find better ways of cutting this brittle hygroscopic material while minimising waste from breakage and dust produced by the saws. All possible methods were considered. This initial investigation was undertaken jointly by staff from the factory and the R&D department.

### 10.3.1    Laser cutting
A low powered laser (60 watts) was found to cut this material. However, the speed of cutting at low power was too slow to make it a viable proposition for a full-scale production plant. A high power 2 kW laser was found and a visit made to test out its cutting speed. To our surprise and horror, it set fire to, rather than cut the magma. The smoky flavour generated was also unsatisfactory!

### 10.3.2    Vibratory saws

Reciprocating saws are used in hospitals to remove plaster casts from limbs that have been fractured and subsequently healed. This device consists of saw like teeth that are moved back and forth at high speed. The rigid plaster is cut by these teeth but the more flexible skin and tissue underneath is not damaged. It is understood that the same principle applies to cutting bone such as the skull. The bone is cut but the soft tissue, in the vicinity of the cut, is undamaged. In the case of the sugar foam, the device worked well for a couple of minutes but the saw generated heat causing the sugar glass to melt and become sticky, clogging it. This technique was quickly abandoned.

### 10.3.3    The next generation of conventional saws

Some form of conventional saw, be it band saw or circular saw, was seen to be the safe option. We should try and benefit from the experience of using saws of varying designs around the world in the last few decades to design a new one for the planned plant.

### 10.3.4    Liquid jet cutting

It was known that liquid jet cutting (using water) was being successfully used for cutting a range of materials. A well-known shoe manufacturer was using it for cutting out patterns. We arranged a visit to see this in action. There was one serious disadvantage. The sugar foam is very hygroscopic. While only a trace of water was left on the cut surface, this was enough to cause the partial collapse of the sugar foam. The technique looked exciting as the cutting rate looked high but we would need to find another cutting medium!

### 10.3.5    The decision

The potential of liquid jet cutting looked so dramatic that it simply could not be ignored. However, the risks were perceived to be very high. There was no in-house experience of this cutting technique or any knowledge of the associated engineering. It was agreed at a meeting of all the parties involved, that the exploration of this technique should be undertaken by the R&D department. They had all the necessary facilities to run extended tests on this equipment. To ensure the viability of the entire project, however, we needed a low risk option. The external consultancy we had employed, were asked to develop the next generation of saws as a fall back position.

## 10.4    Pilot plant development of liquid jet cutting of sugar foam

### 10.4.1    Basics of liquid jet cutting

The first attempt at cutting materials using a high pressure jet was made by Dr Norman Franz in the 1950s. He attempted to cut timber by dropping weights onto columns of water and forcing it through an orifice at the very high pressures generated. This high pressure jet was of course available for only a very short time. He was not successful in making a continuous high pressure liquid jet.

This only became possible when the intensifier was developed in the early 1970s. The basic configuration of an intensifier can be seen in Fig. 10.1. Hydraulic oil at say 200 bar was used to drive a piston or biscuit as it is sometimes referred to, which in turn is linked to yet another piston of say one tenth its surface area which is pressurising the actual cutting fluid. This cutting fluid is in this way pressurised to ten times the 200 bar, i.e. 2000 bar. The maths is fairly simple. However at such extreme pressures there are a number of problems. The high pressure piston needs to be very inflexible, as any distortion would make it difficult to maintain seals. Pistons are now made of tungsten carbide and at one time they were the largest single piece to be made out of this material. The design and development of seals also became crucial, as did the development of check valves and pipe work capable of withstanding the pressures. (More information on liquid jet cutting may be found on the following website www.flowcorp.com/waterjet.)

The other key device was the nozzle itself. It had to withstand the extreme shear forces as the liquid was forced through it. A number of materials are used. The most commonly used one is sapphire. A typical nozzle showing the way the sapphire is held is shown in Fig. 10.2.

It is necessary to understand that the nozzle is a simple hole in a piece of sapphire but that the liquid being forced through it does not touch the sides of the hole. The fluid stream separates itself from the nozzle at the shoulder of the hole and passes through without touching the sides. If it should touch the sides, the jet becomes distorted and ineffective.

Much of this we were yet to learn. So back to the narrative. Once jet cutting had been identified as a possible technique, contact was made with a manufacturer of jet cutting equipment in the USA who had test facilities and who claimed to be able to use food oil as the cutting medium. A visit was arranged.

### 10.4.2    Initial trials

The first cutting trials were conducted using water as the cutting medium. Initially, the cut surface appeared to be satisfactory but over a short period of time the sugar glass began to dissolve and collapse as the liquid jet left a trace of water on the cut surface. What was clear, however, was that the rate of cutting was extraordinarily high. Cutting rates of well in excess of 1 meter per second were clearly possible. We needed to find another medium.

A food oil was substituted as the cutting medium. The cutting rate was equally dramatic and of course the trace of oil on the surface did not cause the glass foam to collapse. It also became clear that as the stress applied to the product was in the vertical plane, there was virtually no horizontal stress, making it possible to cut very thin strips of product just a few millimetres thick. This would never be possible with saws as some vibration is inevitable and the

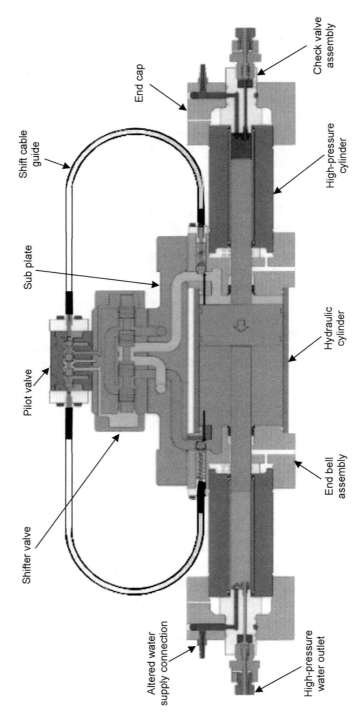

**Fig. 10.1** Mechanical shift intensifier. Reproduced with permission from Flow International Corporation.

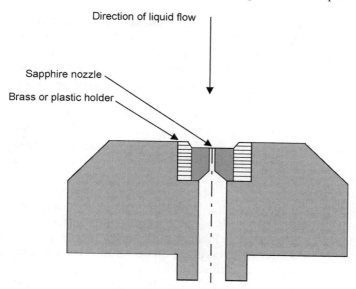

**Fig. 10.2**   A typical arrangement for a sapphire liquid jet nozzle.

product fractures easily. It looked as though we had found a technique that would make it possible to cut this confection and virtually eliminate waste due to product breakage. The commercial advantages were obvious.

It was also very clear that we would be the first in the world to be using a food oil under extreme pressure for a cutting application and that virtually nothing was known about how the oil would behave.

- Would it oxidise or polymerise rapidly?
- Would it degrade the plastic seals in the intensifier?
- Would the nozzle life be adequate? And so on.

There were therefore some very significant risks. These needed to be minimised by an adequate programme of pilot plant testing before we were committed to using this method in a commercial plant.

### 10.4.3   Pilot plant trials: first phase
Pilot plant equipment was constructed to explore all the major parts of the process planned for this new plant. This included syrup cooking, magma mixing, magma cooling, magma height control, cooling belt construction and of course liquid jet cutting using a food oil.

An intensifier capable of generating 2700 bar was hired and the equipment necessary to supply a sapphire nozzle continuously with food oil at this pressure put in place. The oil delivered by the nozzle needs to be collected. The oil as it exits the nozzle is hot, and travelling at something in excess of twice the speed of sound. It needs to be decelerated and cooled. This was done by firing it into a tube in which was maintained a well of water. The emulsion that resulted was then passed to a large vessel where the oil was allowed to separate from the water. This one-G separating vessel was made of glass so we could see what was happening inside. The oil overflowed from the top and the water flowed from the bottom up a pipe and overflowed at a level a little lower than the oil commensurate with the difference in densities of the two materials. The interface of the oil and water was maintained at approximately half way up the separating vessel. As the oil was not being used for cutting under this scenario, there was no need to undertake further cleaning of the oil before returning it to the intensifier.

*The trials began!*

Three members of staff took it in turns to run this plant 24 h a day on 8-hour shifts. All plant parameters such as seal life, nozzle life, temperatures generated, oil quality, etc., were carefully monitored. It quickly became apparent that the life of the jets was initially unacceptably short. An examination of the sapphire jet under light and electron microscopes showed that a deposit was building up in the jet itself on the internal wall. The stream of oil that was passing down through the centre of the jet was being deflected by this deposit and the clean cohesive jet that was otherwise thread-like for 30 to 40 mm became dispersed. This was thought to be caused by gums naturally present in food oils. An oil low in such materials needed to be found. Once a low-gum-containing oil had been selected, the jet life became satisfactory. It also became apparent that some hydration and separation of gum-like materials from the oil was taking place in the separating vessel, as this could be observed through the glass wall. The oil was becoming more satisfactory for use with time. This was an unexpected benefit. The temperature cycling, exposure to high shear forces, and the intimate mixing with air as the jet was collected did not appear to lead to oil deterioration as might have been predicted. The oil quality remained satisfactory.

The results of these tests were presented to the management team and agreement reached that the results were sufficiently encouraging to justify the building of a pilot plant that actually made and cut product. This pilot plant would also be used to explore a number of other unit operations involved in the manufacture of this product, particularly where the sensitivity of certain parameters to scale-up was unknown.

### 10.4.4    Pilot plant trials: second phase

A small plant capable of producing product for sale was built using many of the elements used in the first phase of the development. In this plant, the oil collected after initial separation was further washed with water and passed through a centrifuge to remove any trace of sugar carried over from cutting the confection. The cutting area was cordoned off for safety reasons with interlocks on the doors. The jet looks very innocuous. However, the thread-like stream, sometimes barely visible to the naked eye, was capable of slicing through human flesh just as easily as other materials! Maintaining a safe environment for the staff to work in was always a major consideration.

We needed on this plant to have a travelling jet that cut across the slab of material being made continuously. We caught the jet this time in a trough containing water in the bottom. Deflectors were put in to minimise as much of the splash-back as possible. We found that with time there was some build-up of heat at the slot through which the jet passed. This was overcome by placing a water-cooled tube on each side of the slot. Stationary jets were mounted on each side to trim the edge of the magma sheet.

### 10.4.5   Development of alternative cutting methods for fall back position

Work on a third generation of cutting saws continued in parallel with the work on liquid jet cutting. The use of saws was to be considered as a fall back position if it was found that liquid jet cutting was not commercially viable for any reason. It has already been noted that saws had the inherent disadvantage that they produced a lot of waste in the form of sawdust and broken bars but we knew that they did work! Once the design had been agreed and where possible, tested, a full-sized cutting plant using saws was built. It was therefore available to be slotted into place in the full-sized plant if liquid jet for some reason failed as a commercial cutting technique in this application.

### 10.4.6   The decision

We had reached another crucial decision point. Confidence was now building to the point where we felt sufficiently confident to incorporate this technology into the full-sized commercial plant. All we had learnt on syrup cooking, magma making, height control and cutting, using the pilot plant, was available for incorporation in the full-sized plant. Agreed to go ahead with the liquid jet option but we still had one or two lessons to learn!

## 10.5   The full-scale plant

### 10.5.1   Design of the cutting table

This plant, in addition to the simple slitting of the selvedge on each side and the cross cutting of the slab into blocks, required an XY table to cut the blocks into individual bars. The first two cutting processes had of course been adequately tested on the pilot plant. The XY table was to be designed and put into production without further testing, as there was no real opportunity to do so. It was constructed along the lines of the catch tank for the cross cutter used on the pilot plant. The top was constructed using parallel water-cooled tubes. The tank below, contained baffles to minimise splash up to the cutting area. Six nozzles were to do the cutting. Blocks to be cut were delivered onto the cutting table and the nozzles made six passes.

After each pass, the nozzle was moved sideways the width of a bar before making the next cut. During the sideways move, the nozzle was turned off to avoid damaging the water-cooled tube and incidentally producing a lot of vaporised oil. The valve was mounted just above the nozzle itself so there were six such valves operating together.

### 10.5.2    Testing of the cutting table

One day not long after this part of the plant was commissioned, one of the high-pressure pipes split. We installed a high-pressure transducer so that we could have a look at the pressure that this pipeline was being subjected to. To our absolute surprise we found that each time the valves were turned on, the pressure dropped to 1900 bar. When the valves were closed, the pressure returned to 2600 bar. As these valves were being opened and closed about once every second while product was being cut, the pipe work was being flexed at the same rate. It eventually failed.

Our initial understanding from the liquid jet manufacturers was that viscosity of the cutting fluid was not an issue. However, a search of the literature turned up an obscure Japanese paper that reported that food oils became much more viscous at the pressures we were subjecting them to. We were losing pressure as the oil flowed down the pipe work due to this increased viscosity.

It became necessary to do three things:

- Minimise the length of high pressure pipeline and maximise its diameter.
- Use pipe work that had already been stressed above the pressures we were using.
- Install a large pressure vessel in the pipe work close to the nozzles to dampen the pressure fluctuation.

We also found that while small amounts of atomised oil were produced on the pilot plant, this quantity became significant on the full-scale commercial plant. It was necessary to install large extraction equipment with electrostatic particle collectors to clean up the air before discharging it.

### 10.5.4    Fitting the cutting table into the production line

When we decided to go ahead with jet cutting in preference to the use of saws, the costing included the disposal of the cutting oil after a certain period of use. In practice, however, it was found that as the product picked up about 0.5% of its weight in cutting oil, the turnover was adequate to keep the oil in good condition. It was also possible that the continual washing and centrifuging of the oil was helping to remove any trace of gums and keep it free of degradation products.

The one aspect of the full-sized plant that was not investigated at all was the device for separating and turning over the bars before they were presented to the chocolate enrober. The existing plants had a set of fixed guides that did this job and were thought capable of being scaled up to the increased speed and capacity. A complex set of such guides were constructed and installed.

This did not work. A small blockage rapidly developed into chaos. A rapid investigation and development programme starting with a few pieces of plastic strip bought from a local hardware shop and finishing off with a large profiled table made with machined polythene became necessary. This major plant modification needed to be made as rapidly as possible and the external consultant was asked to take on the solving of this problem, freeing up company staff to progress with plant commissioning.

### 10.5.5   Commissioning the production line

Commissioning a plant this size and with this amount of innovative technology is a big task. This part of product and process development is to be commented on more extensively elsewhere in this book but a few comments about this plant seem appropriate. The person whose original concept it was and who was the team leader, moved to another part of the company before commissioning started. This left a big hole and damaged morale. We were fortunate that by this stage we had built up a robust team of competent people who knew the plant well and furthermore, had access to further human resources if they should need them.

The first job when commissioning, is to check that liquid will pass through all the pipe work and that all the motors turn over (in the right direction in the case of three-phase ones)! On day one we found six plugs that had been left in the pipe work by the building contractors. Not an auspicious start. Then there is all the computer software to be tested. All the control loops to be set up and tuned.

When product is eventually made, it will be found that some devices do not work as well as expected and some do not work at all. The example of the spreading and turn over device has already been commented on. We had trouble with the XY cutting table, too. This equipment was set up with pneumatic controls to move it forwards and sideways using position sensors. In spite of all efforts, we could not get this mechanism to work adequately. The solution was to install an electric stepper motor that has the facility for controlling the number of degrees it rotates and therefore the distance any piece of equipment attached to it moves. The stepper motor uses a computer program to carry out all the movements needed during the cutting operation. Commissioning takes time, dedication and patience. The team may need to call on specialist resources to deal with particular problems as they arise. Commissioning is also likely to be undertaken under extreme time and financial pressures imposed by the commercial environment.

## 10.6   Problems met and lessons learnt

A food technologist or process engineer does not get many opportunities in his or her career to partake in the design, development and construction of an innovative plant of this magnitude. It is worth making some observations about the lessons that were learned.

- Good pilot plant work can be effective in reducing risk. However, in spite of all the effort, there are still likely to be surprises. We may have been naive but we were not expecting such a dramatic increase in viscosity of the food oil under extreme pressure.
- It can be dangerous to assume that some device that works fine at low throughputs will perform equally well when production capacity and speeds are significantly higher. The turnover device was in this category. Take every opportunity to test.
- Small emissions on pilot plants that require no particular devices to deal with them may be very troublesome on production plants. Very finely atomised oil in the air did become a significant problem on scale up.
- You may have the design about right when everybody in the team is tired of talking about it. The design of the XY table was a case in point here. We needed equipment to place five blocks onto the table and then after cutting, remove them a row at a time. The early designs were complicated. A simple but highly effective design emerged only after much tedious discussion involving the draughtsman, the external consultants, production people and the process engineers on the team.
- Some tasks can take very much longer than initially predicted. One particular task on the design of the computer software for the full-sized plant was thought to take an afternoon. It actually took two man years! Be prepared!
- Some parts of the plant will be found not to work adequately or not sufficiently robust to withstand 24-hour, 7-day operation. An example on this plant was the control of the XY table, which turned out to be inadequate. The commissioning team must be given the resources to redesign and rebuild parts of the plant as appropriate.
- Good teamwork and effective communication is essential for designing and building complex plants. This plant was being built while the testing work for the proposed process was being undertaken. This has the advantage that it shortens the time between forming the initial concept and the beginning of commercial production. There are certain financial advantages in following this course. There are financial risks, too, but it demands close co-ordination between the research and development staff, other functions involved, and the factory team responsible for the plant's construction.
- Some increase in costs was inevitable as some parts of the plant become redundant even before commissioning starts. Some were not adequate for the job and needed redesign, and other equipment became necessary to do unforeseen tasks.
- Although some commercial production was achieved quickly, it took longer than expected to fully commission the plant. A balance between the level of innovation being incorporated, the amount of process development needed, the expected gains in plant efficiency, and the timing of the construction of the commercial plant needs to be struck.

## 10.7    Conclusions

The development of the process and the building and commissioning of the plant were long and difficult. It was some years before a satisfactory level of plant efficiency was reached. However, the work on this unique method of cutting made possible the efficient large-scale manufacture of this form of confectionery while producing virtually no waste from broken bars.

Projects like this provide an opportunity for people to develop their technical, managerial and teamworking skills. A number of the team members subsequently took up very senior positions in the company.

# 11

# Process innovation from research and development to production in a large company – development and commercialisation of a low temperature extrusion process

Andrew Russell, UK

*Knowledge has to be established for the individual application, but then needs extension for application to total product development. The last chapter particularly described the development of the technological base through one unit operation in detail. But there are many other stages before a unit operation innovation can be brought from the experimental stage to a product on the market. The design of the unit operation has to be finalised and tested. More important, the product from it is then explored to determine the optimum processing conditions for the product attributes and benefits. From the unit operation development comes a production prototype, and then a complete production line. If the optimum unit operation goes straight into the factory, there often are problems combining the new with the old. But of course it is very expensive to build a new experimental production line, though the smaller scale of the pilot plant reduces the operating costs. This is where the accumulated resource capital, in this instance scientific experience and expensive equipment left over from a quite unrelated product, can become so beneficial. Extensive discussion and even modelling before the plant trials start, is also often worthwhile to prevent chaos in the initial runs.*

*It is important that all the people in the production plant are involved, particularly the engineers and the operators who have the in-depth operational knowledge. And they have to be convinced of the viability of the process, as antagonism can easily multiply problems. But also involvement and enthusiasm have to be built in finance, marketing, distribution and human resources.*

*Finance has to be brought into the discussions because pilot plant development through to production is expensive. There can be major capital costs as well as high operating costs. So this new development has to be compared with other new ventures – what are the predicted financial outcomes and the risks of failure? Predicting some years ahead is always a problem, and relative advantages may shift with time and events.*

*Marketing often has a problem relating to major processing innovations. The timing of the production roll-out can be extended because of technical or supply problems and therefore it is hard to plan the timing of the market launch. But it is important that marketing is involved early so that they can relate the product benefits from the new process with the needs of the consumers. It is easier in the early stages for a process to be modified to meet marketing's identified product needs.*

*Distribution has to know the storage properties of the products, their fragility or other protection needs, and the sizes of production runs and storage capacities at the factory.*

*New process innovations often need staff with different experience, skills and knowledge and if possible they should be recruited early so that they are at the initial trials.*

*There has to be a good project leader or project leaders if personnel change through the long development. Above all there is a need for coordination of the many people involved, if there is to be a successful launch. And a bonus, when all this is well done, it is first class staff training and development.*

*The chapter particularly relates to pages 111–130, 257–316 in* Food Product Development *by Earle, Earle and Anderson.*

## 11.1   Introduction

In the early 1990s Unilever was well established as the largest global manufacturer of ice cream. Successful innovations based on new formats for ice cream, such as chocolate-coated stick products and layered desserts, had cemented this dominance. Within the ice cream category, it was recognised that a more in-depth research approach into the fundamentals of ice cream was necessary to ensure continued innovation in the future. This gave rise to a major research and development (R&D) programme involving exploratory work on ice cream ingredients, microstructure and processing. This initiative has, over the past decade and a half, led to the development of several major technological innovations, of which low temperature extrusion is one, resulting in new products in the market place.

## 11.2   Invention

### 11.2.1   Background

Ice cream is a multiphase material comprising ice crystals, air bubbles and fat droplets dispersed within a viscous solution of sugars, milk proteins and polysaccharides, known as the matrix. The sensory quality of ice cream as a product is highly dependent on its microstructural properties, such as the size and arrangement of each phase, which in turn are sensitive to the processing conditions used during manufacture. The primary objective of the processing research activities was to build understanding of the principles of microstructure formation during ice cream manufacture and to use this knowledge to manipulate and redesign the processing operations in order to optimise the sensory quality of the end product. Much work also focused on understanding the relationship between the complex microstructure of ice cream and the desirable sensory properties, such as creaminess, smoothness and a lack of ice crystal detectability. Broadly speaking, it was shown that these properties improved as the size of the dispersed ice crystals and air bubbles was reduced. Achieving a significant change in the size of these microstructural attributes was therefore a key target of the desired optimised process.

The standard process for ice cream production comprises a series of steps as shown in Fig. 11.1. The ingredients are firstly mixed in a batch vessel and the liquid mix is then homogenised to form an emulsion with fat droplets of around 1 $\mu$m. This is then pasteurised, cooled to around 5 °C and then passed to a storage tank for 'ageing' for between 2 and 24 hours. The ageing period is necessary to allow partial crystallisation of the fat phase, which is important for stabilising the final ice cream structure. Freezing of the aged mix is carried out in a scraped surface heat exchanger, and it is here that the distinctive ice cream structure is created. The freezer performs a number of functions: cooling, ice crystallisation, aeration and mixing of the product. Air is introduced into the barrel together with the premix and whipped to form a stable foam by the action of the rotor. The mixture is cooled to around −5 to −7 °C (depending on formulation) by the evaporation of liquid ammonia (at −20 to −30 °C) in the freezer jacket. Rotor blades continuously scrape the wall of the freezer barrel to prevent build up of frozen material and to maintain high heat transfer rates. A typical freezer barrel can process around 1500 l/hr of ice cream. Once frozen, the ice cream can be shaped and assembled, incorporating inclusions such as sauces or nuts. Stick and bar products may be enrobed with chocolate. The completed product is then packaged and hardened by cooling to around −25 °C in a blast freezer prior to cold storage and distribution. More information on ice cream microstructure and processing is given by Clarke (2004).

The basics of this process had changed little in the 60 or so years that ice cream had been manufactured industrially. It was therefore felt that scope existed to upgrade and improve the process to achieve a step change in ice cream quality, which would enable Unilever to maintain its lead against the ever-increasing competition in the ice cream market.

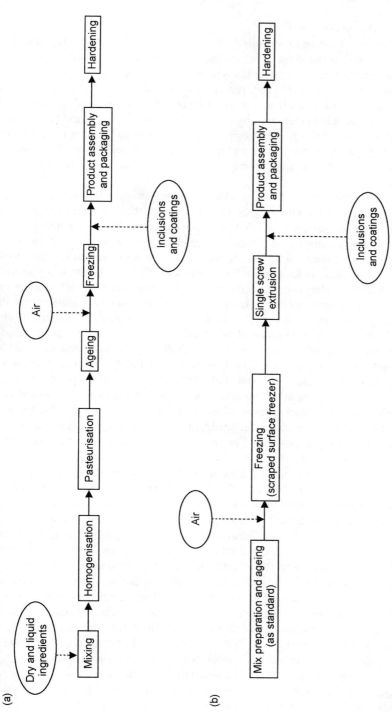

**Fig. 11.1** (a) Sequence of unit operations in a typical ice cream manufacturing process; (b) sequence of unit operations for new low temperature extrusion process.

### 11.2.2  Discovery

Research into ice cream processing concentrated initially on the scraped surface freezer, as this is the stage at which the microstructure is largely created. The early approach was to systematically examine the effect of freezer processing variables, such as scraper speed, residence time and wall temperature on microstructural properties, such as ice crystal size distribution, and ultimately on sensory properties, as measured by a trained taste panel. One of the key findings from this early work was the strong correlation between the temperature of the product at the exit of the freezer and its sensory properties. It was shown that, in general, the ice cream quality improved as the freezer exit temperature was reduced.

Owing to the nature of the ice cream freezing process in a scraped surface freezer, there is a limitation to the extent to which the exit temperature can be reduced. As ice cream is cooled to lower temperatures, the content of ice increases. For a typical formulation, the ice content reaches a level of 40 wt% at a temperature around −5 °C. As the temperature is decreased below this point, the further increase in ice content has a dramatic effect on the product viscosity, which increases almost exponentially with declining temperature (Goff *et al.*, 1995). This in turn has a profound effect on the viscous dissipation in the freezer. Viscous dissipation is the frictional heat generated by movement of a fluid. Figure 11.2 shows measured data for the viscous dissipation rate in a small pilot-scale ice cream freezer, derived from shaft torque measurements. It can be seen how the rate of dissipation increases markedly as the temperature approaches −5 °C, and this trend will continue for lower temperatures. As a result, an exit temperature is quickly reached where the rate of frictional heat generation in the product is more or less equivalent to the rate at which heat is extracted from the product by the refrigerant, and an equilibrium occurs where no further cooling is possible. For most ice cream formulations, this is at a temperature of around −7 °C.

This then posed a limit to the extent to which ice cream freezing could be optimised. There was a clear benefit of freezing to a lower temperature, but the available equipment limited the degree to which this could be done. Various approaches were attempted to reduce the exit temperature from the freezer. One method studied was to significantly lower the rotor speed to reduce viscous dissipation – which allowed a further 1–2 °C drop in exit temperature. This resulted in difficulties with the effectiveness of air incorporation in the freezer, a problem that could be addressed to some extent by pre-aeration with an in-line high shear mixer prior to the freezer barrel. The use of low rotor speeds, surprisingly, resulted in smaller ice crystals in the final product. Current knowledge at the time suggested that high scraping rates were necessary in the freezer to shave dendritic ice crystals from the cold barrel wall before they became too large (the actual mechanism for ice crystal formation was described later by Russell *et al.*, 1999). The results of freezing with low rotor speeds challenged this conventional wisdom and opened up the possibility of using much lower shear devices, with lower dissipation rates, to freeze ice cream.

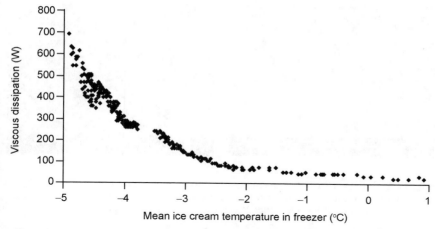

**Fig. 11.2**   Viscous heat dissipation rate in a pilot-scale ice cream freezer (100 kg/hr throughput, 240 rpm rotor speed).

The most obvious low shear process for handling viscous fluids is screw extrusion. Single and twin screw extruders are commonly used in a variety of industries from plastics and ceramics to cereal products and confectionery. The screw action is used to convey a highly viscous, almost solid-like, mass and to force it through a die to achieve a desired shape. In virtually all of these cases, heating of the product occurs during the extrusion process either via an external jacket or from the viscous dissipation arising from the movement of the viscous stream. The concept of cold extrusion did exist at this time, but was used to refer to a system where no external heating was applied. Never before had a screw extrusion system been applied to a product stream that was to be frozen.

The opportunity to test the concept of using a screw extruder to freeze ice cream arose when the existence became known of a small twin screw extruder at a Unilever detergents research facility. This extruder had been designed for extrusion of soap bars and had been specially built for Unilever by a third party. This unit was installed in a research pilot plant and tested for the first time on ice cream in 1994. The unit was continuously fed with ice cream from a standard ice cream freezer, so that the extruder performed the role of a second continuous freezing stage. By using methanol solution as a cooling fluid in the twin screw extruder, exit temperatures as low as −15 °C were achieved. This clearly represented a major step change from the limits of the conventional freezer. From this point on it was clear that 'low temperature extrusion' had the potential to be an exciting new technology for the ice cream business.

### 11.2.3   Product benefits

*Improved sensory properties*
The assumption that decreasing the freezer exit temperature would result in improved sensory properties was born out in the earliest samples produced on

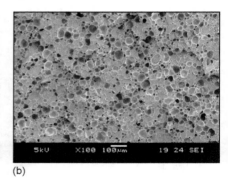

(a)                                    (b)

**Fig. 11.3**   Influence of low temperature extrusion on ice cream microstructure. Cryo-scanning electron micrographs of (a) low temperature extruded ice cream and (b) ice cream produced by conventional processing. Scale bars represent 100 $\mu$m.

the small twin screw extruder. Improvements in the creaminess and smoothness of ice cream produced were immediately evident. This could be related directly to the change in product microstructure in the screw extruded sample (Fig. 11.3). Generally speaking, a finer distribution of both air bubbles and ice crystals was seen in the product that had passed through the twin screw extruder, and this seemed to be the primary source of the improved sensory quality.

*Improved 'stand-up'*

An additional benefit arose from the lower temperature and therefore higher viscosity at which the ice cream left the low temperature extruder compared with a freezer: improved 'stand-up', or the ability of the product to retain its shape, was seen. This offered the possibility of eliminating the hardening step altogether and placing products directly into cold storage after packaging. The normal temperature after hardening is in the order of $-25\,°$C, and one of the primary reasons for performing this cooling step is to ensure products are sufficiently solid to enable palletisation and stacking. The firmness achieved at $-15\,°$C was, in many cases, sufficient to avoid problems in the cold store. Figure 11.4 shows the change in product ice content with temperature. The hardness of the product is directly correlated to the ice content. It can be seen that at the exit of the low temperature extruder, the majority of the final ice content has already been formed, despite the additional temperature drop that would normally occur in a blast freezer.

Another way in which the improved 'stand-up' could be exploited was to allow extrusion of complex and intricately shaped objects, which could survive the subsequent portioning, packaging and conveying operations without collapsing.

*Change in fat state*

A further potential benefit was the change in the fat state of the low temperature extruded ice cream. Normally in the freezer, a certain proportion of the fat

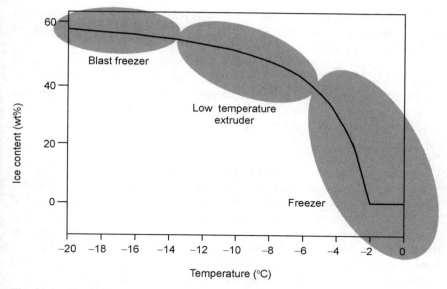

**Fig. 11.4**  Change in ice content with temperature for a typical ice cream formulation.

emulsion becomes 'de-emulsified' and partially coalesces, due to the combined effect of shear action and the presence of a large amount of air interface (Clarke, 2004). As mentioned earlier, the presence of solid fat crystals in the emulsion droplets is important for this process as it prevents full coalescence of the droplets. The creation of this partially coalesced fat phase is an important element of the ice cream structure, as it stabilises the gas cells and improves the product texture. Excessive shear, however, can lead to excessively large fat particles, which are detectable on eating – a situation known as 'buttering'. With low temperature extrusion a noticeable increase in de-emulsified fat was observed, presumably due to the higher shear stresses in the extruder. In many cases this led to improvements in product stability, such as slower melting rates, but for some formulations, buttering was a problem. The latter could be prevented by employing emulsifying agents which improved emulsion stability and suppressed partial coalescence.

## 11.3    Building a business case

Following the discovery of low temperature extrusion as a potential new technology for ice cream processing, and having verified that quantifiable product benefits could be generated, a decision was made in 1995 to begin the process of commercialising the technology. A dedicated team was assigned to the project and their first target was to build a pilot scale low temperature extruder, of larger scale than the small soap extruder and suitable for in-depth experimentation. Based on experience with this machine, the next step would then be to build an optimised prototype production machine for installation in a factory. Operation of the latter would then provide the information necessary to make a decision regarding commercialisation.

Transforming the new concept of low temperature extrusion into a commercially useful technology required a clear idea of the benefit that was to be exploited. Installation of a screw extruder to an ice cream process line was likely to be a large investment and there needed to be a sound business case before such a decision could be made.

*Replacement of blast freezers*
Using low temperature extrusion to replace blast freezers for hardening could be a very attractive option. Blast freezing tunnels are large and very expensive pieces of equipment which require a large factory floor area and a high level of maintenance. Furthermore, they usually require complex process line configurations to convey products to and from the freezer tunnel access points, at the correct height and angle. The use of low temperature extrusion, on the other hand, would allow straight process lines to be realised, with a consequent optimisation of factory layout. Based on these considerations, the use of low temperature extrusion in any newly built factory would be a cost-effective alternative to installation of a blast freezer. At the time, however, all factories had blast freezers already installed, and building of new factories was not planned. So there was not much to be gained in pure financial terms by replacing existing plant with new technology.

*Product textural improvement*
If the product textural improvements could be exploited successfully then the case could be made for investment in low temperature extrusion. Elimination of hardening would then be an option depending on the particular product requirements and factory circumstances. A clear enhancement in the sensory attributes of creamy texture and smoothness had been demonstrated using trained taste panels. The challenge was making a sound business proposition based on this improvement. While there was a general feeling that improving quality of existing products should lead to market growth, it was also known that this was by no means a certainty, as many other factors are involved in growing sales.

*Recipe optimisation*
A more secure way of justifying the capital expenditure was made available by the discovery that use of low temperature extrusion allowed production of ice cream with lower levels of milk solids, but which still had excellent eating quality. This fact enabled the reformulation of ice cream products with no change in quality. Since milk is one of the most expensive ice cream ingredients, reducing the milk solids content gave a significant reduction in raw material costs. This saving could be used to formulate an acceptable business case to justify the initial investment in low temperature extrusion equipment. It was thus

decided to focus, initially, on developing a commercial, low temperature, extrusion process for the purpose of reducing milk solids content with no change in quality. It was later found that a similar approach could be applied to fat reduction in ice cream formulations. This eventually led to perhaps the most important application of low temperature extrusion: high quality low fat ice cream.

## 11.4 Equipment development and scale up

### 11.4.1 Design

The small twin screw extruder performed a useful function in proving the principle of low temperature extrusion of ice cream, and also allowed some limited experimentation in terms of rotor design and freezing rate. However, it was clear that the next step in developing the process was to move to a larger extruder, designed with ice cream processing in mind, rather than soap! Factors that were important for this new unit were:

- use of ammonia evaporation for cooling, rather than the liquid refrigerant system on the small extruder, to enable increased freezing rates and potentially lower exit temperatures;
- hygienic design and construction using food-compatible materials;
- a throughput compatible with pilot-scale freezers (i.e., around 200–300 l/hr compared with the maximum of 100 l/hr possible with the small extruder);
- instrumentation for measurement of temperatures, pressures and shaft torque to enable improved control and process optimisation.

The most important decision to be made at this point was whether to continue with the use of twin screw extrusion or instead use a single screw system. The latter often represents the most attractive configuration for large-scale food applications due to its relative simplicity of construction and low capital cost (around half that of the equivalent twin screw extruder, Ridgway, 1982). It also offered the benefit that a single screw could potentially be accommodated within the existing barrel assembly of an ice cream freezer, enabling the associated ammonia system to be retained. For these reasons, it seemed sensible to test out the single screw approach. It was therefore decided to construct a pilot-scale single screw extruder, by converting an existing ice cream freezer, and use this to benchmark the performance and product quality against the small twin screw extruder. The barrel, refrigeration system and frame of an existing ice cream freezer was utilised and a new drive arrangement and control system were installed. Several alternative screw geometries were produced for evaluation.

### 11.4.2 Commissioning and testing

On completion of this pilot-scale unit in 1996 an intense programme of commissioning, testing and optimisation was carried out. A throughput of 250 l/hr was achieved at a similar exit temperature to that employed with the twin screw unit. A careful evaluation of the product microstructure and sensory quality produced by the two extruders, under similar conditions of mechanical work and exit temperature, showed no significant difference between the two. This was an important conclusion as it demonstrated that there was no justification for employing the more complex and expensive option of twin screw extrusion. Based on this result, the decision was made to pursue single screw extrusion for commercialisation of low temperature extrusion.

### 11.4.3    Design of commercial plant

It was decided to proceed with the next stage of commercialisation of the technology, which was to design and build, from scratch, a production-scale prototype and deliver it to a selected Unilever ice cream factory for evaluation. The factory chosen was a principal centre for the production of tub ice cream in Europe, and this was the type of application that was to be targeted for low temperature extrusion. Some of the requirements for this prototype were:

- a production rate of 1000 l/hr, with the maximum possible degree of cooling;
- all of the learning and optimised design features, such as rotor geometry, from the pilot-scale unit were to be incorporated;
- the mechanical design was to be robust, so that the unit could eventually be reliably used as a production machine.

In the desire to maximise the heat transfer efficiency of this prototype extruder, it was decided to construct the barrel from nickel, as is common for ice cream freezers, since this has a much higher thermal conductivity than stainless steel. Additionally, a very small gap between the flights of screw and the inner barrel surface was targeted, to minimise the thermal resistance due to the unscraped frozen layer at the barrel wall. The challenge of fabricating a nickel tube of the required length (around 2 m) and with a diameter of the specified tolerance required the services of a specialist engineering company. An additional consideration was the potential wear on the inner barrel surface due to the abrasive effect of ice. To counter this, metal spray coating, an approach pioneered within Unilever for margarine equipment, was applied to the inside surface of the barrel. Further specialist companies were contracted to fabricate the screw rotor, the refrigeration system, the sealing systems and to develop the control software. By using different companies to make single components, maintenance of confidentiality was straightforward. Construction of the frame and assembly of the unit were carried out in the Unilever workshops. In early 1998 the prototype unit was delivered to the lead European factory and commissioned for ice cream production.

## 11.5    Building the science base

In parallel with the equipment development, a systematic study of the principles of low temperature extrusion was conducted. This was an essential aspect of the development of the technology as it provided a foundation for optimisation and scale up. The first requirement was to understand the basis for the product quality improvement seen with low temperature extrusion, to know why a finer microstructure was produced. The next step was to relate this to processing and equipment variables so that appropriate design decisions could be made. Building the underlying science base was also an important investment for the future implementation and roll-out of the technology.

The sequence of operations for the new low temperature extrusion process is shown in Fig. 11.1. Careful analysis of the microstructure of products produced by the low temperature extrusion process revealed that by far the biggest change was in the air structure. The mean air bubble size measured (by image analysis of scanning electron micrographs) in a low temperature extruded sample could be less than half that in a product of the same formulation produced by the standard process. The mean ice crystal size typically showed a decrease of around 10%. Four hypotheses for the observed improvement in sensory quality from the new process were proposed and tested:

- reduction in ice crystal networking due to the formation of additional ice phase under shear conditions, rather than static;
- enhanced air structure and stability due to the higher levels of de-emulsified and partially coalesced fat;
- creation of smaller air bubbles in the screw extruder due to more intense working;
- maintenance of small air bubbles due to the lower product temperature prior to hardening.

The effect of low temperature extrusion on the structure of the ice phase was found to be minimal. Three-dimensional tomographic techniques revealed no change in the networking of the ice crystals. Similarly the role of fat partial coalescence on improving sensory properties was discounted, since improvements were still found in low fat products and those where de-emulsification of the fat was kept very low by manipulation of the emulsifier level. The air structure changes, however, were shown to be strongly correlated to sensory attributes, such as creaminess and smoothness – a relationship that has been more recently verified by Goff (2002). It became clear that the benefits of low temperature extrusion derived from the finer air structure that resulted. The question was, whether this was a result of additional bubble comminution or improved air phase stability. The data in Fig. 11.5 give an answer to this question. They show the changes in product temperature and mean air bubble size as the product is hardened in a blast freezer. In the case of the conventional freezing process (Fig. 11.5a) it can be seen that the product starts the hardening process at a warmer temperature and it takes around 50 minutes for it to reach the initial temperature of the product produced by low temperature extrusion (Fig. 11.5b). This period seems to be critical for air phase stability as there is rapid enlargement in mean bubble size in the conventional sample over this period. The temperature of the low temperature extruded product seems to be

(a)

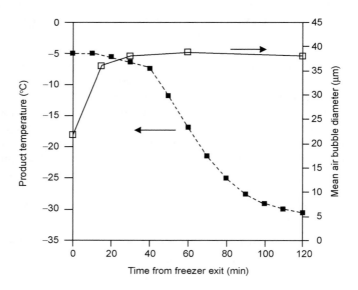

(b)

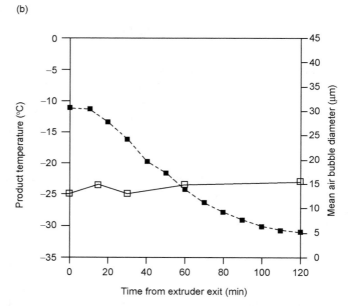

**Fig. 11.5**   Change in product temperature (■) and mean air bubble diameter (□) during blast freezing process: (a) following exit from standard ice cream freezer; (b) following exit from low temperature extruder. Product temperature recorded at centre of 500 ml block of ice cream. Mean bubble size determined by image analysis of scanning electron micrographs (at least 400 bubbles measured for each data point).

sufficiently low that bubble coarsening barely occurs – presumably the product is sufficiently viscous to retard the bubble ripening processes. It is true that the initial mean bubble size is somewhat lower in the low temperature extruded sample, indicating that additional size reduction does occur in the screw extruder. However, by far the more dominant effect is the enhanced stability during the subsequent static blast freezing operation, preventing the coarsening of the air structure.

Following on from this conclusion, it became clear that the key parameter to optimise the sensory improvement was the product viscosity at the exit of the extruder, as measured by the shaft torque. This is basically a function of the product temperature at this point but, as can be gleaned from the steep curve in Fig. 11.2, temperature was not a particularly sensitive measure of this property. By aiming for a product which was as cold as possible, thereby maximising its viscosity, the subsequent stability of the air phase could be enhanced and the sensory quality optimised. Achieving this goal required both a high heat transfer capacity and minimised viscous dissipation. The former was achieved by the design of the barrel and minimisation of the scraping tolerance, as mentioned above. Control of the viscous dissipation was a more complex problem and required an in depth knowledge of the flow characteristics in the single screw extruder.

Owing to its low rotational speed, a screw extruder is a much lower dissipation unit than a scraped surface freezer, hence the reason why such a dramatically lower product temperature could be achieved. However, at the extremely high ice cream viscosities achieved, the heat generated by friction in the extruder was significant. From work on screw extrusion in other industries, it was known that product flow behaviour was highly influenced by the geometric variables of the screw, such as pitch angle, channel depth and the number of flights. The challenge was to optimise the screw design so that as much of the work done by the screw was translated into effective pumping of the product rather than into dissipation of frictional heat. Much work was done on flow visualisation, mathematical modelling and characterisation of different screw geometries to investigate this problem. Ice cream represented a unique system compared with previous applications, due to the fact that the product flow properties changed markedly along the length of the screw: from liquid behaviour at the entry to near solid-like behaviour at the exit. It was eventually found that the pitch angle was a key parameter and an optimised geometry was determined, based on a 40° pitch angle. This geometry was protected in a patent (Bakker *et al.*, 2003) as well as being incorporated in the design of the prototype production unit.

## 11.6   Developing and testing prototype extruder in the factory

Commissioning of the prototype production extruder in the factory proceeded reasonably smoothly. Good product was produced from the beginning, and exit temperatures down to −16 °C were achieved. Some initial problems were experienced with the seals on the barrel jacket, which led to leakage of ammonia. This failure was due to the fact that a late decision had been made to use an ammonia supply that was much colder than had been designed for (in order to try and boost the freezing capacity). The problem was solved by carrying out an extensive search to find a specialist seal material which could cope with the necessary extremes of temperature. Once the seals had been replaced, the commissioning could be completed and the evaluation of the unit for production purposes begun.

The intended application for low temperature extrusion was for tub products of around 1 l portion size. These products often contain inclusions such as fruit pieces, chocolate chips or nuts. In a conventional process, these would be mixed into the product immediately after the ice cream freezer using an in-line mixer. It was realised from an early stage that such a mixing step would be a much more difficult proposition with the low temperature extrusion process, due to the difficulty in mixing the highly viscous product stream exiting the extruder. Adding inclusions prior to the extruder was discounted as the particles were likely to become trapped in the narrow gap between the screw and the barrel wall. Instead a proprietary injection device was developed which ensured uniform inclusion addition. This device was found to work well in the factory.

Products with reduced milk solids contents were produced using the prototype extruder and were benchmarked against standard product. The results were excellent, and no significant difference in quality could be detected between the two product types. The business case for investment in low temperature extrusion units, based on cost savings through reduced milk solids utilisation, was approved and the decision made to implement the process commercially. The route to sourcing low temperature extrusion units with a suitable specification for industrial implementation is detailed in the next section. By 1999, the first commercial units were installed in the factory for production purposes.

It was when these units were connected to a full-scale packaging line that unexpected problems arose. On large tub-filling operations, it is normal to have several lanes of conveyed tubs, each filled by a separate dosing nozzle, which are all fed from a single freezing unit. To split the ice cream flow coming from the freezer (or in this case, the screw extruder) evenly between up to four dosing nozzles requires a complex manifold arrangement with flow control valves on each leg. This normally gives good control of the flow rate through each nozzle. In the case of low temperature extruded ice cream, however, enormous difficulty was experienced in achieving uniform flow splitting. This led to large variation in dosed volumes for each lane, and an unacceptable level of under and over filling of tubs. A systematic analysis of the problem revealed that the lack of flow control was due to a large variation in ice cream temperature across the pipe cross section prior to the flow splitting manifold. The ice cream near the wall of the pipe was measured to be 2–3 °C warmer than that at the centre. As discussed earlier, this range of temperature corresponds to a large difference in viscosity. The warmer, low viscosity, product would always flow through the manifold more readily than the colder, stiffer material, and for this reason balanced flow splitting would never be feasible with such a heterogeneous flow. The temperature variation was found to be due to viscous dissipation and warming of portions of the flow as it exited the screw extruder and as it passed along the transfer pipe to the filling point. The high viscosity of the low temperature extruded ice cream made the extent of this viscous dissipation during transfer much more severe than for standard ice cream. After a lengthy period of experimentation and development, the temperature variation in the transfer line was removed by a combination of modifying the pipe design to minimise dissipation and incorporating an in-line blender which ensured that the temperature was uniform at the point of splitting. This gave a satisfactory solution to the problem and enabled the low temperature extrusion process to be employed for commercial production.

## 11.7  Factory implementation

The prototype 1000 l/hr extruder proved itself in early testing to be a robust and reliable design, and was felt to be close to the sort of unit that would be required for commercial purposes. It was clear, however, that there was scope for cost optimisation of the specification. The cost of the unit would be a key factor in the extent to which the technology would be taken up by Unilever manufacturing operations, and it was important to keep it as low as possible. There was little desire to carry out this cost optimisation in-house, and it was clear that the long-term supply of the equipment would best be done by a third party company, who could subsequently provide the necessary service arrangements. The danger in involving an equipment manufacturer, however, would be the potential for losing the ownership and exclusivity of the single screw technology that had been developed. Minor aspects of the technology had been patented, as mentioned; however, the prior art did not allow patenting of the low temperature extrusion process as a whole. Much of the knowledge that had been developed lay in the equipment design and operation, and this would need to be shared with an equipment manufacturer. It was decided that the best way of protecting the competitive position Unilever had developed was to agree a period of exclusivity with the chosen third party company.

A short list of potential equipment manufacturers was drawn up and the merits of each considered. A suitable company was chosen based on a number of factors:

- the quality of any previous relationship between Unilever and the company;
- the extent of the company's business in ice cream equipment. If this was low, then there was less likelihood that information would be lost to Unilever's ice cream competitors;
- the presence of the necessary skills and experience in hygienic design, ammonia systems and scraped surface heat exchange.

The selected company was approached and after a period of negotiation, a suitable agreement was reached for the company to develop a commercial

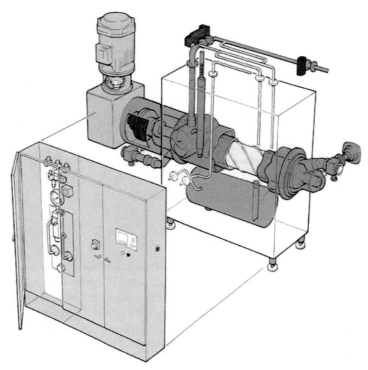

**Fig. 11.6**   Schematic drawing of commercial single screw extruder for low temperature extrusion of ice cream (capacity 1000 l/hr).

specification for the low temperature extrusion units. A one-year period was allowed for industrialisation of the machine design. In optimising the design, a stipulation was that there should be no compromise on the mechanical strength and robustness of the engineering. This was key to avoiding machine failures at a later stage. The manufacturing company made changes to the layout of components, the materials of construction and the mounting of the drive system. The ammonia system was fine tuned and the control system optimised. The basic specification of the unit, however, was largely unchanged. A schematic diagram of the final machine layout is given in Fig. 11.6. The first commercial 1000 l/hr unit was fabricated by the company and delivered to the factory. The unit performed well on testing and was accepted as a suitable unit for roll-out of low temperature extrusion across the Unilever ice cream business.

## 11.8   Roll-out

The first commercial unit was quickly followed by the installation of a second. Once the downstream problems mentioned above had been solved, these units became operational and the first reduced milk solids products manufactured by low temperature extrusion were released on the market in late 1999. As mentioned before, the driver behind this change was cost saving, and the product was designed to be of equivalent quality to standard products, so it was not introduced to the market as a new innovation. However, with the technology in place and proven commercially, the opportunity for using low temperature extrusion for product innovation became of interest. Over the two to three years following the initial installations, a small number of extruders were installed in ice cream factories in Brazil, France, The Netherlands and the UK. These were employed for a variety of reasons, some minor product innovations, but also for the purpose of increasing production capacity – by either eliminating or reducing the hardening requirement for tub products.

A step change in the implementation of low temperature extrusion was prompted by the emergence of low fat ice cream as an important sector of the global market. The improved quality possible with low temperature extrusion enabled new ranges of low fat products with higher eating quality than ever before to be marketed. Unilever began to introduce these new ranges firstly in Europe and then, after installing low temperature extruders in a US factory, in the USA. The latter product line was given the name 'Double Churn' ice cream. Here the new processing method was cleverly used as a marketing message by 'branding' the technology in a consumer-friendly way. This product range has been highly successful in the marketplace.

At about this time, an alternative low temperature ice cream extrusion technology, based on twin screw extrusion, became available to Unilever's competitors. This technology had had been developed independently (Fayard and Groux, 2003) but also enabled improvement in the quality of low fat products. This competitor activity provided additional stimulus to the wider implementation of low temperature extrusion by Unilever.

With the drive to healthier, lower fat products now becoming mainstream in the ice cream market, low temperature extrusion technology had found its niche. The improved quality of products produced by the technology gave Unilever the opportunity to market attractive and high quality offerings using low fat formulations. This has led in recent years to a intensification in the use of low temperature extrusion across the Unilever ice cream business. By the end of 2006, production units were installed in the majority of factories across the world. There is no sign of this roll-out slowing and it only seems a matter of time before low temperature extrusion becomes a standard unit operation in every Unilever ice cream factory.

## 11.9   Problems met and lessons learnt

*Organisational factors*

In a large organisation such as Unilever, transfer of a new technology from an R&D centre to a production environment is always a challenge. Regardless of the technical merits of the new technology, a number of groups of people need to be persuaded of the value in commercialisation before it will take place. In the case of Unilever, the key stakeholders are the marketing, finance and production (or supply chain) organisations. Each of these groups has a different perspective and requires a different strategy to win their support. The case of low temperature extrusion illustrates this very well, and the successful commercialisation of the technology, to a large extent, was a result of the success in convincing each of these organisations of its value.

For the accountants, a robust case could be made by balancing the equipment investment against operating savings on raw materials. This was a much more persuasive argument from a financial management point of view, than basing the justification on the launch of a new product line, where future sale predictions would be far from guaranteed. To gain the support of the supply chain, it was vital that the equipment performed well. Initial problems during commissioning are always to be expected, and this was certainly the case with low temperature extrusion. However, for a production environment, it is vital that equipment is robust enough to withstand continuous operation over the long term, without failures or excessive maintenance problems. Great effort was made during the design stage to build in mechanical strength and to pay attention to aspects such as component wear and robust control systems. These efforts paid off, and over several years of operation, the production extrusion units proved to be very reliable. This fact won the support of the supply chain, and ensured that the technology could be rolled-out with confidence.

Marketers are usually looking to respond to near-term market needs, and therefore often have a shorter perspective than technologists and engineers. An important consideration for the marketing organisation is that any new technology is available and ready to be used. Once this stage is reached, then the business opportunities can be developed more precisely. It was noteworthy, in the case of low temperature extrusion, that the involvement and interest of marketers stepped up once the technology was installed and had been proven to work in practice.

*Championing the technology*

Roll-out of technology within Unilever's European ice cream business was, at the time, coordinated by a central operations group which was involved from an early stage. When it came to applying the technology outside Europe, however, the roll-out mechanism was less clear, since equivalent operations groups did not at that time exist in other regions. The responsibility for championing the technology and finding regional sponsorship therefore fell upon R&D. This is a role that R&D should not shy away from, as it is the technologists and engineers who know first-hand the benefits that can be derived from a new technology, and therefore usually have the greatest belief in it. It is common for businesses to be organised in a manner which does not facilitate easy introduction of new technology; however, it is important that R&D plays its part in overcoming such obstacles.

*Downstream consequences of changes in unit operations*
A key learning from the low temperature extrusion project, which is true of any process innovation, was the importance of considering fully the downstream consequences of new unit operations. The aspect of inclusion mixing had been addressed prior to factory implementation, but the less obvious problem of flow splitting had not been anticipated. This illustrates how a change to a well-established process might have unexpected consequences at a later (or earlier) point in the process line, which may take some effort to resolve. If the time available for commissioning is limited, this might have a catastrophic effect on the success of the transfer of the technology. For this reason, it has now become a standard practice for ice cream process innovation within Unilever to run whole process lines at pilot scale, to ensure all upstream and downstream implications are understood and resolved prior to factory installation.

*Competitor activity*
Finally, the case of low temperature extrusion is an example of how introduction of a new technology can stimulate competitor activity. This is not necessarily a desirable outcome; however, technology leaders should expect to be followed. The market benefits of being first to market with innovative products are well established and this provides ample justification for food technologists and engineers to be proactive in championing appropriate new technology and getting it established in their business. As this case study illustrates, it is only by investing in innovation that a company can be prepared to meet the challenges of today's highly competitive food market.

## 11.10  Further reading

*General sources of information*
Ice cream technology: Information on the general principles of ice cream technology and manufacture can be found in the classic reference by Marshall *et al.* (2003) which also discusses aspects of the global ice cream market. A more scientific perspective on ice cream and its properties is available in the book by Clarke (2004).

Extrusion: More background on the subject of screw extrusion can be found in the extensive literature dealing with applications in the polymer and cereal industries. The works by Tadmor and Klein (1970) and Janssen (1989) are good introductions to the engineering principles of screw extrusion.

*Specific sources of information on low temperature extrusion*
Recently, research on aspects of low temperature extrusion applied to ice cream has appeared in the literature (Eisner *et al.*, 2005; Wildmoser *et al.*, 2005). These studies have been based on the use of a twin screw system.

## 11.11   Acknowledgements

The research team that made low temperature extrusion a reality: Paul Winch, Deryck Cebula, David Cox, Sabina Burmester, Gary Binley, Mark Willmott, Paul Cheney, Loyd Wix, Peter Bongers, Prakash Patel, Dick Luck, Robert Feenstra, Iain Campbell, Jim Crilly, Susie Turan, Tommy D'Agostino, Anna Townley, Ian Burns, Bas Bakker, Nico de Jong, Willie Young, Paul Doehren, Steve Dyks, Wei Wang-Nolan.

## 11.12   References

BAKKER B H, BONGERS P M M, WANG-NOLAN W (2003), US Patent application 20030211192.

CLARKE C (2004), *The Science of Ice Cream*, Cambridge, Royal Society of Chemistry.

EISNER M D, WILDMOSER H, WINDHAB E J (2005), 'Air cell microstructuring in a high viscous ice cream matrix', *Colloids and Surfaces A – Physicochemical and Engineering Aspects*, **263**(1–3), 390–399.

FAYARD G, GROUX M J A (2003), US Patent 6613374.

GOFF H D (2002), 'Formation and stabilisation of structure in ice-cream and related products', *Current Opinion in Colloid & Interface Science*, 7(5–6), 432–437.

GOFF H D, FRESLON M E, SAHAGIAN M E, HAUBER T D, STONE A P, STANLEY D W (1995), 'Structural development in ice cream – dynamic rheological measurements', *J. Texture Studies,* **26**(5), 517–536.

JANSSEN L P B M (1989), 'Engineering aspects of food extrusion' in Mercier C, Linko P, Harper J M, *Extrusion Cooking*, Minnesota, American Association of Cereal Chemists Inc.,17–38.

MARSHALL R T, GOFF H D, HARTEL R W (2003), *Ice Cream,* Sixth edition, New York, Kluwer Academic/Plenum.

RIDGWAY G R (1982), 'Cooking and extruding and its application to the cake and biscuit industry', *Food Trade Review*, **52** (7), 357–363.

RUSSELL A B, CHENEY P E, WANTLING S D (1999), 'Influence of freezing conditions on ice crystallisation in ice cream', *J. Food Engineering*, 39(2), 179–191.

TADMOR Z, KLEIN I (1970), *Engineering principles of plasticating extrusion*, Florida, Krieger Publishing Co.

WILDMOSER H, JEELANI S A K, WINDHAB E J (2005), 'Serum separation in molten ice creams produced by low temperature extrusion processes', *International Dairy J.*, **15**(10), 1074–1085.

# 12

# Up-scaling from development to production by small manufacturers – fishing, baking and sauce industries

Torben Sorensen, New Zealand

*This chapter brings out special problems encountered by the smaller manufacturer in New Product Development (NPD). There are many types of technological knowledge needed in building from the innovative product in the laboratory (or more likely the kitchen for the small manufacturer) to production, but there is also a basic knowledge of procedures and ways of thinking. Someone who has an innovative idea may have formed the new small company, or it may be already operating as a craft or a small technological company. The small manufacturers often lack the knowledge and the resources to build the simple mixers and household pans into the unit operations of a production line, to recognise and test for the qualities of the raw materials and the products, and to preserve and store the products.*

*In the final process and production development, there are several important factors to consider. These include – raw materials and ingredients, production environment, specifications of the product attributes required by the market and the consumer, movement of the project from the laboratory to the production plant, design and development of equipment, internal and external capabilities.*

*Most raw materials used in food products are biological in nature. Their properties vary a great deal according to species, season, handling and storing. There need to be adequate objective testing methods to set up specifications for the raw materials. Often the small company has little control over suppliers and the quality of the raw materials, and can have difficulties in processing and control of the final product qualities.*

*The final product qualities wanted by the market and the final consumer need to be identified, and also turned into objective measurements, so that they can be the standards for the quality assurance programme. The company may not have a laboratory that can build up testing.*

*Processing may be starting from scratch or it may use equipment already in the factory. They will not have pilot plant equipment so they need to identify a university department or a research centre that can develop their pots and pans experiments to a production process. If they have suitable equipment already in the factory, they can use this for the experimental runs. This can cause problems with their present production. There may be the need for an additional piece of equipment, for example a die for an extruder, baking tins or other containers; these have to be bought full size and if the product is a failure can be redundant.*

*The choice of packaging and packaging method is considered along with the process development. The protection needed by the product, the packaging equipment available and the cost often limit the types of packaging considered. Small manufacturers often do not understand how products change during transport and storage and do not have the facilities for storage tests. A number of small companies do not have packaging equipment and use contract packers; in this case the transport between the processing in the company's factory and the packing company needs to be carefully controlled.*

*One of the stumbling blocks in transferring technology from the test bench or laboratory to full-scale production for the smaller manufacturer is simply a lack of expertise. Smaller companies often do not have fully qualified technologists on the staff, and there can be a reluctance to use outside help for reasons of cost, information security, or simply because the help needed is not always available. Small to medium companies often make compromises due to the lack of funds or experience that result in the failure of development or up-scaling projects.*

*The chapter particularly relates to pages 111–117, 165–183 in* Food Product Development *by Earle, Earle and Anderson.*

## 12.1   Introduction

The case studies that follow review the pitfalls in up-scaling from development to production in three small- and medium-sized companies. There are many reasons why the transfer of technology from development to production fails. The three case studies related in this chapter cover some of the conditions that are often found to contribute to the failure of production transfer in small- or medium-sized companies. That is:

1. A lack of understanding of the properties of the raw material used, and an inability to control or monitor these materials.
2. A reluctance to invest in resources not seen to contribute directly to the bottom line, and an inability to control the production environment.
3. Undertaking projects with requirements that exceed the skills of the development and production teams.

## 12.2    Case Study 1: Raw materials that vary

This case study is taken from the fishing industry, where there can be wide variations in fish materials, related to species, season, catching methods and handling between catching and use.

### 12.2.1    Production of surimi from snapper fillet waste

Frozen fish mince and surimi was produced in New Zealand as early as 1970. The aim of the development described was to make use of waste materials recovered from filleting lines.

In the context of this chapter, surimi is a term used to describe a finely ground fish paste used as the basic raw material in a number of traditional Japanese fish products such as 'kamaboko' and 'chikuwa', these being steamed surimi moulded onto small cypress boards or sticks. High quality surimi is prepared from washed white fish mince that may contain phosphates, sugars and salt to assist in preserving the frozen material, and to give the steamed product the rubbery characteristics required.

### 12.2.2    Product developed

The product was developed in conjunction with a Japanese company, to ensure that it conformed to the requirements of the Japanese market. The Japanese were most interested in product produced from snapper. As this was also the main species filleted at the time, it followed that the process was specifically developed to recover fish mince from snapper fillet waste.

The process used for the production of surimi from snapper waste can best be described under the following headings:

- Preparation of the raw material
- Recovery of fish mince
- Washing blood and soluble matter out of fish mince
- Removal of excess moisture
- Additives added
- Packing
- Freezing and storage.

Snapper to be filleted were caught either by trawling or in Danish Seine nets and placed in ice for up to five days before they were off loaded and taken directly to the processing factory. The fish were not gutted at sea, and fish designated for filleting were filleted directly, such that the head, the backbone and belly cavity were left intact.

This material, referred to as 'fish frames' was collected for further processing into fish mince. The first step was the removal of the head. Following this it was necessary to remove the kidney. This was achieved using a specially developed stamping unit that removed the kidneys together with the bones situated along the back of the belly cavity.

Note: The kidney tissue has a deep red colour that cannot be removed from the mince by washing. For this reason it was necessary to ensure that the material was completely removed at this stage of the process.

The removal of the kidney and the associated bones also allowed the gut to be flushed out, as the membranes retaining the gut were destroyed by this process.

The backbone and the belly flap were passed through a meat bone separator fitted with a drum with 5 mm holes. The recovered mince was then washed in fresh iced water to remove blood and soluble enzymes. Following this, the mince was dewatered to a solids content of 80% using a centrifuge. Salt, phosphates and sugars were mixed in and the product packed and frozen.

### 12.2.3   The problem

Although the product developed in the laboratory and produced in pilot trials had been approved and accepted by the Japanese customer, the first production shipment was downgraded. The reduction in value resulted in the process being uneconomic. On reviewing the product, the market identified a number of issues with the product. That is:

- some product failed on taste
- the binding properties were variable
- the level of connective tissue was high.

### 12.2.4   What went wrong?

On reviewing the project it was found that a number of things had gone wrong between the pilot study and production.

In the first instance the project was hampered by language difficulties. The Japanese had identified a high level of connective tissue in some of the earlier samples. Unfortunately, language was still something of a barrier in the early 1970s, and this information did not reach the development team in New Zealand. In time, the problem could have been resolved with the use of a strainer, a machine developed to remove connective tissue and the like, and used at the time in the industry for this purpose.

The more serious problems related to flavour and texture. Both appeared to be the result of changes to the fish at certain times of the year. Over winter and early spring the fish were building up reserves of fat and getting ready to spawn. Under these conditions, the belly wall of the fish was relatively robust, remained intact and untainted on landing, even when the fish had been held on ice for up to five days.

After spawning, the fish lose their condition, and the gonads are depleted, exposing the belly wall. Moreover, after spawning the fish appear to feed at a higher rate, such that the gut is full and the belly cavity more susceptible to becoming tainted. In some instances the gut can rupture, allowing gut contents to flow into the belly cavity whilst the fish is on ice. In fish less affected, changes were not immediately evident. Nevertheless, some product must have been affected and tainted. When the finished (washed) product was tested in New Zealand, the taints were not identified. However, the taints were identified by Japanese experts, who had the product downgraded.

The binding properties were also said to be variable. In the early 1970s the binding properties of surimi were not fully understood outside of Japan, and the testing methods used to test the product tended to be somewhat *ad hoc*. As a consequence it was difficult to find a definitive and reproducible value that could be used to monitor changes, and to use this information to control the conditions that were causing the product to be downgraded.

In the absence of any other information, the problems were put down to seasonal variation, and eventually the project was abandoned, in that it was not possible to produce a consistent product throughout the year.

### 12.2.5   Correcting the mistakes

In the early 1970s there was little work done on the texture of fish pastes or surimi, and it was only after some years that work undertaken in New Zealand (Sorensen, 1973) and Denmark (Sorensen, 1976) identified ways in which these properties could be defined, studied, and ultimately, controlled. The breakthrough can be put down to the application of measuring instruments fitted with 'cells' that could be used to measure textural characteristics such as elasticity and break strength. These techniques are briefly described in the ensuing paragraphs. The work initiated in New Zealand (Sorensen, 1973) was continued in Denmark (Sorensen, 1976), where there was an interest in producing fish mince as a raw material for different products such as Danish fish balls, as well as for surimi production.

Initially, much work was done to describe sensory panel responses in terms that could ultimately be measured using an instrument that could measure properties such as strength and elasticity. A format developed is shown in Fig. 12.1.

Any of the textural characteristics of samples that could be defined in general physical terms could be measured using an Instron texture measuring instrument. Products such as surimi require good strength and a high degree of elasticity. These were measured using a flat cell, such that a cylindrical sample could be squeezed between two flat surfaces. The elastic properties of the sample could be determined whilst the sample was intact. The break point was measured when the sample broke under the pressure exerted, and was used as an expression of the cohesive strength of the sample.

Using a shear press fitted with a chart recorder, it was possible to collect this data on a chart referred to as a 'force deformation curve', as shown in Fig. 12.2.

The ability to measure properties such as elasticity and break point allows for data to be collected in a systematic manner that enables a thorough understanding of the material used, and how it may be affected by criteria such as season, storage or the use of different additives. For example, it was known that the binding properties of fish were related to the level of the salt soluble proteins

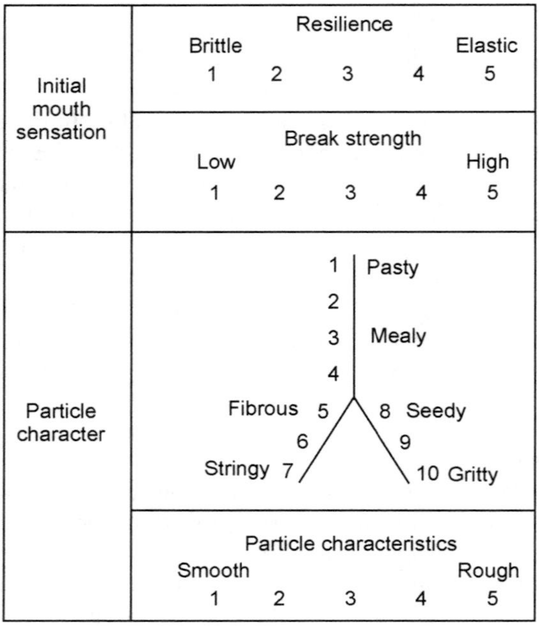

Initial mouth sensation
Resilience
Brittle
Elastic
1 2 3 4 5
Break strength
Low
High
1 2 3 4 5
Particle character
1 Pasty
2
3 Mealy
4
Fibrous 5
6
Stringy 7
8 Seedy
9
10 Gritty
Particle characteristics
Smooth
Rough
1 2 3 4 5

**Fig. 12.1** Texture profile used for taste panel evaluation of cooked fish mince.

Force deformation curve
70.00
60.00
50.00
40.00
30.00
20.00
10.00
0.00
Force (%)
0 5 10 15 20 25 30 35 40 45
Deformation (%)
A
B
C
D

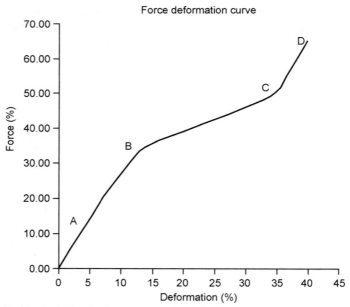

**Fig. 12.2** Typical force/deformation curve of a cooked fish mince sample. The slope AB represents the elasticity of the sample. B is the break point. B is defined as the intersection of the (extrapolated) lines AB and BC. The final resistance of the sample fragments is noted as D.

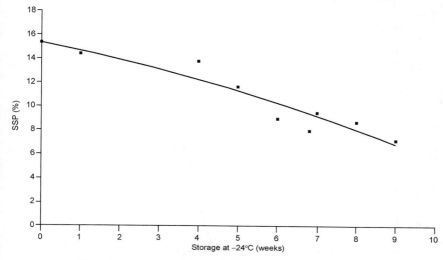

**Fig. 12.3**   Loss of salt soluble protein in frozen cod mince.

(SSP). Work done to find the effect of using frozen ($-24\,°C$) untreated cod mince showed that the SSPs were unstable, and were reduced over time due to denaturation (Fig. 12.3) (Sorensen, 1976).

Using the results from the shear press, it was possible to show just how these changes impacted on the textural characteristics of fish mince products. That is, it was shown that both the elasticity and the break strength was adversely affected as the SSPs were depleted. The effect of break strength is shown in Fig. 12.4.

In the same way it was possible to compare the binding strength of different species of fish, plot the effect of adding different binding agents such as salt and polyphosphate. Data has also been used to find the effect of procedures such as washing, and adding agents such as different sugars to preserve the functional properties of fish mince or surimi. Processing fish during *rigor mortis* has also been found to reduce the ultimate binding effect.

### 12.2.6   Lessons to be learnt

The lesson to be taken from this case study is that raw materials can vary in ways that are not fully understood. The procedures developed to measure the properties important in fish had not been developed in the early 1970s, and in part this was one of the reasons for the failure of the first surimi projects undertaken in New Zealand.

In the early days it was not possible to accurately monitor the changes that occurred through the season. Nor was it possible to measure the effect of excessive levels of connective tissue, or the effect of slight variances in the moisture level. Had this been possible, the reasons for loss of product quality would have been more evident, and it would have been possible to intervene to

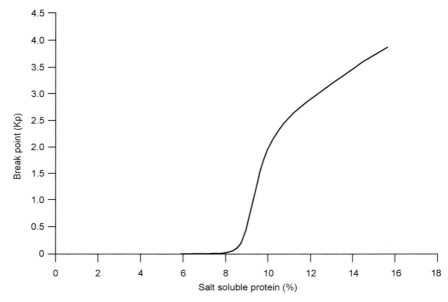

**Fig. 12.4**   Effect of salt soluble proteins on break point of cooked mince samples.

minimise some of the negative properties of the product. In hindsight it might be said that the small company in New Zealand was just too early with this development!

Fish is not the only raw material to change with variety, or through the season. Many crops and food products are subject to similar phenomena. Potatoes, onions and sweetcorn all change through the season, the sugar levels invariably high at the outset of the season, and the starch levels building as the season progresses. Flour and many cereal crops are milled to yield specified levels of protein, often designed for specific purposes. Meats for processing are usually graded by characteristics such as species, cut and fat content. These properties are important when using these materials, and failure to understand how these properties relate to further processing will result in quality shortfalls in the finished product.

Whilst the properties of these types of raw materials are well understood by larger and established operators, there are still small manufacturers who purchase their materials through wholesalers or off the auction floor, without understanding why their productions fail from time to time.

In order to produce a good product, it is necessary to have a thorough understanding of the properties that are important to the product's success. These parameters should be established in the product development phase. Where possible, the parameters identified as being important need to be specified in objective terms.

So, if salt soluble proteins are identified as being important in the manufacture of surimi, the acceptable levels should be specified, together with sampling and testing procedures. Likewise, products selected for sweetness, such as sweetcorn, should be tested for sugar or degrees Brix before they are harvested to ensure they meet the requirements laid down. Care in establishing such specifications may appear tedious at the product development stage, but they will prevent potential failure in the future.

## 12.3  Case Study 2: The importance of correctly designed facilities

One of the difficulties in working in smaller companies is that they are in a state of constant evolution and change, with specifications and production being adapted to accommodate variances in raw material quality, process conditions and orders. In addition, facilities such as temperature control and good ventilation, that may be taken for granted in a larger organisation, may not be available to the smaller manufacturer. This is because finances are invariably overcommitted, and resources such as plant and equipment that are perceived to generate an immediate return are given priority over considerations that are more difficult to quantify in financial terms.

The second case study is taken from the baking industry, or more specifically from pastry manufacture. In New Zealand and Australia pies and pastry products are popular. These products are often produced by smaller operators, of which there are many, who vie for business on the basis of price and quality, with some customers known to drive considerable distances for their favourite pie as a lunch snack.

In the case study cited, a manufacturer had built up a sizeable market, and moved to new premises, and invested in a new production line. The business had continued to grow through the peak of the season, that is, the winter months. During the summer months, however, many customers began to complain about the quality of the product, and a number of retail outlets cancelled their contracts for the company's products.

On investigating the complaints further, it was found that many of the pastry cases were misshapen due to shrinkage of the pastry. There was also a perception that the pie shells were greasy, with lunch bar owners noting that the products left a layer of oil in the hot pie dispensing units. In some instances, there were also complaints that the pastry cases had broken open, allowing the pie filling to spill. As many pies are hand held during consumption, this had resulted in more than one customer returning to work with a soiled shirt or tie.

### 12.3.1  Common problems in pastry manufacture

Meat pies in Australia and New Zealand are usually made with two different types of pastry. A short pastry is used for the bottoms, and a flaky pastry for the tops. This allows the bottom to be more robust, especially when chilled, allowing the pies to be chilled and transported as 'freshly baked' to outlets such as lunch bars and canteens. However the bottom pastry is somewhat uninteresting to eat, and to enhance the eating quality of the product it is usual to provide a flaky top pastry.

Many of the smaller operators producing meat pies rely on skilled bakers to produce their products. Whilst a basic recipe may be used, the baker will often vary ingredients such as the water content to allow for other changes, such as temperatures which vary through the season. Thus much of the industry is still operated as a craft.

This can lead to problems in the manufacture of pastry products, with the most common encountered by industry listed as follows:

Bottom pastry:
• underweight on portioning
• pastry greasy on baking
• shrinkage during baking.

Top pastry:
• underweight portioning
• poor lift
• excessive lift
• breakdown of laminations
• shrinkage during baking.

There are a number of reasons why some of these problems occur. For example, poor lift in the top pastry can be brought about by the use of the flour being too weak (low protein), or the oven being too cold. Poor workmanship can also result in pastries with poor lift. Excessive lift, on the other hand may be due to high oven temperatures.

Whilst this list and the possible solutions may appear daunting, most problems can be controlled or reduced to manageable levels more readily by ensuring the bakery has good environmental temperature control. To understand this it is necessary to review some of the problems experienced by the industry, and how these are dealt with by skilled practitioners.

### 12.3.2   Significance of temperature on pastry quality

The most important factors governing the temperature of pastry are the temperature of the ingredients and the temperature of the dough room or make-up area.

When temperatures are low there is a tendency for margarine and pastry fats to firm up. Under these conditions fats are not readily spread into the flour, and the fat and dough fractions tend to remain separate. As temperatures rise, the fats begin to soften, and to merge with the flour. This reduces the ability of the flour to hold water, and the pastry so prepared tends to be soft. To understand the significance of these changes it is necessary to review the principles of pastry manufacture.

The bottom pastry may be described as a semi-short pastry. That is, the fats are partially blended into the flour to produce a semi-short pastry that is easy to shape into the tin. This style of pastry also sets to a rigid shell when baked, giving the pie a degree of rigidity allowing for the chilled fresh product to be shipped without damage.

It is important that the mixing of the fats is carefully controlled. Under-mixing will result in the fats separating during baking. This causes the pie bottoms to appear greasy and unappetising. Over-mixing, on the other hand, can result in the pastry being soft. This can cause the shell to collapse during baking, causing the product to be misshapen, and detract from the product's appearance.

Mixing of bottom pastries is affected by temperature. If the temperature is low, the pastry fats tend to be hard, making them difficult to mix into the flour. At the correct temperatures the fats are pliable without being excessively soft. This allows for the pastry to be mixed as required. High temperatures can lead to the fats beginning to melt, and for the pastry to break down, being excessively soft and unworkable.

Top pastries are also liable to be damaged if temperatures are not controlled. The top pastry, sometimes referred to as puff pastry, requires layers of fat to be rolled out between the layers of dough. During baking the fats melt, freeing the fat and dough laminations. Some of the moisture in the dough turns to steam, extending the layers of dough, whilst at the same time causing the gluten to set or coagulate. Further baking partially dries the layers, giving the pastry the light flaky structure desired.

Once again the temperature of the raw materials has a significant effect on the outcome. If the temperatures are low, the fats tend to be hard, and break through the dough layers rather than spreading with the dough during the laminating operations. At high temperatures, the fats tend to be worked into the dough, such that the laminations are not fully developed.

### 12.3.3   Underweight pastry portions

Most small- to medium-sized bakeries use manual or semi automatic pastry brakes. These are used to progressively reduce the thickness of the pastry, transforming a block of dough to a sheet that can be rolled up onto a pin, and taken to the production line. The sheet is run out over tins, with self-cutting edges. After the tins are covered with the bottom pastry, the pastry is pressed into the shape of the tin. The pies are filled and a top pastry applied, a rolling pin or an automatic device is used to cut the pastry into the tins.

It follows that the weight of pastry retained in the product will be dependent on the thickness of the pastry. The thickness can vary with the temperature of the pastry. When the pastry is cool ($< 18\,^\circ C$), the pastry has an inherent tendency to bounce back. That is, the pastry will be thicker than indicated by the final gap set on the pastry rollers (referred to as a pastry brake in New Zealand). At higher temperatures the tendency for the pastry to bounce back is reduced, so the thickness of the pastry sheet is altered, even though the settings on the pastry brake are unchanged.

The pastry is recovered from the pastry brake as a long sheet rolled up on a pin or a tube. When temperatures in the bakery are excessive, there can be a tendency for the pastry to be sticky. This can lead to the pastry stretching as it is drawn from the roll of pastry, causing additional reductions in thickness.

Some of the more sophisticated pastry brakes have mechanisms that allow for the final setting to be adjusted to high degrees of accuracy. Most pastry brakes, however, are set up on a ratchet-like system, and it can be very difficult to adjust the final setting. Consequently the final thickness is dependent on the temperature of the pastry at the time it is rolled or sheeted out.

It will be apparent that thick pastry is unacceptable in the product, resulting in a heavy doughy product, and the pastry shell detracting from the eating quality of the product. Pastry that is too thin can also be a problem, especially in products where standards of minimum weight need to be maintained.

### 12.3.4   Means of control

Small- to medium-sized bakeries operating in New Zealand typically experience a range of temperatures varying from 14 °C in the winter to 26 °C in the summer. The range of temperatures in some Australian bakeries is even more variable. To cope with these variances the industry has developed a number of practices, some *ad hoc*, and some based on scientific evaluation. The practitioners tend to make day-to-day alterations based on the feel of the dough, and making progressive adjustments depending on the ongoing production outcomes.

For example, it is common to use iced water during the warmer periods. If this does not give the desired result, the baker may also adjust the water content in the dough. Reducing the water content tends to stiffen the dough. Conversely, increasing the water content allows the pastry to become more pliable.

It will be apparent that changes of this type will affect the final product. With regard to the case study, the soft and greasy pastry bottoms found during the summer months were a result of the pastry margarine working into the flour, such that less water was added. This changed the nature of the pastry, such that the baked pastry cases lost their strength and rigidity. In extreme instances the cases broke apart, spilling the contents.

The problem was partially overcome by using a harder form of pastry margarine during the summer months. This is a common practice in both New Zealand and Australia, and most margarine manufacturers have a 'summer' and 'winter' recipe for pastry margarine. Although the use of different grades of margarine can be used to minimise the extreme effects of temperature, variances in the product are still noticeable. For example, there is a tendency for pastry to be more extensible during the hot months, resulting in lower pastry weights. It is also more difficult to produce a top pastry with the required lift when temperatures are hot, and the practice of reducing the added water results in pastries that are tough and brittle.

The only way in which these variances can be overcome is by providing temperature controls in the raw material stores, and in the dough room. With these in place it is possible to work to set specifications. This leads to a number of benefits that are not possible to achieve without these controls.

In the first instance the quality is consistent through the year. As indicated in the introduction to this section, the industry is very competitive, and a reduction in quality can quickly lead to customers changing their supplier. Companies with variable quality also find it difficult to sign up major customers such as franchised food bars or supermarkets.

Having good controls also allows for tighter standards on pastry weights. Pastry weights can vary by 7% between summer and winter. These variances would be even greater if margarine formulations were not changed for the winter or summer months. It follows that the tighter controls can lead to the reduction in raw material use, as there is not the need to cover high variances with a specification for weight set well over the label declaration.

Finally, better controls allow for the employment of trained semi-skilled staff in the manufacture of pastries. This can be important in New Zealand, where there is a shortage of fully qualified bakers.

### 12.3.5    Lessons taken from the case study

The practice of controlling parameters such as temperature and humidity is usually followed where there are regulatory requirements that are to be met. For example, many meat operators run their production facilities under controlled conditions, with processing rooms maintained at 10 °C. However these same operators may attempt to partially thaw meat to be processed through dicing or mincing machines. The result is, that such operations are almost impossible to control; with meat arriving either completely thawed out, or still frozen to such a degree that machine damage is possible. Other examples abound throughout the food industry.

The case study regarding the pies has been reviewed to show that it may be possible to operate a food business without controlling variables such as environmental temperatures, but that this practice will ultimately limit the company's potential, compromising quality, and preventing the company from trading with some of the better segments of the market, who require high standards and consistency.

## 12.4    Case study 3: The need for qualified and trained people in product and process development

In small businesses, the pressures for survival can override all other considerations, especially during the early stages where capital is stretched and cash flows used to develop the operation. Consequently, formal procedures, such as written quality control programmes, good manufacturing practice (GMP) and Hazard Analysis Critical Control Point (HACCP) programmes that are the norm in larger companies are not given the input or support necessary to ensure that new initiatives do not fail due to unforeseen circumstances.

The third case study in this chapter involves a company that successfully developed a range of tomato sauces packed in sachets for individual use. As an interim measure, the product had been filled by a contract packer. Thus the sauces were produced on one site and packed on another. The product had a pH of 3.8, and was stabilised by heat treatment at the processing plant. The contract manufacturer was a specialist in packaging products into sachets, and also provided a service categorised as an 'aseptic filling line'.

Initially the product was produced in relatively small quantities, and forwarded to the contract packer in 20-litre pails. The product was successfully packed, and over time, customer demand for the product increased to the point where the use of the 20-litre pails was restricting the supply to the contract manufacturer.

A review of the process led to the decision to use 1000 kg bulk pallicons as a means of transferring the product to the contract manufacturer. The pallicons consisted of a large bladder supported by a cage fitted onto a pallet. The bladder had a port that could be used for filling or emptying the unit. The plant was modified to enable the change, and a pallicon of sauce was sent off to be packed into sachets. The contract manufacturer also made changes necessary to accept the bulk pallicons, and the first production with the modifications was run soon after this. A number of samples were retained to ensure that the product was stable as a special precaution, and that the change had had no appreciable effect on the product. In time it was established that the retained samples were stable, and random tests found no significant difference between the old and the new process.

The new process was established as the benchmark, and production continued using the bulk pallicons. Not long after this the company began to get customer complaints regarding blown product.

### 12.4.1    The problem

The causative agent was found to be an osmophilic yeast identified as *Zygosaccharomyces bailii*. Whilst this may be the technical explanation, the real issue is how the problem arose after the company had undertaken steps to ensure that the change in the scale of production did not affect product quality.

The short answer is that the company did not have the technical resources to undertake a risk assessment of the changes planned and ultimately put in place. The tests that were undertaken were considered to be sufficient, and the results finite. The mistake made was that the results from one production run were extrapolated into the notion that the system of production was sound. When a professional risk assessment was undertaken, it was found that there was a risk of the valve of the pallicons being contaminated. It was also found that the cleaning and sanitation steps taken could not be guaranteed to remove all traces of micro-organisms that might contaminate the product at this point. As with all microbiological hazards, the risk was a matter of probabilities rather than finite, with the organism that caused the problem being one that could grow in the product, should it survive processing, or be introduced after the heat processing step.

The case study demonstrates the need to thoroughly examine all aspects of proposed changes. In addition to undertaking a thorough desktop evaluation of the proposed changes, the final product should be tested before product is produced for release onto the market.

Tests should include shelf life testing. In certain instances it may also be necessary to undertake challenge tests, using micro-organisms known to be a potential risk to the operation.

### 12.4.2    Applying professional expertise

There are a number of tools that the professional food technologist can use to prevent the outcome described. These are applied to ensure that the possibility of overlooking an area of concern is minimised, and shortfalls found are corrected. Procedures or evaluations that may be considered include:

- Product specification
- Process evaluation
- HACCP (Hazard Analysis Critical Control Point) food safety programme
- Shelf life testing of final product
- Challenge testing of operation or product.

### 12.4.3    Product specification

The product specification will include a breakdown of the ingredients used. These should be assessed to determine the possibility of contaminants or risks that they could bring to the operation. Special handling requirements and restrictions such as the shelf life of the material should also be noted. Any allergens or food additives used should be highlighted, and segregated by time or space as appropriate.

Special characteristics of the product should be included in the specification. pH and water activity ($A_w$) are two that would be included if these contribute to product safety or shelf life. Others would be special packing conditions such as gasses used in product packed in modified atmosphere packs, or reagents such as preservatives applied to packaging film.

The specification may include a microbiological specification. This normally applies to chilled products with a relatively short shelf life, but may also be included in products such as mayonnaise, baby foods, and products that are designed to be eaten without further cooking. The specifications will include a nutritional panel, set out in the correct format, and any special parameters used in safeguarding the product be specified together with acceptable limits. Packaging specifications and storage conditions should also be included in the product specification, together with the shelf life of the product and factors that may affect this.

### 12.4.4  Process evaluation

The process evaluation will invariably begin with a process flow chart. In operations where significant quantities of waste material are generated, it may also be prudent to undertake a mass balance, and set out a summary of all services used to supply or remove material to and from the operation.

The process flow chart is used to identify where unit operations are applied. Process objectives are identified, and process requirements specified. Areas of particular focus would be operations affecting process stability, such as the temperature/time requirements of a pasteurising operation. Transfer operations should also be scrutinised. In undertaking this evaluation, 'what if' questions should be applied. It is also useful to question the effect of deviations from the stated specifications, and to question the validity of stated process specifications in relation to what is known about the capability of the process.

All this information is used to determine the product type, and to evaluate factors that could put the product at risk. Sometimes a decision tree can be used to assist in this process. An example is set out as in Fig. 12.5.

### 12.4.5  Food safety programme

Operations used to produce foods that have been classified as being 'high risk' will need to be evaluated to ensure that the control procedures applied will be sufficient to achieve the required level of product safety.

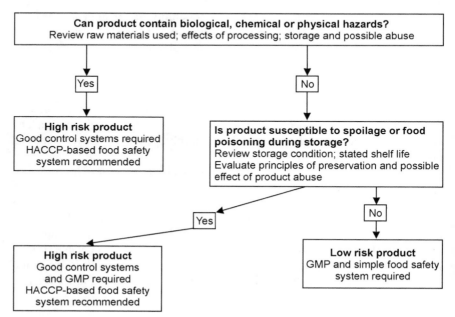

**Fig. 12.5**  Food safety decision tree. Adapted from *Food Industry Guide to Good Manufacturing Practice* (1999).

Food safety programmes give focus to the question of product safety. In recent times the trend has been to use HACCP (Hazard Analysis Critical Control Point)-based programmes. These are well documented, and will not be reviewed in this chapter (see *Food Industry Guide to Good Manufacturing Practice*, 1999; Untermann, 2000; Mortimore and Wallace, 1994).

At this point it is sufficient to note that HACCP is a tool that should be used where a high risk product has been identified. By definition, the focus of HACCP is food safety. However, the discipline of identifying hazards involved in the preparation of a food product, and systematically setting up the monitoring, control and supervision of such controls does have a flow over effect that can identify other issues that are relevant to product quality and shelf life.

### 12.4.6    Shelf life testing

The shelf life of a product should never be assumed. Even different varieties of the same product can be affected in different ways. Product being tested for shelf life should be stored under conditions that are similar to those in which they are to be marketed.

In some circumstances it may be possible to undertake accelerated trials. Quite often there is pressure to place the product into the market before shelf life trials have been completed, and accelerated trials are undertaken. This is usually done by raising the storage temperatures. This approach should be given careful consideration, however, as changes in temperature may effect parameters that are not immediately obvious. For example, raising the temperature may change the status of fats in a product, or affect the vapour pressure in a pack. Such effects can give results that will not be relevant in the product were it stored at normal handling and storage temperatures.

When setting up shelf life testing, it is important to establish the standards for acceptance and rejection. Measures used in shelf life testing will include both objective and subjective measures. The objective measures will cover properties such as moisture content and microbiological criteria. It is relatively straight-forward to measure and compare these to the standards. Subjective measures such as flavour and texture can be more difficult to evaluate or measure. For example, the flavour of a product may change gradually over time. Evaluating these changes in a systematic way will be difficult without a reference to a comparative sample that can be used as a standard.

Often freshly made product is used for comparative purposes. However, this approach can introduce variables that make it difficult to focus on the changes that are actually occurring. Another approach is to store the 'standard' under conditions known to minimise changes in the product. Invariably this will involve storing the product at low temperatures. For example, 'standards' for frozen product ($-18\,°C$) would be stored at $-60\,°C$, and 'standards' for chilled or shelf stable product would be held under frozen conditions, or at temperatures just below $0\,°C$ for product that may be affected by freezing.

It is important that the shelf life tests include conditions under which the product will ultimately be handled. In this regard, consideration should be given to warehousing, transport and consumer use. In scientific terms this will involve subjecting the product to conditions of temperature, humidity and light to which the final product will be exposed.

### 12.4.7   Challenge testing

Challenge tests are normally carried out to ensure that potential pathogens are removed by a given process, or prevented from growth in a prepared product. The approach can also be used to ensure spoilage organisms do not grow in a given product. Organisms considered to be likely to spoil the product are added to the product, and tested to find out if they survive the process, or if they die out as a consequence of a specific operation, or are overcome in the product during storage.

Challenge testing will require the input of a qualified and experienced microbiologist, as it is important that the test is set up correctly with regard to the experimental design. Criteria that must be considered include:

- the selection of micro-organisms
- level of inoculation
- point of inoculation
- incubation temperatures
- reading and interpretation of results.

### 12.4.8   Application to case study

By applying the evaluation techniques to the production of the sachets of tomato sauce it was shown that there were a number of points that were overlooked in the decision to proceed with the operation.

The first stage of the review was to identify the cause of the packs blowing. The causative organism was found to be *Zygosaccharomyces bailii*, a yeast that is characterised by the ability to produce large quantities of gas, even when the organism is present in low numbers. A review of the product specification showed that the product is of a type that is susceptible to spoilage by yeasts such as zygosaccharomyces. Although the pH and the $A_w$ of the product would have prevented the growth of most bacteria, there was no real protection against acid tolerant yeasts. Thus the product was not stable on the basis of formulation, but required a heat processing step to eliminate the possibility of yeast contamination.

The next step was to review the process. An abbreviated process flow chart is shown in Fig. 12.6. The operation can generally be classified as 'heat treatment followed by aseptic filling'. The manufacture of aseptic products using this approach requires the application of a series of complex technologies. Essentially it involves the production of a commercially sterile sealed package of food, by pre-sterilising the food, followed by cooling, filling and finally sealing

Part 1 (Processing factory)

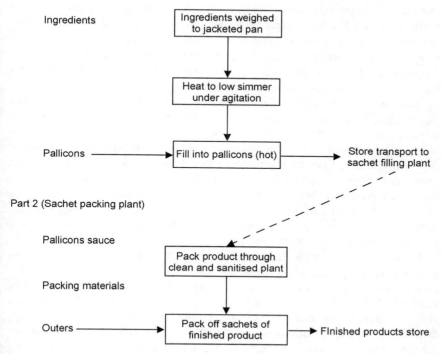

**Fig. 12.6**   Process flow diagram (abbreviated form) for the production of sauce sachets.

in sterile containers. It is usual that there be no break between the respective operations of pasteurisation, cooling and filling.

One of the mistakes made was that the complexity of the process was not fully understood by the manufacturer. On evaluating the process it became clear that the process could have failed at any time, and that the successful running of the operation prior to the use of the pallicons was the result of a very vigilant cleaning and sanitising programme, and a certain level of luck.

On reviewing the process flow diagram, the break between the two plants stands out as a potential cause of the problem. The other point that was scrutinised in the process review was the heating step. On questioning the operators, it was discovered that they had not been simmering the product during the heating stage. The reason given was that the product took a long time to cool in the pallicons (due to their large size), resulting in an unacceptable dark colour.

This illustrates the need to talk to the people who are actually doing the work. It is not unheard of to find that the operator on the floor has made changes to a procedure, believing the change has been done in the company's interest. This point also illustrates the dangers that occur, when formal quality control procedures are not monitored on a routine and ongoing basis.

The final finding of the review concerned the shelf life trials. It was found that the sachets put away for evaluation had been examined over a period of four weeks after they had been prepared. At this time it was decided to continue with the altered form of the production to meet market demand. When the samples were examined some months later, as a consequence of the customer complaints regarding the blown packs, a number of sachets were found to be blown. This outcome shows the need to continue and conclude shelf life trials.

There was one other issue identified in reviewing the shelf life trial. The trial had been initiated during winter, when temperatures in the warehouse were relatively cool (14–18 °C). With the onset of summer, the temperature in the warehouse increased to 20–30 °C. It is not uncommon to find that low levels of organisms such as yeast lie dormant in products until conditions become more favourable, allowing apparently stable products to spoil some time after they have been produced. The problem was solved by attending to all the issues identified in the product and process review. That is, a sachet filling line was brought on site. Most importantly, changes were made to the process to include a post-packaging pasteurisation step.

In order to include the final pasteurisation step, it was first necessary to ensure that the sachet seals would not be compromised. This required the product to be filled directly after cooking and re-cooling. Essentially this required a heat exchanger to be installed, to enable the product to be raised to a standard temperature, then cooled to allow filling into the sachets. The heating, cooling and filling operations were conducted in line, with no breaks between the operations. The cooling step was necessary, as the sachet could not be correctly sealed if the product was above temperatures that would affect the ability of the film to 'set' after the sealing operation.

Once sealed, the sachets were passed through a steam tunnel, to pasteurize the product after filling. As the product specification stipulated a pH of 3.8, it follows that the temperature/time relationships to achieve this were well within the capability of a tunnel operating at atmospheric pressure. The sachets were partially cooled, allowing them to dry off before they were packed into cardboard outers.

This process, sometimes referred to as post-packaging pasteurisation, does not allow for the product to become contaminated after the heat processing step. No incidents have been reported since the application of this process.

### 12.4.9   Lessons taken from the study

This case study highlights a number of typical issues that can cause problems in the transfer of technology from the laboratory to production, particularly in smaller operations, where there can be a lack of technical expertise. Whilst a number of problems were identified, the most important point to be taken from this study is that some operations require certain levels of skill, knowledge and experience to ensure that they are completed successfully.

The lack of this skill resulted in shortfalls in the original design of the process. The company also made assumptions that could not be sustained in regard to the way in which the shelf life study was undertaken. In addition, quality control procedures were not followed through, with the loss of control over operations that were essential for the successful operation of the line.

Most small food manufacturing operations will have had similar experiences at some time in their history. This is because of the evolving nature of such companies. What the study highlights is the importance of growing beyond this point, and ensuring that the company has access to professional expertise as is required. Whilst professional food technologists can assist in the general development of new products and give an input into operation and quality control procedures, they will not necessarily have all the skills needed for technically demanding operations. It is also unrealistic to expect a person working in isolation to cover a wide range of highly specialised skills. For example, correctly undertaking a challenge test will require specialised microbiological input. To cover such areas it is important that companies who do not have all the expertise in house form good liaisons with reputable consulting agencies, universities or companies.

## 12.5   References

*Food Industry Guide to Good Manufacturing Practice*, 2nd edn (1999) New Zealand Institute of Food Technology Inc. p. 15.

MORTIMORE, S. and WALLACE, C. (1994) *HACCP: A Practical Approach*. Chapman and Hall, London.

SORENSEN, T. (1973) Restructuring of Fish Mince. Masters Thesis. Department of Food Technology, Massey University, Palmerston North, New Zealand.

SORENSEN, T. (1976) Effect of Frozen Storage on the functional Properties of Separated Fish Mince. Proceeding from *The Production and Utilization of Mechanically Recovered Fish Flesh (Minced Fish)*. Ed. Keay, J.N. Torry Research Station, Aberdeen.

UNTERMANN, F. (2000) *Hazard Appraisal (HACCP): Encyclopaedia of Food Microbiology*. Eds. Robinson, K.R., Batt, C.A., Patel, P.D. Academic Press, London.

# 13

# Spirit of entrepreneurship in commercialisation – product and process development of coconut beverage mix

**Choon Nghee Gwee, Malaysia**

*This chapter shows how the company strategy itself can become a 'product' for development, becoming a basis for strategic enhancement of food company decision making. Both product development knowledge and process development can fall short for commercial success if the adequate human skills are not added in, and added with the enthusiasm that comes from total commitment.*

*An important word is missing from the index of the book* Food Product Development *by Earle, Earle and Anderson – entrepreneur! In the food industry, the entrepreneur who starts a business from their innovation is an important person in food product development. They may notice a new trend in consumer behaviour, have developed a new processing technology (maybe from an old craft or new knowledge), or invented a new product. Often the new products are original. The products may fail because they are too unusual. But more usually the entrepreneur fails because of lack of money and resources or lack of know-how or because of competition from large companies. But they are also a source of new products for the large companies and can be taken over by them and developed further.*

*Basically the Product Development Process (PD Process) is the same for the entrepreneurs, but they have to build up an external team to give them the expertise they need. They do not have internal knowledge and people as in the big company. Also, because of their lack of resources, they usually take greater risks by leaving out the activities seen as necessary by the big companies. For example, they may not do the market research necessary to prove if there is a market and how big it could be, or the research to optimise the process, and more often, they may go ahead without sufficient money to last through the initial launch and the following months.*

*If they have no experience in product development, they may not identify the*
*activities in the first part of the PD project – defining the project, developing the*
*product concept, identification of processes, distribution and marketing,*
*development of product design specifications, planning of the project, and*
*predictions of project cost and financial outcomes. They often go ahead with*
*only part of this knowledge, contract a processor to make the product without*
*specifications and then take it to the supermarket buying staff without the*
*knowledge to persuade them to put it on their shelves. They have to cut corners*
*but they need to know which are the safe ones to cut.*

*The chapter particularly relates to pages 96–130, 165–176 in* Food Product
Development *by Earle, Earle and Anderson.*


## 13.1   Introduction

The aim of this case study was to set up a company to develop, produce and
market a drink based on coconut cream.


## 13.2   Entrepreneur's dream

There comes a time in the lives of most people when they begin to think 'Why
am I working for someone, when I could be doing my own thing?' The difference between the entrepreneur and everyone else is that the successful entrepreneur gets up and does their own thing, and is not content to dream about it.

After having successfully developed several new products and created
successful businesses through new innovations in the specialised field of tropical
products, during more than ten years of working life, it was always my dream to
become an entrepreneur. Make my own unique product, and run my own
business, making and selling product of my own brand.

This interest in setting up my own business made me enrol in all sorts of
courses related to starting a business, read all sorts of books and pamphlets and
even to attend a MBA course to prepare myself for the challenge. So much for
the theory, in reality many successful businessmen do not attend such courses,
but I was determined to succeed and thought that, well, every little helps.

To get a business started very much depends on how one makes use of
opportunities at the right place and time and to recognise these essential key
points when they occur. My substantial expertise in product development
acquired through my years in the food industry, particularly in the field of
coconut products, opened opportunities for me to explore further the
development of new beverage- and dessert-related products.

After successfully stabilising the production and quality systems of a fully
integrated coconut processing factory in Indonesia in the early 1990s, I received
again and again enquiries about developing a version of the same product, a
canned coconut juice drink, similar to or better than the existing products.

After some studies and involvement in determining the actual market potential, I was convinced to take up the challenge of developing a version of this same product myself, but using a different approach.

I was on the way to realising my dream.

## 13.3    Product idea

### 13.3.1    Background

Coconut juice drinks began to be popular in the late 1970s and early 1980s in Hong Kong. During that time, these products were only available from one or two street stalls, and were prepared from extracts of desiccated coconut together with sugar. Consumers liked these exotic tropical drinks. They had a pleasant taste, and were also popular because in traditional Chinese medicine, they are considered to have some beneficial nutrient or health value. Hence coconut juice drinks are marketed as health drinks in both Hong Kong and mainland China.

With advances in technology, coconut juice drinks became available in cans and aseptic paper composite packs. The world's largest producer of coconut juice drink is Haikou Canned Food Factory in China. Their Coconut Palm Brand has been produced since 1984. It had become a nationally available drink and was selling better than some of the multinational beverage companies' products. There were more than 10 to 15 factories in operation in Hainan producing more than 100,000 tonnes of coconut-based foodstuffs and drinks per year.

Coconut product manufacturers in China buy coconut raw materials from Indonesia, Malaysia, Vietnam and Philippines to supplement local coconut production. Coconuts are very demanding in their climatic requirements and can only be grown commercially on Hainan Island in China. The acute shortage of coconuts in Hainan Province makes imports of coconuts, coconut cream or coconut raw materials essential to enable the drink producers to cope with the enormous demand in China. Sensory evaluation of coconut juice drinks by a trained panel of flavourists showed that the products were extremely variable in quality. This was not surprising, considering the reliance on imports, I knew just how poor quality control could be in some of the supplying factories.

Besides China, coconut beverages have also gained popularity in Taiwan, Japan, Australia, Europe, USA and other parts of the world. A range of coconut juice drinks, with added pineapple juice or other juice flavours are the choice product of the new generation. Some restaurants and coffee houses serve coffee and tea with coconut cream in place of dairy products or traditional creamer.

I was then convinced that it would be possible to prepare a coconut juice drink of higher and more consistent quality than anything available in China. China was the logical initial target market: it was big, the product type, coconut juice drink, was known and accepted, and there was room for a better quality version. The product I was to develop had to be competitive in price and in a format easily transported and distributed.

### 13.3.2   New product concept

Coconut cream, being the key ingredient for making coconut juice drinks, is easily perishable due to its high water content, high fat and low acid nature. Another natural phenomenon in coconut cream is the rapid fat separation which makes the product look less than appealing on standing. The common processes used to produce safe and long shelf life drink products involve heating or pH adjustment, both of which can affect the flavour, colour and texture of coconut cream. Much research was required to optimise process requirements in order to retain a nice taste and fresh aroma and yet produce a product with a long shelf life and pleasant appearance.

Canned and paper packaged, sterilised, ready to drink products made from diluted coconut cream already exist and are particularly popular in S.E. Asia and China. The high water content of these makes transportation from Malaysia prohibitively expensive. Spray drying of coconut cream powder was the other alternative method of food preservation. During spray drying, the water content of a liquid foodstuff is reduced to very low level so as to minimise micro-biological and enzymatic activities in the product. The trick is to effect drying without destroying the delicate flavour and damaging the reconstitution properties of the material. With the right process conditions, a coconut cream powder with superior and stable quality can be prepared. This product would have the additional advantage of being concentrated during the drying process, and taking up far less room. This was the desired breakthrough. Several companies began production of coconut cream powder, with more or less satisfactory end products.

Coconut cream powder of improved quality would be the basis of a whole range of branded new powdered drinks, easy to prepare, cheap to transport. Lengthy product development resulted in a basic series of drink mix bases, containing specially treated coconut cream powder, sugars, flavours and essences, and other ingredients to ensure smooth consistency, rapid and complete dissolution.

## 13.4   Product design qualities

The specifications for the product design were as follows.

*Raw material availability*
The leading coconut cream powder producers are all located in South East Asia. The major exporters are Malaysia, the Philippines and Indonesia, with relatively minor players in Sri Lanka and India. My supplies were to be sourced where I was most confident of quality, but a wide geographical range was necessary to ensure supplies (in terms of both quantity and quality) were available throughout the year.

*Superior quality*
Recent advances in coconut process technology contributed to the improved natural coconut taste in the coconut cream powder, which has earned recognition

in terms of its consistent quality and better aroma compared to products processed in cans. With strict quality control, good selection of nuts, and proper pre-treatment of the fresh coconut extract, a premium quality coconut cream powder can be prepared with very low moisture content. The product quality can be further enhanced with high barrier packaging to prevent moisture and oxygen absorption during storage resulting in minimum microbiological, enzymatic and oxidative changes.

*Product flexibility*
The instant powder drink that resulted from my initial trials was a product for all seasons. It could be reconstituted and served hot during winter and cold during summer. It is a good base for smoothies, mocktails, cocktails and other exotic beverage concoctions.

*Ready markets*
The range of commercial products in liquid form (cans or UHT pack) available in the market were already in demand in China, Hong Kong, Taiwan, Japan, Australia, USA, etc., and so we needed a product which was superior so that we could eventually take on these markets, too.

*Exciting varieties*
A variety of Asian favourite tastes would be developed to make the product even more attractive and exotic. For the beverage range, the most popular flavours elected for development and launch initially were:

- natural coconut
- honeydew melon
- pineapple
- coffee.

Later, coconut slush and ice bases were also *to be* developed through the use of slush machine, ice blender even soft-serve ice cream machines. Five exotic flavours were introduced:

- natural coconut
- mango
- honeydew melon
- strawberry
- coffee.

*Convenience and easy availability*
For a product to succeed in the market place it needs to be easily made available in all forms of retail stores and food service outlets. Hence it helps if it needs no refrigeration, and is compact, taking up little space. This was another advantage of the powdered beverage.

*Long shelf life*

For export markets and in areas where distribution is less developed, it was desirable to have a product shelf life of 2–3 years at ambient room temperature, or about twice the realistic shelf life of canned, UHT or bottled products.

*No added preservative*

A powder product with a low moisture level and protected with good packaging does not require addition of a chemical preservative.

*Product know-how expertise*

Our strong R&D team has in-depth product and market knowledge in the related industry internationally. This was an absolute essential to innovate and to achieve our intended objectives.

## 13.5   Process development: improving the solubility of coconut cream powder

Further research was needed to improve the quality of coconut cream powder regarding its solubility and stability. Since my personal expertise at that time was very much focused on liquid coconut cream process technology, a food powder expert was then invited to assist in the development of this new product. Most coconut cream powders available are mainly suitable for cooking purposes, especially for use in curries, or for sale to international confectioners for further processing. When choosing coconut cream powder for powder beverage applications, there were various additional quality aspects to consider. Most of the existing products were not easy to dissolve either at room temperature or in hot water. Good solubility was of paramount importance to my product. There were two possible approaches in solving this problem: direct spray drying and the surfactant method.

### 13.5.1   Direct spray drying method

Coconut cream powder is difficult to dissolve and too rich to serve as a drink base but could be used as the start of an entirely new process. The dryers could prepare a powder, not of coconut cream but of my fully formulated drink base. This would be simply packaged and easily reconstituted into a drink. Gradually my product took shape, and I arranged trial drying. The seed had germinated.

Solubility of products is very much affected by ingredients, formulation and process conditions during spray drying. Pilot trial drying was carried out in a private R&D institute with the most modern spray drying technology, to establish optimum process conditions.

Our early research showed that by the addition of small amounts of soluble potassium and sodium salts, themselves naturally occurring in coconut cream, the delicate protein complexes of the cream, which are largely responsible for the characteristic flavour and aroma, could be made to survive the demanding pasteurisation and spray drying processes (Chee *et al.*, 1997). We had a product that had a good shelf life and could be reconstituted to resemble the original fluid. Reconstitution was, however, not easy.

The powder formed needed a lot of hard work to get it to disperse and then dissolve in water. It was not a good characteristic. The busy housewife has more to do than to mix coconut drinks. We therefore sought solutions that might make our base powder easier to dissolve.

Initially we attempted to increase the powder particle size, by altering the drying temperature and dryer airflow, and ensuring individual particles adhered to others while still damp, and formed agglomerates, large clusters of many particles joined together. The idea was that water would be drawn into the centres of the cluster by capillary action, once the powder was added to water. While this certainly occurred, the results were not as good as we had hoped.

Microscopic examination of the various powders produced in our early trials, showed that the tiny capillaries between individual particles within the agglomerate cluster were blocked with fat, preventing water ingress.

This was a surprise. The large oil content of coconut milk has been carefully prepared by nature in the form of tiny droplets of oil contained in thin protein envelopes or bags, a little like balloons, where the air is the oil and then rubber the protein outer. No free oil should have been present. Looking at cream samples from the many parts of the process, from picking nuts through to the packaging of our drink recipe, soon showed that handling the coconut cream at each stage resulted in damage occurring to the delicate protein membrane around the fat globules, and some oil or fat escaping. The more processing that occurred, the higher was the free, or unencapsulated, oil content. Free oil would clog agglomerate particles and make dissolution difficult, and sadly free oil in contact with air would result in oxidation, which could affect the shelf life of the product.

We experimented by adding protein from other sources, both of vegetable and milk origin, and found that a good part of the oil could be made to be present in the encapsulated form in carefully produced powder; expensive, but essential. Sadly, this was not the complete answer. Try as we might, there was always in the region of a few per cent of the coconut cream oil content present in the free state, just enough to give the powder poor solubility. We had to look for some physical/chemical solution.

### 13.5.2  Surfactant method

Not all the existing coconut cream powder dryers were able to produce instant powder as required. As we had no intention of investing in a spray drying plant, we looked for an alternative and yet cheaper means of production.

In the process of washing dishes or clothing, oils and fats are removed by adding detergents to the water. More properly termed surfactants, detergents reduce the surface tension of water and make the oil deposits more easily soluble.

Surfactants in this context are complex molecules of long, thin forms, like sticks. One end is hydrophilic or water loving, the other lipophilic, or fat loving. When added to water and dirty plates, the surfactant covers the fat in the lipophilic ends of its molecular chains, producing something like a pincushion. The water is presented, not with a blob of oil or grease, but a hedgehog of water loving molecules which it lifts and keeps in suspension. It emulsifies the grease or oil.

So, we needed an edible surfactant. Surfactants occur naturally in our stomachs, but these, although they can be synthesised easily, taste awful. We had to look elsewhere. Research into heart and circulatory diseases had shown that deposits of cholesterol in human arterial systems could be removed if the lecithin level in the blood was increased. Could it be that lecithin was a surfactant?

Lecithin occurs naturally in many plants and animal products. It is responsible in part for the creaminess of egg yolk and for the richness of soya bean products. It had been used for many years in the chocolate industry as an emulsifier, to keep the cocoa butter (fat) properly mixed with the water-based constituents of chocolate. We took refined lecithin from soya beans and experimented and experimented. It worked. If we could cover the surface of our powder particle agglomerates with lecithin, some would dissolve in the free oil present, and render the powder water soluble. Treated powders dissolved easily with minimum stirring. We were on the way to a breakthrough.

There were still problems to be solved. Lecithin is very, very expensive. Even the refined product has a pronounced beany taste. We needed to develop a means of dispersing tiny quantities of lecithin intimately over our powder particles and to minimise the flavour impact.

Fluid refined lecithin is a very gooey viscous material, not easily handled or dispersed. It could be made less viscous by heating, but raising the temperature meant that the surfactant properties were destroyed. An experienced flavour house was brought in to attempt to produce a flavour additive that might lessen the beany note of our surfactant.

Both activities are the results of expensive research and cannot be disclosed in detail, but suffice it to say that the problems of application on powder particles and its strong flavour were eventually overcome, and our drink base was rendered soluble, even in cool water.

We felt that we had won the day. As the discovery of this form of surfactant addition was new and has good potential for other food powder applications, we patented the invention.

## 13.6    Process development: other coconut cream powder problems

### 13.6.1    Fat separation

This occurs naturally in coconut cream but is not acceptable or appealing in powder used for making a beverage base. Formulation with the right stabiliser and emulsifier and use of the correct drying process parameters were found to be the solution. Reconstituted as a drink, our product would stay homogenous for more than an hour, a big improvement over the natural liquid.

### 13.6.2    Consistent quality
Selection of nuts and good practices in post-harvest handling contribute to initial aroma freshness in the raw material. Fresh coconut aroma is delicate and sensitive to high temperature. Research was done on optimum parameters at various process stages and minor differences in process were evaluated. The results were observed and samples tested. Eventually we arrived at a process that is close to the original flavour and freshness. Protected by good packaging, the product can be stored without loss of aroma for at least two years. Close supervision and a good quality control implementation system are important to ensure consistent quality of the product. The key ingredient is the in-depth raw material and product know-how which enables us to work closely with the manufacturer to supply ingredients of consistent good quality.

### 13.6.3    Consistent supply
Coconut supplies are much affected by weather and season and sometimes for political reasons. Consistent supply is always a worry. As predicted in the sales projection, once the volume starts to move, the company needed to plan to secure more supplies to ensure year round availability. With better understanding of the supply patterns, the advantage of the long shelf life of the product allowed more flexibility in the inventory control and delivery planning.

## 13.7    Testing of prototype product

Prototype products were prepared in plain single-serving packaging. Samples were distributed to various potential distributors for quality feedback. The positive response from potential customers encouraged me to proceed with my business venture dream.

## 13.8    Sourcing capital

### 13.8.1    Capital requirements
The initial approach was to start with absolute minimum capital until the sales started to pick-up. We decided to use a contract manufacturer initially, avoiding the huge amount of cash needed to purchase processing and packaging equipment. A company was formed with the founder being the major shareholder. The potential market targeted was mainland China. Predicted from costing and project evaluation, the start up, excluding initial development costs, was

estimated to be RM 500,000 with a further requirement of RM 400,000 needed for commercialisation of the new products.

### 13.8.2    Funding

The company intended to borrow from the bank(s) but was turned down because of having no track record in business at that time. A typical 'catch 22' situation frequently faced by entrepreneurs.

Another option was to invite investors to join us. Potential investors were selected from various industries such as electronic, construction, venture capitalists and food companies. There were various considerations in choosing the 'right partner', including background, co-workers' comments, track record with similar projects and good standing in the financial world.

### 13.8.3    Right partner

After assessing the pros and cons of various possibilities, I was very fortunate to get financial assistance from friends who are also from the food industry. At this initial period of the new business, the company benefited a lot from the partners' existing network and supports, not just financially.

Selecting the correct partner is not easy and sound advice from those who have trodden the path before is invaluable. So often investors have a different set of aims than the entrepreneur ... they may expect a quick return on capital, rather than sustained business growth, and it is important to establish mutual respect and for each partner to give credence to the other's area of expertise or input. Genuine good partners with commitments to the same objectives are important for the success of the business.

## 13.9    Marketing development

### 13.9.1    Branding

Preparing the optimum presentation and pack size often necessitates a specialist from the target market, as does giving a name to the product. These are the first steps to creating branding, essential when the product is new. Branding costs money and can be exceptionally expensive. Co-branding with an existing popular brand was decided upon, to take advantage of existing distribution and marketing channels.

### 13.9.2    Promotion through trade exhibition

Simplicity was the keyword for the initial product launch, and the three most popular flavours were selected: natural coconut, honeydew melon and pineapple. The products were packed in a cheerful display box filled with single serving sachets (see Fig. 13.1). Brochures and postal flyers were printed for promotion purposes.

**Fig. 13.1**   Initial 'Coconut Express' packaging.

The first exposure of our products was at an international food exhibition in Hong Kong attended by trade visitors from the food industry in Hong Kong, China and South East Asia. We also invited potential distributors to visit our booth. We were serving the drinks through juice dispensers and at the same time selling our products. The response was overwhelming. On the spot, we sold a few hundred cartons of our products. We realised that wet sampling in this way was a very effective means of convincing potential customers. Coconut products had been considered very traditional and nothing very exciting. Besides promotional materials and other advertising methods, on the spot product tasting lets the product speak for itself.

In subsequent years, the company participated in as many trade exhibitions as possible, in order to promote our products, as it is one of the most effective ways of introducing new products and getting new potential distributors interested.

### 13.9.3    Choosing potential distributors

We were now ready to investigate distributors. We had received enquiries from several interested parties from China during trade exhibitions, and visited their operations to understand their background.

After fairly lengthy evaluation and consideration, we confirmed distributorship with a foreign joint venture company in China. At that time, this company had been distributing cereal drinks nationwide for at least six years and was looking for a product suitable for summer. We had also invited the potential distributor to visit our exhibition booth in Hong Kong to further enhance their confidence in our products. Since they already had established their brand in China, we decided on a co-branding collaboration.

'ACES' brand was in the top three in terms of turnover in the 3 in 1 cereal market, between Goldroast and Super, with a turnover of SGD 70 million at that time (1998). The company liked our initial packaging presentation and 'Coconut Express' brand (see Fig. 13.2). So, we decided to have a co-branding of Aces

**Fig. 13.2**    Initial 'Coconut Express' presentation.

**Fig. 13.3**   Aces 'Coconut Express' presentation.

Coconut Express as shown in Fig. 13.3. (Coconut Express was printed in Chinese as it is for the Chinese market.) The products were introduced to the major hypermarkets in the major affluent cities around Shanghai, Hangzhou, Nanjing, Suzhou, etc.

### 13.9.4   My first order
Soon after our negotiations were completed, occurred the happiest single event in my first business venture. We received our first order, valued at USD 500,000. Within the first year, we managed to export over thirty 40-foot containers of goods.

## 13.10   Post-launch activities

### 13.10.1   The ups and downs of business
We decided that we should best concentrate on the problems of sourcing ingredients, stabilising product quality and the production system, and ensure

that deliveries occurred on schedule, and allow our distributor, who had vast experience in the area of distribution, marketing and sales, to take care of the product launch.

Exports to China through the same distributor went on for 1–2 years. Owing to unclear legal issues on distributorship, the product launch was interrupted and delayed. We had many lessons to learn, not the least of which was the best approach to product familiarisation. At the initial launch, the distributor did not follow our intended method of promotion by giving on spot free tasting. Instead, our products were given as free sample in the cereal packs. We thought this would not be so effective because customers might not try the samples or prepare them in the right way.

Techniques often used in this region, such as giving away free samples, were not effective for new product introduction in China. China is a huge market, and many containers of free goods can easily be absorbed for little return. Careful target marketing in specially selected key areas is vital. This we learned the hard way. It is better to target and focus retail outlets in one city initially.

We continued participating in food exhibitions in Guongzhou, Shanghai and also worked together with MATRADE (Malaysia Trade Development Board) to develop new potential distributor(s) for the retail market. We found some other potential distributor(s) but we realised huge capital is needed for selling in retail outlets as China is a huge market. We then called for further investment for the business expansion.

At the same time we further developed a new range of products: ice blend, granite (Slush) with more varieties of exotic flavours for the food service market segment. Coconut Express powder mix was then accepted by a large local fast food chain in Hong Kong. It is used as a beverage base with tropical fruits combination. It has become one of the most popular beverages in the outlets.

### 13.10.2    Product line extension
As our product slowly became accepted, and sales developed, we decided to expand our product range. Besides coconut drinks, new trendy products were developed projected for the youth market. Exotic tropical flavours, granita and ice blend beverages with coconut, mango, strawberry, melon and coffee flavourings were to be launched.

### 13.10.3    Continuous promotion, expansion
The company continued to take part in fairs and trade shows and exhibitions, throughout China, but with particular emphasis on target cities, Guangzhou, Shanghai and Beijing. The intention was continuous promotion to increase sales potential and allow further expansion of the business.

## 13.11    Problems met and lessons learnt

When bringing new products into new market places, we often face unexpected problems and obstacles. Strong beliefs and lots of passion for the business are the main driving force.

In the foregoing I hope I have shown that the would-be entrepreneur cannot go it alone except in very exceptional circumstances. Successful businesses stem from a simple idea, the founder's concept or vision. To convert this concept into a viable and profitable business needs the co-operation of many minds and skills, and lots of decision making. The successful entrepreneur knows who to ask for advice, and how to sift the valid bits from such advice. He or she knows their limitations and is not too proud to seek expert help. They look at the track record of those advising or carrying out part of the operation. They keep a shrewd eye on costs. To be an entrepreneur it is not enough to have a great idea. He or she needs this of course, it is the acorn that will grow into a flourishing oak, but he or she needs to cultivate the ability to develop relationships with other areas of expertise, to be able to listen to others' experiences, to determine what competitors are doing well or badly, to evaluate expert advice. All of these attributes form the essence of entrepreneurship.

## 13.12    Acknowledgements

Sincere thanks to those who contributed to this chapter, especially to Terry Green who is one of the key important partners who assisted in the initial development of the project.

## 13.13    References

CHEE, C. SEOW and GWEE, C.N. (1997) Coconut milk: chemistry and technology, *International Journal of Food Science & Technology*, **32**, 189–201.

EARLE, M., EARLE, R. and ANDERSON, A. (2001) *Food Product Development*. Cambridge: Woodhead Publishing Ltd.

# Part V

# Consumer and market research

# 14

# Sensory testing in the Product Development Process – the sensory researcher as a trusted advisor

Anne Goldman and Elizabeth O'Neil, Canada

*This chapter shows how consumer strategy in consulting companies needs to broaden from sensory testing to knowledge of the consumers' needs, wants and behaviour. The importance of sensory properties in acceptance and rejection of foods has been recognized for a long time but there are many other factors that also affect the consumers' attitudes to food products. Over the last few years, it has been recognized by food companies and also sensory consulting companies (called 'suppliers' in this chapter) that sensory testing is needed to broaden to the total product assessment to gain a deeper understanding of the reaction in the market place to a new product.*

*There has been a growing sentiment in both academic and business circles that there is a need to develop new research strategies and relationships between food companies (called 'clients' in this chapter) and their consumer product research suppliers. Today, perhaps more than ever before, business is increasingly adopting a more fact-based, decision-making strategy. Clients are looking to their market and sensory research supplier to provide more than the delivery of data tables and significance levels. They now want to obtain market and consumer insights for their products. What the consumer thinks about their products or services is key. Today, gut feel and intuition, without a clearly defined problem and a supporting argument, are not acceptable. How does this impact any market research supplier and sensory researcher specifically in terms of their client relationships?*

*The focus is now on the consumer and their evaluations and relationships with the brand and product category. One way to respond to this evolving landscape is to expand on the traditional market research function and to develop a more collegial relationship with clients. This new type of relationship pertains just as much for the external research supplier as the internal supplier. This involves open communication between the researcher and the client about what is driving a particular research need or request, and interpreting how the findings of the research will impact their particular business. This type of effort and skill set brings new dimensions to the competencies of the current typical practising market researchers and sensory researchers. It changes the consultancy from 'technicians' (limited scope, limited range, limited responsibilities, limited ethical constraints) to 'professional' – unlimited. This means transforming not only the activities but also the attitudes. However, this role, often defined as that of the trusted advisor, is attainable and necessary to address the evolving and growing needs of the client. It could open gates to potential growth, bringing in new expertise, but also vulnerabilities.*

*A useful text for the total food development process that relates to this chapter is Chapters 4 and 5 in* Food Product Development *by Earle, Earle and Anderson.*

## 14.1   Introduction

The Dunmark Company was founded as a sensory evaluation research company in the mid 1980s. The research services available in the market at the time did not adequately address the different needs of marketing and research and development (R&D). Dunmark viewed this as an opportunity to supply a service that was unique in nature, by positioning their product performance research service to provide direction to marketing and R&D. Over the next 20 years, Dunmark recognized that the business environment in which it was operating was changing. Who was the change agent? Was it Dunmark? Their clients? Or both? Clearly the sensory research function, primarily that of a diagnostic technician, was shifting with some clients, to a more consultative role. This meant that the output of product research was not solely focused on the sensorial dimensions. It also encompassed marketplace variables that impacted the product. This shift created both a philosophical and organizational challenge and change for Dunmark. However, the ability to accommodate varying types of research projects was necessary in order to meet the evolving and differing needs of the clients. Dunmark needed to be able to meet the client needs in one of three ways:

- as a diagnostic technician
- as a research coach
- as a trusted advisor.

The method of choice was a function of the level of client/supplier trust and the scope and application of the results.

## 14.2    Scenario 1 – The diagnostic technician

In past years, the client viewed the sensory researcher primarily as an order taker who would provide data with a basic interpretation (Fig. 14.1 Scenario 1). In this instance the sensory researcher might be weak on industry and business knowledge and could be seen more as a white coat 'techie'. While an inherent assumption might suggest that this no longer is the operating paradigm, in fact, there continues to be instances where this is the reality. The relationship is efficient, yet perfunctory. There is a distinct lack of a dynamism that fuels a more creative and insightful research product. The point that is being made here is that this used to be the norm, whereas today it is possibly more the exception. However, the road to becoming a trusted advisor (as depicted in Fig 14.1 Scenario 3) is not easy, predictable or straightforward.

Part of the reason that the sensory researcher was not perceived as providing much more than data is that the typical research undertaken usually involved difference tests such as blind, triangle taste tests. The triangle test methodology was often the only sensory method taught in academic programmes other than sensory training programmes. This type of research was mostly done in-house using employees to test the products. R&D would likely have been the initiator or the 'client' of this type of research and in some cases still is today.

The research process was focused on the task at hand, such as product evaluation for quality purposes. The objectives of the research were primarily focused on how to discriminate between products with the report being read or filed by R&D. General consumer research, in this model, was the domain of market research alone, where sensory dimensions were not well understood.

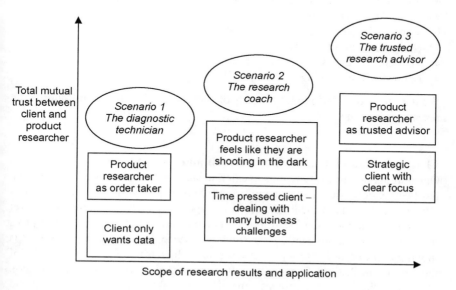

**Fig. 14.1**    Model of trust.

However, marketplace factors started to increasingly influence the consumers' choice process. Along with the proliferation of choice options available within most categories, this model of product evaluation was simply not 'real' enough to ensure successful product entry to the market by food manufacturers.

## 14.3    The changing business environment

It is fair to say that the environment in which sensory research is now being conducted has changed. It is now more closely linked with traditional market research and is recognized as an important dimension of the overall market plan for a product.

Nothing occurs in a vacuum. Oftentimes, changes in business or how business is conducted come about as a result of macro level initiatives that have no direct bearing on specific processes, but impact them nevertheless. The research process has also been affected in this manner.

## 14.4    The effects of the changing business environment in the food company

The complexity of the current business environment leads to tremendous pressures on all parties (client side and supplier side). The pressure comes from many places including:

- Outsourcing of market research and sensory testing by the food company
- Retail demands on the food company
- Brand manager movement and involvement
- The need to be innovative
- Payment by results
- Team decision making.

What is the potential impact of each of these variables in general and to the research function?

### 14.4.1    Outsourcing of market research and sensory testing

Outsourcing of various functions to external suppliers, as a cost savings solution, has been widely used across many departments of a food manufacturer such as HR, accounting, IT, product development and ingredient procurement. Market research, and in particular, sensory evaluation (consumer and trained descriptive panels) is also predominantly an outsourced service. In some organizations, it is moving in the direction of a procured service, which often means the supplier of the service is dealing financially with someone else within the organization other than their primary project contact person. While it is not a widespread phenomenon currently, it does potentially alter the dynamics of the research relationship. The supplier incurs costs, and procurement procedures are often inflexible and not time sensitive.

### 14.4.2   Retail demands on the manufacturer

The retailer is also putting pressure on the food manufacturer by demanding that they prove that their products have merit to be on the shelf. Churn rates are monitored and slow moving products will quickly be de-listed by the retailer, at a high cost to the food manufacturer. Slotting or listing fees in the North American marketplace are generally very expensive to obtain and maintain. Thus one of the key objectives for product research might be how well the submitted test product compares to the in-market leader rather than the research having an optimization focus. Additionally, branded products face increasing competition from the retailers 'own label' brands.

### 14.4.3   Brand management movement and involvement

Whereas years ago R&D might have mostly been involved with the sensory research process, the primary contact today will likely be a brand management team. These people are not necessarily as knowledgeable in the R& D aspects of product development and in addition, may not be that familiar with specialized market research methods. As soon as a key client contact becomes conversant in the product evaluation process and appreciates the merits product research can offer in gaining knowledge on consumer behaviour, that person will be moved off the business and replaced by someone new. Thus, the process of educating brand management begins again. Increasingly, the research supplier is expected to offer guidance and education in this area. Some of the efficiencies that can be gained are then lost as new brand managers come onto the business ... one step forward, then possibly two steps backward.

### 14.4.4   The need to be innovative

The pressure on brand managers and product developers for product innovation is coming at the same time that the marketplace is saturated with all possible variations on the product theme. Is there really more money out there to be spent or are consumers simply switching to a more innovative product at the expense of another 'tried and true'? An exaggerated example of this might be attempting to sell garlic-flavoured ice cream as a consumer commodity. Here, the primary focus of the product evaluation or research process is around the idea of 'new and different' purchase interest rather than 'instead of' or 'in addition to', etc.

### 14.4.5   Payments by results

In the larger food manufacturing companies, the prospect of payment by results can impact the research process. Bonuses are usually tied to a combination of performance, completion of required tasks, sales, innovative products, and efficiencies, to name a few. Clearly, research results, which support the prospect of a bonus, are enticing. What this can mean is that at times judgment might be clouded by head office demands to rush to market ahead of the competition with a product, regardless of research findings.

### 14.4.6   Team decision making

Another pressure on business is the movement to cross-functional teams making decisions, rather than the decision making resting with one individual. In the product research area the team might include the consumer insight manager, the brand manager, R&D, a finance person and possibly the Vice President Marketing of the category. While this provides an excellent check and balance process, it often has implications on the efficiency of executing the research project.

## 14.5   Some results of the changing business environment

### 14.5.1   Incremental degradation of the product

The ability to 'do more with less' is age old. However, the focus of corporations is more than ever on profit margins and shareholder demands. As a result, they have reduced costs in other areas of the market mix. Is there room to remove some of the expensive ingredients from the product? Here the research objective might be to determine if consumers notice a significant difference between the current product and the one with a reduced amount of the more costly ingredient in it. In this situation, the impact on the role of the research supplier must be more proactive than the Scenario 1 position in Fig. 14.1.

It is no accident that many products in today's marketplace are poor representations of the formulations that existed several years ago. For example, chocolate that has lost its chocolate profile as a result of increased levels of vegetable fat replacing cocoa butter, and jams and preserves that have lost their fruit integrity and resulting fruit flavour character to allow better handling in an automated processing environment. All are the result of situations where cost driven changes were made at the expense of retaining the product's original sensory standards. Whereas at year 1 and at subsequent time points thereafter (Fig. 14.2) a product change did not result in a significant difference in liking for the current product versus the new formula. When the changes occurred incrementally over a 7-year period or more, the change from year 1 to year 7 resulted in significant product degradation. This situation can occur when the sensory researcher does not fulfill a trusted advisor role and where reducing product costs takes priority over product quality. In another situation the history for the product may be deficient. For example, when a brand is sold and is passed to another corporate home, the result is a lack of knowledge of the product's sensory standards. Degradation can also occur when test action standards are too lenient and allow a new formula to be approved over the current formula.

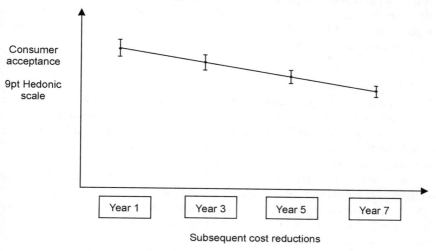

Fig. 14.2    Incremental degradation.

### 14.5.2    Lack of knowledge of the marketplace and the market environment

Clients often want to initiate surveys that tend to be packed with too many attribute ratings, poorly worded purchase interest questions. etc.... . 'while we have the consumer ... let's find out what they also think about x, y, z and get the most out of our research spend'. This raises validity and reliability concerns.

Alternatively, a client has decided to adapt a new research tool, 'the magic bullet', such as segmentation studies or consumer satisfaction studies. The segments they identify can then be difficult to recruit for research and may not be relevant in the marketplace. Often the client is going from putting out one fire to the next without a macro or strategic perspective of the overall marketing strategy associated with the brand (internally) or the category (externally). An overall consistent strategy for research action standards is often missing.

What does this mean in terms of research supplier and client relations? There is still a need for specialized knowledge such as food science, engineering, etc. At the same time, there is an increasing need for the supplier and the client to have a broader understanding of the impact other aspects of the marketing mix have on the product under study. This could include an appreciation of the consumer behaviour role.

We have observed that because the client is so close to their product and/or category, there can be an inability or lack of experience to see the macro environment. In other words, they have difficulty in appreciating how the consumer of their brands actually sees the marketplace. There is a real need to ask the tough question of 'what will you do if you get "x" results?' and 'do consumers even think in those ways?' This can be an issue when so much stands to be lost or gained on the results of a research project. The ability to review the results from a realistic perspective, both in terms of how the product really performs and how the consumer probably approaches the product in the market-place, is critical. What a client might see as a major change to their package or product may hardly be noticed by the consumer.

### 14.5.3    The client's reality

More and more for clients there is little or no time to think. Clients are looking for suppliers who can do a lot of the thinking for them. This means the supplier has to have an understanding of the real drivers of the research and much more insight on the business background and objectives. In some cases the client may not even know what the real underlying business objectives are.

The place of marketing research in the hierarchy varies from company to company. If the client is part of the internal research or consumer insights team and is subordinate to marketing, then how are the results really viewed by marketers? The market research function, in some companies, has changed in both title and function. These people were always responsible for many facets of market research, which could include some or all of the following; sales tracking and volume data, advertising tracking, consumer complaints, marketing com-munication, packaging, retailer tracking studies, sensory and product evaluation. In some cases the title has been revamped to that of consumer insights or consumer intelligence. While the core function is generally still the same, the role has been expanded to incorporate consumer marketplace trends and infor-mation. They often look to the various research suppliers to provide input at all of these levels.

### 14.5.4    Keeping up-to-date with rapidly changing technologies

Clients are faced with an inundation of information due to advanced techno-logies. Even the sources of marketing research information are expanding to include consumer-generated media such as blogs, online concept evaluations, virtual focus groups and ethnographic research. The reality is that clients are overwhelmed. Are they able to process the vast quantity of information and do they have time to vet the sources of this information?

### 14.5.5    Sensory positioning within the corporate structure

Experience suggests that sensory research resides in a number of places within the client organization. It may be the responsibility of different groups either internal to the organization or with one or more external suppliers. In the cor-porate world, sensory research may be a dedicated department reporting to R&D or technology guidance. In another model, it may be integrated within a con-sumer insight or market research services group. If an external service provider is being used, the organization may opt for a separate sensory-only supplier or general market research supplier who also has consumer sensory science capabilities.

## 14.6   Scenario 2 – The research coach

In thinking about the challenges previously outlined, that are taking place in the business environment, how might the research supplier respond? Much of what influences the client also impacts the supplier. As these factors become part of the business reality, there is an opportunity, driven by a need, to alter the service offering of the research supplier. How can the researcher accommodate the growing pressures that the client is facing? Do they have the capability to handle these changing needs? These challenges may impact all or other aspects of the marketing mix. For example, advertising has had to do a complete about face to accommodate the explosive growth of new media of consumer communication. Gone are the 30-second television spots that were once the cornerstone of advertising. Enter the customized message delivered via the cell phone, iPod and internet.

Scenario 2, as represented in Fig. 14.1, indicates an improved level of understanding on the part of the product/sensory researcher. However, the client (while increasing their trust of the supplier) might be distracted with the day-to-day challenges of the job. In addition, the client may not really appreciate how product research can serve their business needs. The danger here is that the product researcher may not be able to rely on the client to articulate the business and research objectives for a project. Sometimes, in an attempt to be efficient, the research supplier may attempt to draft those needs. However, in the end, this process might be likened to simply 'shooting in the dark', without the complete knowledge of the business issues of the client, who in turn does not understand what the research supplier can do for them. While a client may not know what they want at the beginning of a project, they are able to know what they don't want once they see the results. This presents potential issues of reliability and validity.

### 14.6.1   Scenario 2 example

If the client does not fully appreciate the importance of product research and in light of time pressures to 'get to market', there might be the desire to simply 'check off' the box next to the question of 'have you undertaken product research?'. In addition, research is often denied adequate time lines in the scheduling process. The client might feel that they know what the consumer wants more than the consumer himself. Sometimes the client acts on personal interest, bias and liking rather than reality.

A good example of this comes from the snack food sector, which has seen remarkable flavour innovation over the past few years. Some exotic flavour profiles are just that 'exotic'. While they may clearly be innovative, they are still being applied to a commodity-based product. In the salty snack category, there is only so much room for niche-flavoured products. Profit comes from the big movers and crowd pleasers. In one case, a management team became quite enamored with a particular flavour profile that had been made popular as a result of growing interest in sushi. The sensory results from a consumer research study identified flaws in the flavour variant and the recommendation was not to proceed to market. It was only during the presentation of results that the external research supplier learned that the decision had been made to adopt the new flavour regardless of the consumer results. Previous success stories, based on gut feel, provided the necessary rationale for the brand team. However, months later, it was revealed that the exotic or innovative flavour profile had not produced the anticipated results. The product had been pulled from some of the retail channels. This was a costly gamble on the part of the brand team.

Similar to Scenario 1, you will always have some clients who will fall into this category. In some cases, the supplier might fluctuate between clients under the conditions of Scenario 1 and others in Scenario 2. This usually happens when new researchers on the client side join the team and have to be 'educated' on the role and function of product research. The challenge is to have these clients progress to the next stage (Scenario 3) of the Model of Trust.

## 14.7    Scenario 3 – The trusted advisor

In Fig. 14.1 (Scenario 3), you have the situation where there is close to total trust between the client and the researcher. Both parties have business knowledge as well as product research knowledge. Each party augments the other and there is a sharing of information. The researcher can behave and serve as a trusted advisor and the client also accepts him/her in this role. This really is the most efficient paradigm for the client and supplier to operate in for maximum effectiveness.

It is at this stage that honest and frank discussions about reliability and validity issues occur. The relationship has been elevated to a stage where challenging questions can be asked of the client and responses discussed in a collegial rather than adversarial manner. The client will be looking to the supplier to provide an experiential perspective of the category. While the primary research objective may have been a product evaluation, with a focus on sensory aspects, attaining general consumer insights and behaviour are also becoming increasingly important objectives. We have chosen an example from the food service sector to demonstrate the effectiveness of Scenario 3.

### 14.7.1    Scenario 3 example

An existing food service client had historically developed their core products based on qualitative research insights. However, when sourcing new products from their suppliers, they were being challenged by these suppliers to show supporting evidence for product acceptance or rejection by the target consumer. The research supplier suggested a quantitative consumer sensory approach to measure consumer acceptance of the new product prototypes submitted by their suppliers and to determine the consumer driven product strengths and weaknesses. The initial research resulted in the selection of a formula for a new beverage introduction that became the most successful product launch in the client's history with consistent year-to-year sales increases. The product also set the gold standard for other competitors and potential new entries into the market. The research also set a standard for all product research going forward with the client. They embraced the concept of sensory driven quantitative research benchmarks as part of their business model for all new product introductions and formulation changes. Thus, the result was that of a client developing research knowledge under the guidance of the researcher.

## 14.8    Establishment of mutual trust in the food company

### 14.8.1    Features and benefits of being a trusted research advisor
The advantages of being a trusted research advisor are as follows:

- the business environment in which both parties operate is proactive rather than reactive
- insights are shared between client and supplier
- less competitive bidding and more supplier agreements
- ongoing relationships with products and brands
- opportunity to extend research service package.

However, with these advantages come increased responsibilities for the trusted research advisor. These include:

- being more available to discuss or address client needs that are either directly or indirectly related to a project
- being accountable to meeting the diverse and changing needs of a client
- being subjected to reviews by the client and also participating in reviewing the client as part of their internal review process
- dealing with more bureaucratic red tape pertaining to supplier agreements.

What initiated the transition from 'order taker' to 'trusted advisor'? It comes from a combination of better service from suppliers and greater need from clients. While this symbiotic relationship is generally ideal, there are some aspects to increased client needs that must be discussed. Oftentimes, marketers are reluctant to make decisions on their own without market research support. At times they rely on market research too much. The result is that they rush to research too early, possibly not carefully addressing which of the marketing levers really is the driver of the research. For example, on branded product/sensory research, a 'value' question is often asked. Is value a product issue? A pricing issue? A retail management issue? Regardless of your research specialty, it is clear that you need to expand your understanding of the client's business as much as possible in order to transition from the order taker mode to that of trusted research advisor. There is a strong indication that this role yields improved business results and prevents poor marketing decisions from being made.

While it would appear that the trusted advisor role is mutually beneficial, there are some risks. For the external research supplier, this collegial relationship may take you closer toward an exclusive arrangement or discussion with clients. Can the external sensory research supplier really afford that type of arrangement? Clearly, some of the strengths that you bring to the role are there because of your experience in the general category with other competitive companies.

There are two types of risks for the client in moving to the trusted advisor relationship with an external research supplier – knowledge risk and personal risk. Does the trusted advisor know too much about proprietary results and issues? Is there a potential problem with confidential information? Could the external research supplier upstage the client? It is important to always be aware that you only win the client over in their game, not yours. It is then as an external research supplier, you might then be offered a seat at the table (Grover and Vriens, 2005). The trusted advisor role is not a particularly new concept to business. It is one that has already been established for a number of years amongst other parts of the business such as finance, legal and advertising but is less common with the research function.

There are four component parts to the trusted advisor role. Some of the characteristics are objective, while others are subjective and may be perceived differently from person to person. Overall, it is the ability to see the bigger picture that encompasses the following:

- Business knowledge (objective) – This extends to include a general understanding of organizational behaviour, how corporate decisions are made and what role, if any, internal politics play. This is a key component when dealing on the client side of the business. Suppliers, who can demonstrate a solid appreciation of what it might be like to walk in the shoes of their clients, stand to win support and respect.
- Industry and product knowledge (objective) – The client often seeks out industry and product knowledge. They anticipate and ask suppliers to provide this type of information for specific projects or for their own internal training sessions with new company hires.
- Your own professional and personal trustworthiness (subjective) – Professionalism is generally demonstrated as timeliness, congeniality, tactfulness (Zaltman and Moorman, 1998). This can be boosted by professional certification. Membership in various professional organizations within your research specialty, as well as broader business organizations, also provides the type of credentials that clients respect. Personal integrity is something that one earns over time. The trusted advisor role is generally reserved for those suppliers who have demonstrated their commitment to excellence over a succession of projects and in some instances beyond the call of duty (Zand, 1972).
- Your own personality, ability to be self-reliant and your entrepreneurial capabilities (subjective) – Personality is something that can be very subjective in nature. Rigid or warm, friendly or stoic, all of these variations have their merit. What is important here, though, is that the personality of the

trusted advisor plays a key role. It is important to match the right researcher to the right client with personalities that complement rather than clash. (Grover and Vriens, 2005)

### 14.8.2  Managing and maintaining trust

Now that there is an understanding of what the trusted advisor role is, how does the researcher actually manage this relationship in practical terms? Grover and Vriens (2005) suggested that the operational process could be likened to a patient/doctor relationship.

The researcher should strive to have a relationship with the client similar to that which a physician has with a patient. The client articulates the symptoms of the problem to the researcher, who then interacts with the client to determine the problem and the research approach. Just as the physician would resist and refuse a request for a particular drug by the patient, the researcher should resist taking orders about the type of data or methodology of data collection that should be used to solve the client's problem.

Here is how the process might work.

*Obtain background information*
Can the product be changed? What are the constraints – cost, supplier, regulatory, legal? What are the brand characteristics (penetration of the brand or category, historical brand information)? Who is the consumer target market (demographics, screening, regional differences and possible impacts)? Which department is driving the research? Are there sales performance or volume statistics available? Is there any previous research that might be useful such as BASES or any qualitative front-end work? In fact is there sufficient background information for the researcher to propose a research plan with appropriate methodology?

*Purpose and key information needs*
What is the key question about the consumer that will help the client achieve their business objectives? A mutual understanding allows for more realistic boundaries to be established in terms of expectations and outputs. What specific information do you want the research to provide? What will you do with the information once you have it? What will the client do if they do not get the results they either expect or want? It is important that the action standards are clearly articulated. These are required in order to evaluate the data and to 'diagnose' the outcome.

*How to report the findings?*
You have to be a truth teller. Honesty is what is expected from the trusted advisor. It is important to make a believer out of your client. You can do that by managing the results and presenting them in a manner that will maximize their level of understanding. For example, for some clients you might present

the big picture comprising insights and an executive summary with recommendations, instead of a litany of numbers or charts. If the data produces results that are unclear, it is the role of the trusted advisor to manage this situation. Knowing your audience when you report the findings and what their information needs are, is critical to sustaining the trusted advisor role. Of course not all research results are positive for the client. Managing bad news requires diplomacy and skill. It is always best to give the client advance notice of a bad news story.

## 14.9    Problems met and lessons learnt in developing and maintaining a model of trust

### 14.9.1    The trusted research advisor's role

The transition to a trusted advisor role requires the researcher to be able to articulate research data into insights that bring recognizable value to the client's business needs. It is imperative that the researcher speaks in a research language that the client can understand and not in a mix of research jargon (hedonic, incomplete block design, top box numbers, etc.). Data and the processes by which the data have been collected may have low interest or even value to the business. However, insights and outcomes from the research data generally have much more value to offer to the business decision process. Several measures of the potential value that research insights may have are:

- a factual insight that represents a profound truth
- an insight that is clearly relevant to the brand or business
- an insight that resonates with consumers
- an insight that meets an articulated consumer need
- an insight that will change behaviour and lead to a competitively differentiated idea.

### 14.9.2    Going forward – the challenges

In the course of this chapter reference has been made to the various challenges to be overcome by the researcher in order to progress to a trusted advisor role with their client. In the view of the authors, there are several issues that will continue to be challenges.

In order to equip future generations with the diversity of skill sets that are required to become effective researchers and consultants, some post secondary academic institutions may wish to re-evaluate their current programmes. Co-op and work term programmes where students gain practical working knowledge are often seen as the domain of community colleges and not part of university level programmes. It will be critical that the business community works with academic institutions at all levels to realize better work/study opportunities for the next generation of researchers. This will enable new graduates to gain a better understanding of the business environment. In the authors' experience, clearly one of the keys to attaining the trusted research advisor status is to develop the ability to understand business objectives and determine how and when sensory consumer research can be utilized to fulfil these objectives. Academic grounding, self-reliance, entrepreneurship, and analytical skills for problem solving and project implementation will be requirements for this role (Does, 2006).

Increasingly, it will be necessary to have the ability to deal with change. Just as the advertising industry found that their business model underwent a dramatic shift, it is likely that the research industry will also be challenged to remodel the way they currently conduct their business. As technologies advance, researchers will be challenged to adopt new ways of obtaining information from their consumer target groups.

Businesses will continue to grow and globalization will continue to drive the decisions that corporations make for their products. There will be an increasing need to bring the same level of research expertise to those parts of the company that are both far removed from corporate headquarters and lack the dedicated manpower or resources for research. Global corporations will have to meet the financial obligations of extensive and ongoing intra-corporate training programmes in order to bring all parts of the global company to the same level. Without that level of commitment there will continue to be too many inconsistencies between the 'richer' and 'poorer' parts of the company for effective progress.

Additional complexities will be the continuation of company mergers and acquisitions both at the client and supplier side of the business, as well as outsourcing internal company services (accounts payable, for example) to outside suppliers who may even be situated in other parts of the globe.

With reference to the future of sensory research, there is a need to develop a better understanding of where it fits in the corporate model. For sensory research to truly come of age, it needs to reside in a dedicated domain, well understood and recognized by all parts of the business model. Communicating the value of sensory research in the Product Development Process will continue to be a challenge for all researchers. However with more individuals progressing to the trusted advisor role, this will change the overall status of the discipline.

As a result, there will be plenty of opportunities for individuals to advance from the diagnostic technician role through research coach to the ultimate trusted advisor role.

End Note: The objective of this chapter was to provide a strategic perspective on the research process and the relationship between the research supplier and the client. It must be made quite clear that this was not a chapter about research methodologies as they pertain to sensory and product evaluation. If you are interested in sensory research methodologies, you might want to obtain a copy of Lawless and Heymann's (1998) *Sensory Evaluation – Principles and Practices*.

## 14.10    References

DOES, R. (2006). Researchers vs. Consultants. *Research World*, ESOMAR March, 38–39.

EARLE, M., EARLE, R. and ANDERSON, A. (2001). *Food Product Development* (Cambridge, UK: Woodhead Publishing).

GROVER, R. and VRIENS, M. (2005). *The Handbook of Market Research: Do's & Don'ts* (London: Sage Publications).

LAWLESS, H.T. and HEYMANN, H. (1998). *Sensory Evaluation of Food: Principles and Practices*. Published for ARF (Advertising Research Foundation) by the World Advertising Research Centre (Washington, DC, USA).

ZALTMAN, G. and MOORMAN, C. (1998). The importance of personal trust in the use of research. *Journal of Advertising Research* October/November 16–24.

ZAND, D.E. (1972). Trust and managerial problem solving. *Administrative Science Quarterly* 17, 229–239.

# 15

# Consumer research in the early stages of New Product Development – market-oriented development of meal complement beverages

Joe Bogue and Douglas Sorenson, Ireland

*This chapter illustrates the identification of the appropriate priorities amongst the many possibilities for consumer research and the integration of the results of this research into the Product Development Process (PD Process). The early stages of New Product Development (NPD) are often called the 'fuzzy' front-end stages because the direction for the project is not clear. For incremental NPD such as line extensions or product improvements, the market, consumers, processing and distribution systems are already set. In contrast, radically innovative new product projects can have nothing defined. In this case, the direction may be from the company strategy, which just states that the company wants to enter a new market, develop a new technology, or launch a new product platform. The problem for the product developer is, therefore, to decide what knowledge is necessary and then to create this knowledge with the people, money and other resources available to the company.*

*In the early stages, a clear project aim must be identified and developed, which should include all aspects of the technological and marketing development: What are the products? Can they be made? Can they be sold? Who wants them? What do they need? The research at this stage is using knowledge within the company, and desk market research. to find already published information and data. In the large company, information on raw materials, ingredients, processing, consumers, competing products, distribution methods, market size, can be stored for ready access; and there is also a great deal of tacit knowledge from previous projects and from producing and marketing other products. In the small company, it is usually impossible to collect and analyse all this knowledge and this is where risk-taking starts. When the aim, projected outcomes and constraints are presented to management and accepted, the next stages can proceed. Product concepts are developed, researched and the final product concept is then built into product design specifications for the product design and process development. This involves a multidisciplinary team (or in the small company, a multidisciplinary person!) so that a total product concept can be described and developed into product design specifications. This must include a study of consumer needs, so that the product attributes and benefits can be identified and combined in a product profile for the product design specifications.*

*Integrating technical and consumer research often presents a problem in large companies because the two types of research are in several different departments, often not in the same location, and a significant amount of the consumer research is contracted outside the company to consultants. This can present real difficulties in co-ordination and can end in the extremes of marketing telling research and development (R&D) the product concept they have to use, or R&D designing a product prototype and presenting it to marketing! The medium-sized company is likely to be the most successful in co-ordinating the knowledge to produce a total product concept.*

*The consumer research techniques to study consumer needs and wants are numerous and the problem is to select the most suitable for the present project. In using consultants, the company may not play any part in choosing the techniques, because of their lack of experience, and the expensive research may not produce the knowledge they need. Therefore, consultants need to be brought into the project at the early stages so that they understand what knowledge is needed. Sometimes, the same techniques can be used time and time again, and there is a need to keep up-to-date with the research in universities and research centres.*

*This chapter relates particularly to pages 200–236 in* Food Product Development *by Earle, Earle and Anderson.*

## 15.1    Introduction

Health and convenience have been important drivers of innovation in the food and beverage industry over the last twenty years. In particular, there has been a significant increase in the number of new product introductions targeted at health-oriented, 'time-poor' consumers, such as meal replacement and meal complement beverages. A meal complement beverage is a portion-controlled, pre-packaged nutriceutical drink designed to compensate for a nutritionally inadequate meal. However, the development of innovative meal complement beverages presents two major problems for product developers. The first major problem relates to beverage manufacturers successfully repositioning ready-to-drink (RTD) juices and diet soft drinks as meal complement beverages.

Therefore, from a product or category substitution strategy perspective, will consumers substitute an innovative meal complement beverage for conventional drinks currently on the market? The second major problem relates to trade-offs that consumer groups would be expected to make between health benefits and price. The challenges are: identifying the optimal premium that specific consumer groups would be willing to pay for a functional meal complement drink; and identifying product design and marketing strategies that would facilitate higher levels of 'adoption' of meal complement beverages by these consumer groups.

This case study investigates Irish consumers' needs and wants for innovative meal complement beverages, and focuses on one specific element of the NPD process: new product concepts and new product design as it relates to the early stages of the NPD process. The research presented in this case study formed part of a larger multi-disciplinary NPD project conducted by researchers from the Departments of Food Business and Development, Process Engineering, and Food and Nutritional Sciences, at University College, Cork, Ireland, and led to the development of innovative meal complement beverages with high levels of intellectual property value. The case study outlined in this chapter presents a framework by which firms can more effectively manage consumer knowledge at the early concept stages of the NPD process, leading to the technical development and marketing of more consumer-led new foods and beverages.

## 15.2    Market dynamics and New Product Development trends: meal complement beverages

Successful new product launches require an in-depth understanding of markets and trends in order to anticipate changing consumer needs and preferences. For example, the European Commission (2003) reported a strong trend towards significantly smaller households, where the increase in single-person households could be attributed to increased divorce rates, increased life expectancy with more 'surviving singles', and also independent young people moving out of home. These socio-demographic trends have led to dramatic changes in consumers' eating habits. Specifically, Cullen (1994) noted that single-person and single-parent households were more likely to eat out, where significant increases in value-added, such as health and convenience, were increasingly being sought by consumers.

Overall, this growing trend towards health and convenience-oriented foods and beverages can be attributed to an additional number of factors including: an increase in both the number of food service outlets and variety of meal solutions available (Datamonitor, 2005); socio-demographic changes, the increasing prevalence of individualistic lifestyles, and consumers' perceived time constraints (Mintel, 2004); the broadening of distribution channels away from specialist dietetic outlets and into the mainstream food and beverages market (Leatherhead Food Research Association, 2003); rising consumer interest in diet and health (Boyle and Emerton, 2002); an acceleration in the decline of traditional fixed meal times (Groves, 2002); and the increasing use of single-serve packs and individual portion formats (Eurofood, 2001).

From a review of the published literature it can be concluded that product development activities to-date in the health and convenience markets can be viewed as a continuum ranging from the repositioning and adaptation of existing product offerings to the development of new products, particularly in relation to 'nutritional' quality, portion size and portability. There has been a noticeable shift away from carbonated soft drinks towards a range of still beverages and fruit juices in recent years (Retail Intelligence, 2002). The decline in carbonated soft drink sales, and especially cola-flavoured drinks, has been attributed to lifestyle changes as consumers sought alternatives that were natural and healthy (Beverage Industry, 2003). For example, Weisberg (2001) highlighted the growing popularity of smoothies and juices 'in lieu' of breakfast or lunch, which were targeted towards both dieters and busy consumers that wished to eat or drink healthily 'on-the-go'. As Perlik (2004: 33) noted: 'people tend to skip breakfast because they don't like traditional breakfasts, and don't have time to prepare them. They're used to drinking juice, but want more of a meal replacement'.

Indeed, the interest that juice manufacturers have shown in the smoothies category has grown in tandem with the growing trend towards convenience and the consumption of meals-on-the-go (Leatherhead Food Research Association, 2003). For example, Croft (2005) and Beverage Industry (2001) reported that Tropicana launched a range of yoghurt and fruit juice smoothies in the US in response to this growing trend towards on-the-go beverage consumption. Concurrently, PepsiCo purchased PJ Smoothies, the leading smoothie brand in the UK, in 2004 in order to consolidate its dominant presence in the UK smoothie and premium juice market (Centaur, 2005).

Similarly, dairy companies, manufacturers of weight loss foods and beverages such as the Slim-Fast Company, and manufacturers of meal replacement foods and beverages such as Abbot Laboratories, have repositioned their existing products to broaden their respective brand's appeal to the healthy lifestyle sector (Roberts, 2003). For example, Yoplait introduced the *Nouriche* and *Go-Gurt* yoghurt food and drinks range to meet new consumers' needs for healthy drinks on-the-go, while Campina introduced the *NutriStart* range of dairy-based drinks as an alternative to a sit-down breakfast (Food Production Daily, 2003; Cooksey, 2000). In contrast, other food and beverage manufacturers have been more proactive in terms of their NPD activities in the health and convenience food and beverage markets. For example, some manufacturers have been engaged in incremental innovations such as line extensions through the adaptation of existing products such as Kellogg's *Special K* bar, while other firms have pursued the development of more innovative nutrient-dense 'on-the-go' products such as Sanitarium's *Up and Go* and Otsuka's *Energen* beverage ranges (Leatherhead Food Research Association, 2004). However, the ability to predict which product categories will experience strongest growth, and where customer demand can be expected in terms of market and product trends are significant considerations for manufacturers of meal complement beverages. So how can firms become more consumer-led when bringing new products to market? Integrating the consumer with the new food PD Process can help overcome product design and strategic marketing issues associated with emerging new product categories when consumers' needs and values are less well understood by product developers.

## 15.3    Concept ideation and development during the early stages of the New Product Development process

Organisations require information from both internal and external sources to evaluate and monitor business activities as well as make informed business decisions. Consequently, knowledge is widely considered one of the most important intangible resources that firms can possess, and is considered essential to the development of organisations (Grant, 1997). Product development is considered a knowledge-intensive process where the generation of new ideas and concepts requires detailed knowledge of both products and consumers. Therefore, firms that effectively manage knowledge throughout the NPD process create more evident values in a firm's offering in order to effectively meet consumers' needs. In order to manage consumer knowledge more effectively throughout the NPD process, a market-oriented approach to innovation is considered a key factor for new product success. In market-oriented organisations, consumers are viewed as co-designers in the NPD process since they can make an effective contribution to new food product design.

The integration of the consumer with the NPD process can best be achieved at the pre-development stages of concept ideation, concept screening and optimisation (Sorenson and Bogue, 2006; Bogue, 2001; Cooper, 1993). The incorporation of consumers' value-creation at the pre-development or concept stages of the NPD process make organisations better able to adapt to changes in consumers' needs, reduce uncertainty in NPD, and can ultimately lead to higher quality and consumer satisfaction. Concept ideation and concept generation represent the earliest steps in the NPD process, and entail the generation of new product ideas from various sources including: internal brainstorming including marketing, production and technical R&D; external sourcing through suppliers, wholesalers, distributors and retailers; tradeshows and food fairs; market analysis and retail audits; gap analysis and product tracking; and most logically, desk research and consumer research. So how can a multi-disciplinary NPD team integrate the consumer with the new food PD Process? Consumer research techniques, which focus on the early stages of the NPD process, lead to a more systematic and multi-disciplinary approach to product design.

## 15.4   Selection of methods for consumer research on meal complement beverages

There exists a family of consumer research techniques, which can utilise both technical R&D and marketing information, which promote closer integration between the marketing and technical functions for food product optimisation. These include: qualitative methods such as ethnography, in-depth interviews and focus groups; and quantitative techniques such as Quality Function Deployment (QFD), sensory analysis and conjoint analysis.

### 15.4.1   Qualitative consumer research methods

Ethnography and in-depth interviews generate rich data on life experiences and reveal a wealth of information on informants (McDaniels and Gates, 1991). As Hill (1993: 258) comments: 'living through the highs and lows of informants allows the researcher to know the phenomenon under investigation in a way that few other methodologies permit'. It can be difficult to truly participate with consumers in a situation without disturbing the authenticity of their behaviour. In that sense, ethnography can be used to make sense of human behaviour as it occurs in a natural setting; explain activities that engage people for a significant amount of time; and view a situation through the eyes of those involved in that situation. Ethnography therefore becomes an extremely valuable tool at the earliest stages of the NPD process in terms of observing and then understanding consumer behaviour. From a new food product development perspective ethnography can be used to: explore emerging and unmet consumer needs; determine how well products fit in with consumers' lifestyles; and importantly, understand how consumers might use an innovative product such as a meal complement beverage in their everyday lives. On the other hand, in-depth interviews are considered most beneficial in terms of understanding the rationale for consumers' individual choice motives (Stewart and Shamdasani, 1990). In that sense, emerging behaviour patterns can be more easily recognised earlier through ethnography and in-depth interviews, and provide deeper insights into multi-faceted behaviour and motivations of respondents.

Unlike ethnography and in-depth interviews, focus groups are considered most beneficial in eliciting insights into group human behaviour (Krueger and Casey, 2000). A focus group can be defined as 'a carefully planned discussion designed to obtain perceptions on a defined area of interest in a permissive and non-threatening environment' (Krueger, 1994: 6). Focus groups collect information from small numbers of interacting people, and are most frequently described as 'brainstorming sessions' with potential consumers (Trenkner and Achterberg, 1991). In terms of understanding consumers' needs for product design and concept evaluation, focus groups are an obvious initial starting point (Sultan and Barczak, 1999; van Trijp and Steenkamp, 1998). The focus group technique has a unique strength in that it is well suited to determining innovation possibilities; exposing consumers to new technologies and products; providing perceptions on different products; stimulating new ideas; and gathering consumers' impressions of new concepts. In particular, focus groups are considered especially valuable to product developers for the purpose of 'fast-tracking' the NPD process, where consumers' needs are poorly understood, and where new products are radically innovative such as the meal complement beverage concept (Sorenson and Bogue, 2005a). A pertinent question thereafter arises: what quantitative consumer research techniques can be used at the concept optimisation and refinement stages of the NPD process that promote a multi-disciplinary approach to new product design?

### 15.4.2   Quantitative methods: product optimisation research techniques

Research has highlighted the competitive, rather than the co-operative element, within organisations for developing new products, and the lack of communication between marketing and technical R&D departments, where the co-ordination of different functional groups adds to the complexity of NPD (van Trijp and Steenkamp, 1998; Crawford, 1994). Traditionally, consumer and technical R&D research were kept separate during the NPD process. However, Earle and Earle (2000) advocate the need to develop a more holistic approach to NPD and argue that marketing and R&D research techniques need to be integrated when applied to product attributes, and importantly, related to consumers' needs and preferences. There is a range of contemporary scientific consumer research techniques, such as conjoint analysis, sensory analysis and QFD, which bring marketing and technical R&D together to 'assess the potential for innovation in the marketplace' (Cardello, 1995: 878).

*Conjoint analysis*
Conjoint analysis is a multivariate technique that models purchase decision-making processes through an analysis of consumer trade-offs among hypothetical multi-attribute products. The conjoint analysis technique views a product as a combination of a set of attribute levels, where varying these attribute levels, according to a statistically determined design, facilitates the estimation of utility values that quantify the value that consumers place on each attribute level. This determines consumers' total utility or overall judgement of a product (Green and Srinivasan, 1978). The conjoint analysis technique therefore assumes that consumers' perceptions of product attributes control purchasing patterns, and that the attributes represent the most suitable determinants for conceiving marketing and product design activities (Fornell, 1992). In that sense, conjoint analysis is believed to mimic real choice situations where respondents are required to simultaneously consider many dimensions of alternatives.

Conjoint analysis has a number of commercial applications of relevance to both marketers and R&D personnel. From a marketing perspective, conjoint analysis has been used extensively to: estimate the value that consumers associate with particular value-added product features; segment markets based upon the differing benefits sought out by consumers; and design effective product pricing and positioning strategies based on consumers' trade-off decisions among alternative design features (Sorenson and Bogue, 2005b; Herrmann *et al.*, 2000; Green and Krieger, 1991). From an R&D perspective, conjoint analysis is becoming increasingly important in terms of: defining consumer-led new product concepts with the optimal combination of features; predicting consumers' preferences for new concept features; and identifying viable market opportunities for new products not presently on the market (Sorenson and Bogue, 2006; Hair *et al.*, 1998; Kamakura, 1998). Conjoint analysis is therefore an extremely effective consumer research technique, which can bring R&D personnel closer to understanding the voice of the consumer and facilitate more effective integration of marketing and R&D personnel, to aid the new product design process for innovative meal complement beverages.

*Sensory analysis and quality function deployment*
Sensory analysis is important for the development of food products due to the role sensory perception plays in food choice, and in increasingly competitive markets, provides firms with '*actionable*' information on their products and those of their competitors. Sensory analysis has naturally developed from its traditional role in quality control, to being used in product development, by contributing to an understanding of consumers' sensory preferences and the evaluation of competing products from a sensory perspective (Bogue *et al.*, 1999). Sensory analysis can also be used to evaluate how consumers rate product prototypes and to see how they measure up against ideal products (Earle and Earle, 2000). Sensory analysis provides marketers with an understanding of product quality, direction for product quality, evaluations of new product concepts, and profiles of competitors' products from a consumer perspective (Bogue *et al.*, 1999; Shukla, 1994; Moskowitz, 1991).

Another specific example of advances in consumer food research, in relation to concept optimisation and refinement, is offered by Bech *et al.*'s (1994) application of QFD to the food industry, where they translated consumer requirements into marketing and technical R&D product specifications. QFD has been described in detail by various authors (Schonberger, 1990; Hauser and Clausing, 1988), since its origination in 1972 at Mitsubishi's Kobe shipyard site in Japan. QFD integrates information on consumer demands with technical details for the development of new products (Hauser and Clausing, 1988). Schonberger (1990) describes the process as the product developer reaching out to consumers to define and rate their needs. A competitive product analysis is also an integral element of the QFD consumer research technique.

QFD is a market-oriented structuring tool that utilises the consumer to describe preferences based on desired attributes. Six Sigma (2007) define QFD as a systematic process for motivating a business to focus on its customers. It is used by cross-functional teams to identify and resolve issues involved in providing products, processes, services and strategies which will more than satisfy their customers. Bech *et al.* (1994) modified the process to include a central sensory element. This adaptation acknowledged the strong influence sensory attributes have on food preferences. Bech *et al.* (1994) also proposed that market and sensory analysis be linked with the voice of the consumer to help product developers identify product attributes desired by the consumer. The process, they suggest, further facilitates the evaluation of competitive food products. QFD provides a good starting point for the development of market-oriented food products through an examination of consumer demands, consumers' perceptions of products, and the sensory analysis of specific products, and provides product developers with a means of utilising both sets of information in combination, rather than competitively, in a food development environment (Bech et al., 1994; Hofmeister, 1991).

The challenge is to identify consumer demand for quality, to identify possible product improvements, and to test and launch new products. The adapted model illustrates how marketing and sensory information can be utilised to target segments with particular products (Bech et al., 1994). In the following sections of this case study, the concept of managing consumer knowledge at the early stages of the NPD process is explored, through applying it to the development of a range of innovative meal complement beverages.

### 15.4.3    Planning of the research

In terms of new product and process improvement, the approach adopted in this case study provides a framework by which firms can use consumer research more effectively, and manage consumer knowledge more efficiently, at the early stage of the NPD process. Therefore, the development of innovative meal complement beverages is presented in two distinct sections, which reflect the pre-development stages of the NPD process, and the different research methods used at each stage of the process. In-depth interviews were used to understand consumers' beverage purchase motives, and to identify the most important product design attributes in terms of choice alternatives for meal complement beverages. Focus groups were then conducted to illicit consumers' group attitudes towards meal complement beverages, at the concept screening and development stage of the process. Following this, the conjoint analysis technique was selected to generate possible new product design concepts for meal complement beverages. The conjoint analysis technique made it then possible to understand the trade-offs which consumers might be expected to make, in terms of product design attributes, for the purpose of product optimisation and refinement in NPD. Although QFD is also an extremely significant technique in terms of identifying consumer-relevant trade-offs in product design specifications, conjoint analysis is especially beneficial when consumers are required to evaluate intangible product concepts before product prototypes are developed. In addition, given the importance of sensory perceptions to consumer food choice and acceptance, sensory scientists constituted key members of the multi-disciplinary NPD team, and subsequently profiled and evaluated consumers' sensory acceptance of the meal complement beverages arising from this consumer research study.

## 15.5    Concept screening and development for meal complement beverages

In-depth interviews and focus groups were used to generate consumer-led knowledge that would provide direction in terms of concept screening, product design and formulation, and product positioning strategies for meal complement beverages. Overall, forty-two respondents were recruited by means of intercept and convenience sampling to participate in the in-depth interview and focus group discussions. The three focus groups each with eight people were held between March and June 2005, and 18 in-depth interviews were conducted between January and March 2006 (see Table 15.1). An experienced moderator designed the in-depth interview and focus group guides in line with best practice, and conducted the qualitative research sessions, which were audiotape recorded and lasted approximately one hour and thirty minutes. All respondents were rewarded with a small payment of €40 for their participation in the study. The qualitative data was transcribed and then coded using the computer package N6$^{TM}$ (QSR International, 2002).

### 15.5.1    Consumer diets and meal complement beverages

During the interviews and focus group discussions, participants were introduced, by means of product prompts and information on a flipchart, to the meal complement beverage concept. Participants were told that a meal complement beverage was a portion-controlled, pre-packaged nutriceutical drink designed to compensate for a nutritionally inadequate meal (Bogue *et al.*, 2005). The in-depth interviews and focus group discussions revealed that respondents generally held mixed attitudes towards the meal complement concept. A number of younger interviewees and focus group participants were receptive towards the meal complement concept, and were willing to trial purchase a meal complement beverage. The concept of a meal complement beverage was considered most beneficial in terms of compensating for a nutritionally inadequate fast food meal: 'Well because if I felt I was eating too much junk food, then at least if I had the nutrition coming from a drink it may make up for it' (FG1). In contrast, older interviewees and focus group participants appeared more satisfied with their dietary habits, and seemed less likely to trial purchase a meal complement beverage than younger respondents. In particular, dietary supplement consumption appeared to exert a strong influence as to whether older respondents would consume a meal complement beverage: 'Not very regularly [consume a meal complement beverage] as I don't eat very much junk foods or fast foods, and I take a supplement every morning' (FG3).

### 15.5.2    Consumers' lifestyles and meal complement beverages

The majority of interviewees and focus group participants reportedly skipped breakfast most frequently, although a number of participants frequently missed lunch also: 'Because I am rushed and too busy to stop, I often work through

**Table 15.1**   Socio-demographic profiles across in-depth interviews and focus groups

| Socio-demographic variables | Interviews | FG1 | FG2 | FG3 |
|---|---|---|---|---|
| Participant numbers | 18 | 8 | 8 | 8 |
| Gender | | | | |
| Male | 9 | 4 | 3 | 4 |
| Female | 9 | 4 | 5 | 4 |
| Age Group (years) | | | | |
| 18–24 | 5 | 4 | 0 | 2 |
| 25–34 | 5 | 4 | 2 | 2 |
| 35–44 | 4 | 0 | 3 | 4 |
| 45+ | 4 | 0 | 3 | 0 |
| Marital status | | | | |
| Single | 7 | 4 | 2 | 1 |
| Married | 8 | 2 | 3 | 4 |
| Separated/divorced | 0 | 0 | 1 | 1 |
| Cohabiting | 2 | 2 | 1 | 2 |
| Widowed | 1 | 0 | 1 | 0 |
| Education | | | | |
| Completed primary level | 1 | 0 | 0 | 1 |
| Completed early secondary level | 3 | 0 | 2 | 1 |
| Completed late secondary level | 6 | 3 | 3 | 4 |
| Completed tertiary level | 8 | 5 | 3 | 2 |
| Location | Cork | Cork | Limerick | Dublin |

lunch' (Interviewee 8, Male aged 35–44 years). Consequently, these respondents often snacked on convenience and on-the-go foods such as breakfast rolls and potato chips, which were perceived by them as unhealthy: 'I usually focus on making it into work on time, and if I then have time I will grab something to eat' (FG2). On that basis, the majority of respondents felt that they would most likely consume a meal complement drink with either breakfast or lunch: 'I suppose breakfast [the meal occasion] because I may not have time to make something or just be lazy and eat a breakfast roll or something fatty and convenient' (FG3). Significantly, interviewees and focus group participants expected a meal complement beverage consumed at breakfast, or lunch time, to be low in fat and calories and rich in nutrients.

### 15.5.3   Design attributes of meal complement beverages

Participants were introduced, by means of product prompts and information on a flipchart, to specific meal complement beverage concepts, in order to identify key product design attributes that would lead to high levels of consumer acceptance. The meal complement carriers or base products investigated in this study were rice milk, green tea and fruit juice. These were chosen based upon their satisfactory micronutrient and macronutrient profiles, and textural and satiety attributes. An important product design issue investigated in this study is whether consumers would accept novel carriers and flavour profiles for meal complement beverages. Consumers' perceptions of all three carriers were positive, and were considered healthier alternatives to other beverages, such as carbonated soft drinks. Although all three carriers were generally perceived as healthy and natural, only the rice milk and fruit juice-based meal complement beverage concepts were deemed most appropriate for consumption in the morning or lunch time: 'I think that it [fruit juice] would go with breakfast, and maybe lunch. I would consider juice to be the healthiest [of the three]' (FG1).

Interestingly, the in-depth interview and focus group research revealed that certain product design-related characteristics represented 'health' or 'wholesomeness' indicators from the consumers' perspective, and similar observations have been reported elsewhere (Bogue et al., 2005; Sorenson and Bogue, 2005a). On that basis, it appeared that consumers used these product design-related features to discriminate between alternative meal complement beverage concepts on the basis of healthiness and naturalness. For example, a textured meal complement beverage was considered healthier, more natural, and more satiating, than smooth-style meal complement beverages. Likewise, consumers preferred still to carbonated meal complement beverages. In addition, respondents felt that still drinks were more suitable for consumption in the morning, and were viewed as healthier than carbonated drinks, even by respondents that generally preferred carbonated drinks. Interestingly, the in-depth interviews and focus groups revealed a high level of inertia amongst interviewees and focus group participants towards both the concept of adding functional ingredients to foods or beverages, and the functional meal complement beverage concepts evaluated in this study. Specifically, it seemed that the health benefits offered were considered a 'bonus' rather than a primary purchase motive factor for meal complement beverages by consumers. Overall, these respondents most preferred a meal complement beverage that could either 'lower cholesterol' or 'boost the immune system'.

The final step outlined in this part of the meal complement case study involved using the information generated from the in-depth interviews and focus groups to design more consumer-led meal complement beverage concepts through conjoint analysis.

## 15.6    Product concept optimisation and refinement

The conjoint analysis technique was used in this meal complement study to gain a better understanding of consumers' choice motives for the purpose of new product design, and to guide positioning and pricing strategies. The product attribute alternatives, and associated attribute levels, used in this conjoint-based study were derived from a combination of the in-depth interview and focus group results, and also from discussions with the technical R&D personnel involved in this multi-disciplinary NPD research project (see Table 15.2). The Orthogonal Design Procedure in SPSS generated 22 hypothetical meal

**Table 15.2**   Attributes and attribute levels used in the conjoint-based survey

| Attribute | Level |
|---|---|
| Flavour | Strawberry |
| | Raspberry |
| | Plum |
| Carrier | Rice milk |
| | Green tea |
| | Fruit juice |
| Health benefit | Lower cholesterol |
| | Boost immune system |
| | None |
| Pack size | 200 ml |
| | 500 ml |
| | 1 litre |
| Packaging | Carton |
| | Glass bottle |
| | Plastic bottle |
| Price | €1.50 |
| | €2.50 |
| | €3.50 |

complement beverages, four of which were holdout beverage profiles, based upon the selected attributes and associated attribute levels. A holdout profile is a hypothetical product description that is rated by consumers, and is included only to determine the reliably of the subsequent conjoint models in terms of predicting consumers' choice preferences. Therefore, in this study, the four holdout beverage profiles would be rated by consumers, but not used in the estimation of utility values. This made it possible to then determine how consistently the conjoint models could predict consumers' preferences for hypothetical meal complement beverage concepts that would not be evaluated by consumers (Hair *et al.*, 1998). A conjoint-based survey was designed which presented 22 hypothetical meal complement beverages for consumer evaluation using a nine-point Likert scale. Additional questions which related to consumers' eating habits, lifestyles and their socio-demographic information were also included in the conjoint-based survey.

A total of 297 consumers participated in the conjoint-based survey that was conducted during June and July 2006 (see Table 15.3). The questionnaires were analysed using SPSS v11 (SPSS, 2003). The individual level conjoint analysis procedure in SPSS determined the importance of each attribute, and attribute level, to consumers' preferences for innovative meal complement beverages. The conjoint analysis revealed that the carrier (29.55 out of 100), health benefit (17.93 out of 100), flavour (16.57 out of 100) and price (12.71 out of 100) attributes were most important in terms of choosing between alternative meal complement beverage concepts. The Pearson's $R$ (0.991) and Kendall's tau

**Table 15.3**    Socio-demographic profiles in conjoint analysis

| Socio-demographic variable | Category | Sample (N) | Sample (%) |
|---|---|---|---|
| Gender | Male | 126 | 42.4 |
| | Female | 171 | 57.6 |
| Age group (years) | 18–24 | 72 | 24.2 |
| | 25–34 | 113 | 38.0 |
| | 35–44 | 63 | 21.3 |
| | 45+ | 49 | 16.5 |
| Marital status | Single | 150 | 50.5 |
| | Married | 84 | 28.3 |
| | Cohabiting | 51 | 17.2 |
| | Separated/divorced | 8 | 2.7 |
| | Widowed | 4 | 1.3 |
| Educational status | Completed primary level | 6 | 2.1 |
| | Completed early secondary level | 17 | 5.7 |
| | Completed late secondary level | 106 | 35.9 |
| | Completed tertiary level | 168 | 56.3 |
| Employment status | Employed full time | 194 | 65.3 |
| | Employed part time | 48 | 16.2 |
| | Student | 39 | 13.1 |
| | Disabled | 2 | 0.7 |
| | Retired | 3 | 1.0 |
| | Unemployed | 11 | 3.7 |

(0.931) values were high and indicated strong agreement between the averaged product ratings and the predicted utilities from the conjoint analysis model.

### 15.6.1    Meal complement cluster profiles

Agglomerative hierarchical cluster analysis was employed initially to determine the desired number of clusters. This preliminary segmentation process suggested that three to five clusters existed for meal complement beverages in this case study. A five-cluster solution was finally chosen to reflect the variation in consumers' preferences that might exist in the marketplace. K-Means cluster analysis was then used to group respondents into five clusters of consumers with similar preferences for meal complement beverages based on attribute utility patterns (see Table 15.4). In Table 15.4 the highest utility values are in bold and the lowest utility values are in italics. The socio-demographic profile of each cluster is also presented in Table 15.4.

Cluster 1 contained 45 respondents, of which 26.7% were male and 73.3% were female. The majority of these respondents were 45 years of age and over (37.8%), single (44.4%) and in full-time employment (57.8%) (see Table 15.4). This cluster was most influenced by the health benefit attribute when choosing between alternative meal complement beverage concepts, and could therefore be considered health-driven in terms of purchase preferences. Interestingly, this health-driven cluster was the only cluster that most preferred a functional meal

**Table 15.4**  Averaged utility values and socio-demographic profiles across clusters

| Attribute | Factor/attribute level | Cluster 1 | Cluster 2 | Cluster 3 | Cluster 4 | Cluster 5 |
|---|---|---|---|---|---|---|
| Flavour | Strawberry | 0.02 | **0.33** | **0.08** | **0.78** | **0.15** |
| | Raspberry | **0.23** | 0.01 | −0.01 | 0.45 | −0.14 |
| | Plum | −0.25 | −0.34 | −0.07 | −1.23 | −0.01 |
| Carrier | Rice milk | −0.51 | 0.93 | −1.99 | −0.23 | −0.05 |
| | Green tea | −0.07 | −2.52 | −0.60 | −0.52 | −0.03 |
| | Fruit juice | **0.58** | **1.60** | **2.59** | **0.76** | **0.08** |
| Health benefit | Lower cholesterol | **1.24** | −0.03 | 0.10 | −0.08 | 0.13 |
| | Boost immune system | 0.80 | **0.22** | **0.33** | **0.46** | **0.18** |
| | None | −2.04 | −0.19 | −0.43 | −0.38 | −0.30 |
| Pack size | 200 ml | −0.41 | −0.27 | −0.19 | −0.32 | −0.12 |
| | 500 ml | 0.05 | −0.01 | 0.07 | **0.18** | −0.03 |
| | 1 litre | **0.36** | **0.28** | **0.12** | 0.13 | **0.15** |
| Packaging | Carton | 0.04 | **0.08** | **0.02** | **0.09** | **0.08** |
| | Glass bottle | −0.10 | −0.13 | −0.03 | 0.00 | −0.06 |
| | Plastic bottle | **0.06** | 0.05 | **0.02** | −0.09 | −0.01 |
| Price | €1.50 | **0.58** | **0.13** | **0.24** | 0.15 | **0.28** |
| | €2.50 | 0.01 | 0.03 | 0.02 | **0.16** | 0.06 |
| | €3.50 | −0.59 | −0.16 | −0.26 | −0.31 | −0.34 |
| Cluster size | | 45 | 23 | 60 | 68 | 101 |

| Socio-demographic variable | Factor | Cluster 1 (%) | Cluster 2 (%) | Cluster 3 (%) | Cluster 4 (%) | Cluster 5 (%) |
|---|---|---|---|---|---|---|
| Gender | Male | 26.7 | 60.9 | 46.7 | 42.6 | 43.8 |
| | Female | 73.3 | 39.1 | 53.3 | 57.4 | 56.2 |
| Age group (years) | 18–24 | 15.6 | 21.7 | 20.0 | 27.9 | 29.2 |
| | 25–34 | 33.3 | 43.5 | 43.3 | 45.6 | 28.1 |
| | 35–44 | 13.3 | 21.7 | 15.0 | 16.2 | 34.8 |
| | 45+ | 37.8 | 13.0 | 21.7 | 10.3 | 7.9 |
| Marital status | Single | 44.4 | 47.8 | 38.3 | 64.7 | 51.7 |
| | Married | 42.2 | 17.4 | 41.7 | 17.6 | 24.7 |
| | Cohabiting | 11.1 | 30.4 | 18.3 | 16.2 | 15.7 |
| | Separated/divorced | 2.2 | 4.3 | – | 1.5 | 5.6 |
| | Widowed | – | – | 1.7 | – | 2.2 |
| Educational status | Completed primary level | 2.2 | – | 1.7 | 2.9 | – |
| | Completed early secondary level | 11.1 | 8.7 | 5.0 | 5.9 | 3.4 |
| | Completed late secondary level | 28.9 | 52.2 | 31.7 | 22.1 | 48.3 |
| | Completed tertiary level | 57.8 | 39.1 | 61.7 | 69.1 | 48.3 |
| Employment status | Employed full time | 57.8 | 60.9 | 76.7 | 63.2 | 65.2 |
| | Employed part time | 13.3 | 17.4 | 10.0 | 23.5 | 15.7 |
| | Unemployed | 6.7 | 4.3 | 1.7 | 4.4 | 2.2 |
| | Student | 20.0 | 13.0 | 11.7 | 8.8 | 13.5 |
| | Retired | 2.2 | – | – | – | 2.2 |
| | Disabled | – | 4.3 | – | – | 1.1 |

complement beverage that could lower cholesterol (1.24). The price and carrier attributes were equally important to this cluster, and these consumers most preferred low priced (€1.50), fruit juice-based (0.58) meal complement beverages. Interestingly, from a marketing perspective, although this group appeared most receptive towards functional beverages, this cluster was also most price sensitive across clusters, and similar findings have been reported elsewhere (Sorenson and Bogue, 2005b). Cluster 1 was also the only cluster that preferred raspberry- to strawberry-flavoured meal complement beverages (see Table 15.4).

Cluster 2 was the smallest cluster identified in this study and contained 23 respondents. These respondents considered the carrier attribute most significant in terms of choosing between alternative drinks, and this group most preferred fruit juice-based (1.60) meal complement beverages. However, unlike the other four clusters, this group of consumers gave a positive utility value for the rice milk carrier (0.93). The flavour and pack size attributes were also significant to this cluster. Cluster 2 most preferred strawberry-flavoured (0.33) meal complement beverages, and most liked the 1-litre pack size format (0.28). Cluster 2 was biased towards males that were single (47.8%), had completed secondary education (52.2%) and in full-time employment (60.9%).

Cluster 3 consisted of 60 respondents of which 53.3% were female. The majority of consumers in Cluster 3 were in the 25–34 years age group (43.3%), married (41.7%), had completed tertiary level education (61.7%), and were in full-time employment (76.7%). Cluster 3 exhibited relatively similar preferences for meal complement beverages as Cluster 2. Specifically, Cluster 3 also considered the carrier attribute most important in terms of choosing between alternative beverages, and this group of consumers most preferred fruit juice-based (2.59) meal complement beverages. However, unlike Cluster 2, the health benefit attribute was important to this group of consumers and Cluster 3 most preferred a functional meal complement beverage that could boost the immune system (0.33).

Cluster 4 scored the highest utility value for the flavour attribute across clusters, and therefore, could be considered flavour-driven in terms of purchase preferences (see Table 15.4). Specifically, Cluster 4 most preferred strawberry (0.78) and raspberry (0.45) flavoured meal complement beverages. The carrier was the second most important attribute to Cluster 4, and this cluster expressed the greatest preference for fruit juice-based meal complement beverages (0.76). The health benefit attribute was also important and Cluster 4 most preferred a functional meal complement beverage that could boost the immune system (0.46). Cluster 4, the second largest group of consumers, contained 68 respondents, and the composition of this cluster was biased towards adults aged 18–34 years (73.5%) (see Table 15.4).

Cluster 5 was the largest cluster identified in this study, and this cluster considered the price attribute most important in terms of its purchase preferences. However, similar to the other four groups, Cluster 5 was also receptive towards medium priced (€2.50) meal complement beverages. The health benefit and flavour attributes were also important to this cluster, and Cluster 5 most preferred a strawberry-flavoured functional meal complement beverage that could boost the immune system. The socio-demographic membership of Cluster 5 was biased towards adults aged between 25 and 44 years (62.9%) (see Table 15.4).

### 15.6.2    Designing the optimal meal complement beverage

The next step in this NPD process was to design and evaluate alternative new product concepts for meal complement beverages using the group level simulation analysis procedure in SPSS. The group level simulation analysis procedure in SPSS requires the generation of hypothetical product concepts that are not evaluated in the original survey. These hypothetical new product concepts can represent new market entrants, alternative marketing strategies, or in this case, new product offerings that a firm may wish to commercialise. In that sense, the group level simulation analysis technique represents a powerful tool which can assist product development personnel predict consumers' preferences for new hypothetical product concepts at the early or concept stages of the NPD process.

In this study the hypothetical meal complement beverage concepts were generated following rigorous analysis of the qualitative results and cluster analysis data, and from discussions with the technical partners involved in this project. In particular, interpreting the cluster analysis results for the purpose of designing the simulation analysis research must be approached carefully. For example, the group level simulation analysis procedure in SPSS could be used to identify meal complement beverages specifically targeted at each group of consumers identified in this study. This strategy is most appropriate when consumers' preferences differ markedly across clusters, and in competitive markets where a firm needs to segment selectively in order to gain a superior competitive advantage in the marketplace.

The group level simulation analysis technique was used in this meal complement case study to identify a limited number of meal complement beverages that would appeal to a number of consumer groupings. This strategy is most appropriate in emerging markets or where consumers' preferences are relatively similar across clusters. In fact, it appeared from Table 15.4 that all five clusters most preferred the fruit juice-based carrier. In addition, Clusters 1 to 4 exhibited relatively similar preferences for strawberry-flavoured functional meal complement beverages to boost the immune system. Similarly, Clusters 1 to 4 most preferred the carton-packaging format.

Overall, a Kendall's tau value of 1 for the four holdouts was obtained which suggested perfect agreement between the holdout ratings and the model predictions. It was therefore possible to analyse consumers' preferences for alternative meal complement beverage concepts using choice simulators, both maximum and probability (BTL and Logit) modelling, across clusters. These models were used to estimate preference scores associated with each hypothetical meal complement beverage concept included in the simulation analyses. Although the maximum utility model assumes respondents only choose beverage concepts with the highest predicted utility scores, the probability models assume respondents rarely make decisions using such precise notions of utility (Hair et al., 1998). Importantly, the group level simulation analysis across clusters provided for a more market-oriented approach to NPD whereby the preferences of each consumer group were taken into account when optimising the product design formulation for meal complement beverages.

Five hypothetical meal complement beverage concepts (MCOMPL 1–MCOMPL 5) were generated for the group level simulation analysis across clusters (see Table 15.5). MCOMPL 1 was chosen for inclusion in the group level simulation analysis as it represented the hypothetical meal complement beverage concept, which according to the cluster analysis results (Table 15.4) would yield high predicted preference scores across groups. However, new product concepts that combine the optimal product design attributes may not represent commercially feasible new products. This simplistic approach to new product design neglects the multi-faceted nature of consumer food choice, where the interplay between market-related factors such as price, and product-related factors such as sensory perception, health benefits and user benefit, ultimately influence consumers' cognitive food choice motives. Therefore, four further hypothetical meal complement beverage concepts (MCOMPL 2–MCOMPL 5), which were slight variants of MCOMPL 1, were included in the simulation analysis (see Table 15.5). This made it possible to identify which consumer clusters would be expected to make trade-offs between inter-related attributes, namely health benefits and price, when evaluating alternative meal complement beverage concepts.

**Table 15.5**  Results of the group level simulation analysis for meal complement beverage concepts across clusters

| Attributes/preference scores | MCOMPL 1 | MCOMPL 2 | MCOMPL 3 | MCOMPL 4 | MCOMPL 5 |
|---|---|---|---|---|---|
| Flavour | Strawberry | Strawberry | Strawberry | Strawberry | Strawberry |
| Carrier | Fruit juice | Fruit juice | Fruit juice | Fruit juice | Fruit juice |
| Health benefit | Boost immune system | Boost immune system | Boost immune system | None | None |
| Pack size | 1 litre | 1 litre | 500 ml | 500 ml | 1 litre |
| Packaging | Carton | Carton | Carton | Carton | Carton |
| Price | €1.50 | €3.50 | €2.50 | €1.50 | €2.50 |
| Cluster 1 (pref. score) | 7.7 out of 9 | 6.4 out of 9 | 6.9 out of 9 | 5.5 out of 9 | 5.5 out of 9 |
| Cluster 2 (pref. score) | 8.8 out of 9 | 7.7 out of 9 | 7.9 out of 9 | 6.7 out of 9 | 6.6 out of 9 |
| Cluster 3 (pref. score) | 8.7 out of 9 | 7.3 out of 9 | 8.0 out of 9 | 6.9 out of 9 | 6.6 out of 9 |
| Cluster 4 (pref. score) | 8.3 out of 9 | 7.5 out of 9 | 8.8 out of 9 | 7.5 out of 9 | 7.9 out of 9 |
| Cluster 5 (pref. score) | 8.8 out of 9 | 7.3 out of 9 | 8.0 out of 9 | 8.2 out of 9 | 7.5 out of 9 |

The conjoint models predicted that Clusters 1, 2, 3 and 4 would not make trade-offs between the health benefit and price attributes when evaluating alternative meal complement beverage concepts. Specifically, Clusters 1 to 4 would be expected to be more receptive towards the functional meal complement beverage concepts MCOMPL 2 and MCOMPL 3, than the non-functional meal complement beverage concepts MCOMPL 4 and MCOMPL 5, according to the predicted preference scores and probability (BTL and Logit) models (see Table 15.5). In contrast, the group level simulation analysis revealed that Cluster 5, the largest consumer group, would make trade-offs between the health benefit and price attributes. Specifically, if a competitor's beverage MCOMPL 4, a low priced non-functional variant of MCOMPL 3, was launched on the market in the future then the conjoint models predicted that Cluster 5 would be expected to choose MCOMPL 4 (mean score 8.2 out of 9) over MCOMPL 3 (mean score 8.0 out of 9) (see Table 15.5).

## 15.7    Final product concept and its use in product design

The attraction of the functional food and beverages market lies in adding value to otherwise conventional foods and beverages in reaction to the downward pressure on prices, where consumers are increasingly seeking value for money in their food and beverage choices. The functional food and beverages category has indeed come to represent an important strategic and operational orientation for food and beverage, biotechnology and pharmaceutical firms (Sunley, 2000). Specifically, the functional food and beverages category has proved attractive to firms with an average growth rate ranging from 15–20% per annum, in comparison to growth rates of 2–4% per annum for both the general foods market and lighter food and beverages market (Weststrate et al., 2002). In increasingly competitive markets, food and beverage manufacturers have therefore targeted functionality, vis-à-vis the health benefits offered, as an extremely significant marketing tool in creating value and competitive advantage in order to differentiate their product offerings from their competitors (Heasman and Mellentin, 2001).

However, the conjoint-based survey in this NPD project confirmed the findings of the qualitative consumer research, which showed that the health benefits offered were not as important to consumers, in terms of purchase motivations or value systems, as manufacturers and retailers have been led to believe. In fact most consumer groups would be expected to make trade-offs in terms of health benefits at the high price level (€3.50). These findings were congruent with Wennström and Mellentin's (2003) strategic analysis of the healthy foods market, and explained the niche market appeal of functional foods and beverages presently. However, as can be seen from Table 15.5, most consumers would be expected to choose a medium priced (€2.50) functional meal complement beverage (MCOMPL 3) over a non-functional meal complement beverage at the low price level (€1.50) (MCOMPL 4). The market-oriented approach to NPD presented in this case study can help firms better understand the interactions and relationships driving consumers' choice motives. This in turn can assist food and beverage manufacturers to identify the optimal price or premium that consumers would be willing to pay for functional ingredients added to foods and beverages.

The consumer-led approach to NPD presented in this case study made it possible to identify the optimal combination of product design attributes for a meal complement beverage, targeted at a number of potential consumer clusters, using information generated from consumer research. This consumer-led approach to NPD also provided guidance to marketers in terms of relevant positioning and pricing strategies for meal complement beverages.

The final outcome of the consumer-led NPD case study was the identification of a commercially feasible meal complement beverage concept: MCOMPL 3. MCOMPL 3 was described as a low-fat, low-calorie, non-carbonated, strawberry-flavoured, fruit juice-based meal complement beverage that contained selected functional ingredients to boost the immune system, and retailed at €2.50 per 500 ml carton (see Table 15.5). Generally, these findings suggested that the technical development and market positioning of functional meal complement beverages on a 'healthy' and 'natural' platform would be an effective strategy decision for firms. This product design information can now be used by the multi-disciplinary NPD team to guide the technical development of a consumer-led functional meal complement beverage with relatively high predicted levels of consumer acceptance.

## 15.8    Problems met and lessons learnt

1. Although qualitative methods such as ethnography, in-depth interviews and focus groups are very effective market-oriented NPD research techniques, the process of data transcription and analysis is extremely time consuming. This can have the effect of slowing down the strategic decision-making process with a resultant increase in NPD lead-times.
2. Full profile conjoint analysis is an excellent technique for understanding consumer trade-offs in relation to product design but participant fatigue becomes an important issue when consumers are asked to evaluate a relatively large number of hypothetical product concepts.
3. A methodological critique of the full profile conjoint analysis approach is the limited number of attributes and associated attribute levels that can be included for evaluation by consumers. Although qualitative research techniques can be used to identify the most important characteristics of a product, often consumers cannot express or articulate their needs in terms of product design attributes.
4. Integrating sensory analysis data into the conjoint analysis technique would generate more realistic product design information given the importance of sensory acceptance to consumers' purchase motivations.

5. The most preferred product concept that emanates from the simulation element of the conjoint analysis research may not be the most commercially feasible product concept.
6. Consumers were not as price sensitive as one might expect. This was revealed through the conjoint analysis procedure, but has also been reported elsewhere for other innovative impulse purchases. This would suggest that combining health and convenience might offer a means of leveraging higher premiums in certain consumer markets. However, consumer tolerance for premium pricing strategies will be dependent on the specific product category, the level of value perceived by consumers, and whether they can make direct price comparisons with potential competitor products.
7. Although this case study presented a systematic approach for designing more consumer-led meal complement beverages, the conjoint analysis research design used in this study, by its nature, did not address the central issue: 'would consumers substitute their regular beverage choice for a functional meal complement beverage?'.

## 15.9   Summary

New food product development is a multi-disciplinary knowledge-intensive process, which necessitates the generation, dissemination and management of consumer knowledge across all functions involved in the development of new foods and beverages. This case study explored the concept of managing consumer knowledge at the early stages of the NPD process, and applied it to the development of meal complement beverages, through the use of consumer research techniques, namely in-depth interviews, focus groups, and conjoint analysis. This market-oriented approach to NPD facilitated the integration of the consumer as 'co-designer' at the early stages of the NPD process. The integration of the consumer during the concept stage of the NPD process, through in-depth interviews and focus groups, provided a valuable insight into consumers' attitudes and perceptions towards the concept of meal complement beverages. The conjoint analysis technique complemented the qualitative findings as it provided a clearer understanding of consumers' cognitive choice motives, and specifically the trade-offs consumers would be expected to make between alternative functional and non-functional meal complement beverages.

This case study illustrated how an understanding of consumers' choice motives and value systems could provide guidance to marketers, in terms of segmentation, pricing and positioning strategies, and to R&D personnel, in terms of concept development and product design, when bringing innovative products, such as meal complement beverages, to the market. Overall, the consumer-led approach to product development presented in this case study promotes a multi-disciplinary approach to NPD, and importantly, provides a systematic framework for managing consumer knowledge during the new food product development process. Finally, given the significance of NPD to organisational performance and long-term profitability, consumer research methodologies that advance both a firm's understanding of consumers' choice motives and value systems, and its knowledge management processes, can improve the chances of new product success.

## 15.10   References

BECH A C, ENGELUND E, JUHL H J, KRISTENSEN K and POULSEN C S (1994), 'QFD – optimal design of food products', *MAPP Working Paper*, No. 19, March.

BEVERAGE INDUSTRY (2001), 'Juicy stuff', *Beverage Industry*, 92 (6), 38–39.

BEVERAGE INDUSTRY (2003), 'The 2003 soft drink report', *Beverage Industry,* 94 (3), 15–19.

BOGUE J (2001), 'New product development and the Irish food sector: a qualitative study of activities and processes', *The Irish Journal of Management, incorporating IBAR*, 22 (1), 171–191.

BOGUE J, DELAHUNTY C, HENRY M and MURRAY J (1999), 'Market-oriented methodologies to optimise consumer acceptability of Cheddar-type cheeses', *British Food Journal*, 101 (4), 201–316.

BOGUE J, SEYMOUR C and SORENSON D (2005), 'Market-oriented new product development of meal replacement and meal complement beverages', *Journal of Food Products Marketing*, 12 (3), 1–18.

BOYLE C and EMERTON V (2002), *Food and Drinks through the Lifecycle*, Surrey: Leatherhead International.

CARDELLO A V (1995), 'Sensory evaluation and consumer food choice', *Cereal Foods World*, 40 (11), 876–878.

CENTAUR (2005), 'News roundup', *Brand Strategy*, 190 (6), 2–3.

COOKSEY K (2000), 'Portable Packaging', *Food Product Design*, 10 (3), 137–144.

COOPER R G (1993), *Winning at New Products*, 2nd edn, Reading, MA: Addison-Wesley.

CRAWFORD C M (1994), *New Products Management*, 4th edn, Homewood, IL: Irwin.

CROFT M (2005), 'Innocent launches children's website', *Marketing Week UK*, 28 (14), 15.

CULLEN P (1994), 'Time, tastes and technology: the economic evolution of eating out', *British Food Journal,* 96 (10), 4–9.

DATAMONITOR (2005), 'Consumer beat: is low-carb Kaput?', *Restaurants and Institutions*, 115 (2), 14.

EARLE M and EARLE R (2000), *'Building the Future on New Products'*, Management Series, Leatherhead Publishing.

EUROFOOD (2001), 'Food On-The-Go Increasing as Habits Change', *Eurofood*, 10 May.

EUROPEAN COMMISSION (2003), *The Social Situation in the European Union*, Brussels: Office for Official Publications of the European Communities.

FOOD PRODUCTION DAILY (2003), 'Breaking news in food processing and packaging', *Food Production Daily*, 13 June.

FORNELL C (1992), 'A national customer satisfaction barometer: the Swedish experience', *Journal of Marketing*, 56 (1), 6–21.

GRANT R M (1997), 'The knowledge-based view of the firm: implications for management practice', *Long Range Planning*, 30 (3), 450–454.

GREEN P E and KRIEGER A M (1991), 'Product design strategies for target-market positioning', *Journal of Product Innovation Management*, 55 (1), 20–31.

GREEN P E and SRINIVASAN V (1978), 'Conjoint analysis in consumer research: issues and outlook', *Journal of Consumer Research*, 5 (2), 103–123.

GROVES A M (2002), *Food Consumption 2002*, Watford: IGD Business Publication.

HAIR J F, ANDERSON R E, TATHAM R L and BLACK W C (1998), *Multivariate Data Analysis*, 5th edn. Englewood Cliffs, NJ: Prentice-Hall.

HAUSER J R and CLAUSING D (1988), 'The house of quality', *Harvard Business Review*, 66 (3), 63–73.

HEASMAN M and MELLENTIN J (2001), *The Functional Foods Revolution. Healthy People, Healthy Profits?*, Surrey: Leatherhead International.

HERRMANN A, HUBER F and BRAUNSTEIN C (2000), 'Market-driven product and service design: bridging the gap between customer needs, quality management and customer satisfaction', *International Journal of Production Economics*, 66 (2), 7–96.

HILL R (1993), *Ethnography and Marketing Research: A Post-modern Perspective*, Chicago: American Marketing Association.

HOFMEISTER K R (1991), 'Quality function deployment: market success through customer-driven products', *Food Product Development from Concept to the Marketplace* (Graf, E. and Saguy, I.S., eds.), New York: Van Nostrand Reinhold.

KAMAKURA W (1998), 'A least squares procedure for benefit segmentation with conjoint experiments', *Journal of Marketing Research*, 25 (3), 157–167.

KRUEGER R A (1994), *Focus Groups: A Practical Guide for Applied Research*, 2nd edn, Thousand Oaks, CA: Sage Publications.

KRUEGER R A and CASEY M A (2000), *Focus Groups: A Practical Guide for Applied Research*, Thousand Oaks, CA: Sage Publications.

LEATHERHEAD FOOD RESEARCH ASSOCIATION (2003), *Drinks On The Go – International Trends and Developments,* Surrey: Leatherhead International.

LEATHERHEAD FOOD RESEARCH ASSOCIATION (2004), *Low and Light Food and Drinks. International Trends and Developments in Weight Control*, Surrey: Leatherhead International.

MCDANIELS C and GATES R (1991), *Contemporary Marketing Research*. Minnesota: West.

MINTEL (2004), *Healthy Eating – Ireland*, London: Mintel International Group Ltd.

MOSKOWITZ H R (1991), 'Optimizing consumer product acceptance and perceived sensory quality', in Graf E and Saguy I S (eds), *Food Product Development from Concept to the Marketplace*, New York: Van Nostrand Reinhold.

PERLIK A (2004), 'Breakfast yoghurt blends', *Restaurants and Institutions*, 114 (25), 33.

QSR INTERNATIONAL (2002), *N6 (Non-numerical Unstructured Data Indexing Searching & Theorizing) Qualitative Data Analysis Program*, Melbourne, Australia: QSR International Pty Ltd. Version 6.0, 2002.

RETAIL INTELLIGENCE (2002), 'Water puts fizz into Coke and Pepsi', *Retail Intelligence*, June.

ROBERTS W A (2003), 'Bar None', *Prepared Foods*, 172 (4), 18–20.

SCHONBERGER R J (1990), *Building a Chain of Customers*, London: Hutchinson Books Ltd.

SHUKLA T P (1994), 'Sensory research in food development and marketing', *Cereal Foods World*, 39 (11/12), 876.

SIX SIGMA (2007). Quality Function Deployment, www.isixsigma.com

SORENSON D and BOGUE J (2005a), 'Market-oriented new product design of functional orange juice beverages: a qualitative approach', *Journal of Food Products Marketing*, 11 (1), 57–73.

SORENSON D and BOGUE J (2005b), 'A conjoint-based approach to product optimisation: probiotic beverages', *British Food Journal*, 107 (11), 870–883.

SORENSON D and BOGUE J (2006), 'Modelling soft drink purchasers' preferences for stimulant beverages', *International Journal of Food Science and Technology*, 41 (6), 704–711.

SPSS (2003), *Statistical Package for Social Sciences v 11*, SPSS Ins., 444 North Michigan Avenue, Chicago, IL 60611, USA.

STEWART D W and SHAMDASANI P N (1990), *Focus Groups: Theory and Practice*, Thousand Oaks, CA: Sage Publications.

SULTAN F and BARCZAK G (1999), 'Turning marketing research high-tech', *Marketing Management*, Winter.

SUNLEY N C (2000), 'Functional foods in an emerging market', *Food Australia*, 52 (9), 400–402.

TRENKNER L L and ACHTERBERG C L (1991), 'Use of focus groups in evaluating nutrition education materials', *Journal of the American Dietetic Association*, 91 (12), 1577–1581.

VAN TRIJP J C M and STEENKAMP J E B M (1998), 'Consumer-oriented new product development: principles and practice', in Jongen W M F and Meulenberg M T G, *Innovation of Food Production Systems*, Wageningen: Wageningen Pers.

WEISBERG K (2001), 'More than a pick-me-up: functional beverages', *Food Service Directory*, 14 (8), 84.

WENNSTRÖM P and MELLENTIN J (2003), *The Food and Health Marketing Handbook*, London: New Nutrition Business.

WESTSTRATE J A, VAN POPPEL G and VERSCHUREN P M (2002), 'Functional foods, trends and future', *British Journal of Nutrition*, 88 (2), 233–235.

# 16

# Consumer research in product design – market-oriented development of healthy vegetable-based food for children

Helle Alsted Søndergaard and Merete Edelenbos, Denmark

*This chapter shows how particular techniques of consumer research were applied to refine and improve product design. In the total approach to New Product Development (NPD), the aim is from the beginning of the product development project to integrate the consumer researchers, market researchers, food engineers and technologists, and food product designers in developing the product ideas, product concepts and the product design specifications. In food product design, there is a need to understand the social and cultural environment for the new product and from consumer research to build up knowledge of both the basic product qualities (composition, nutritional value, microbiological counts, physical form, sensory properties) and the differentiating qualities (aesthetic, health, safety, fun, convenience).*

*The aim in developing from the product idea to the product concept to the product design specifications is to focus what may be initially a vague description into identification of the consumers, specific product attributes and product benefits, gradually building to a product profile. The product profile, which details the group of product attributes, which is the unique identification of the product, has to be related to both the processing and marketing in the product design specifications.*

*Creativity is important and therefore there is a constant cycling between consumer research, product idea generation, product idea screening, product concept, and product prototypes, before the product concept is developed to a stage to build into the product design specifications. The product concept is still not finalised; either it may not be achievable or it may change on consumer testing of the product prototypes. Small food companies may not have the finance, people, knowledge, time, to do much consumer research and they often take risks using the information they can obtain from the market to identify the product concept. But they should still be systematic in their development and be aware of their lack of consumer knowledge.*

*The selection of the consumer techniques is very important at this stage of the NPD project. The outcomes need to be related to the knowledge required by the product designer. Very often, the research starts with focus groups in order to identify the general needs of the consumers and to start product idea generation. But there needs to be more in-depth and quantitative research, which can identify specific product attributes and their importance. In the last chapter, conjoint analysis was used, and in this chapter means-end chain (MEC) analysis. It is important that product designers know what consumer research technique is being used and to understand the outcomes they can expect and how to use them in developing the product concept and finally the product design specifications.*

*This chapter particularly relates to pages 101–111, 224–240 in* Food Product Development *by Earle, Earle and Anderson.*

## 16.1    Introduction

This chapter reports part of the results from a Danish collaboration project on developing healthy vegetable-based food for children. The motivation for the project is to supply parents with healthy vegetable-based meal alternatives for their children. There has been, for some years now, a tremendous focus on higher intake of vegetables for all age groups including children, since a diet rich in fruit and vegetables has a significant protective effect on human health. The Danish health authorities recommend that children above the age of 10 years eat at least 600 g fruit and vegetables a day and at least 400 g a day under the age of 10 years. It can be difficult for parents to reach these recommendations since children tend to dislike vegetable-based foods and prefer foods with a high content of fats and sugars. Additionally, time is a constraining factor in most modern families, which means there is a need for nutritious and healthy convenience food for children that parents can trust.

The Danish food industry, in general, and more so for smaller companies, does not have a strong tradition of market-oriented product development. Therefore, apart from the objective of developing healthy vegetable-based food for children, the project aims at developing a market-oriented development process where market information is channelled into the Product Development Process.

This chapter primarily deals with the development of vegetable-based product concepts based on consumer market information. The specific health aspects of vegetable-based foods, i.e. nutritional value, were investigated by other project partners and will not be reported in this chapter. The chapter describes the theoretical foundations of the project and gives details of the methods used in the collection of consumer market information. Results from the different consumer market information analyses are reported as well as the product ideas and prototypes developed by the company during the PD Process. Results regarding the effects of the market information on the PD Process are also described. Recommendations for further reading are given at the end of the chapter.

## 16.2    The background

Since the early 1990s and the two major contributions to market orientation (Kohli and Jaworski, 1990; Narver and Slater, 1990) there has been much debate in the NPD community about the benefits of this approach. This case study shows the results of a real life PD Process where valuable market information about consumers is generated and introduced into the PD Process.

One of the primary goals of the collection and use of market information in NPD is the identification of customer preferences. Empirical studies show that customer focus is central to the success of innovation (Han *et al.*, 1998; Lukas and Ferrell, 2000). However, there is still much to know about the role of market information in product development (Hart *et al.*, 1999). An empirical study shows that the use of market information has a direct positive effect on new product success, and the importance of using market information effectively is stressed (Ottum and Moore, 1997). Market information may be used instrumentally, i.e. the use of information for decision making, implementation and evaluation of decisions, or *conceptually*, i.e. the information is valued and the company has processes for converting information to knowledge (Moorman, 1995). The barriers inhibiting market orientation are many and complex. Adams *et al.* (1998) found that some barriers can be overcome by broadening the functional participation in acquiring, interpreting and using market data. Furthermore, collected data should be vivid and in a language understandable for the users (Adams *et al.*, 1998). We therefore needed an approach that would be suitable and comprehensible for both the technical and marketing people involved in the NPD process.

The company under investigation is a small Danish manufacturer of vegetable foods. All their products are produced without the use of artificial additives and GMO and products are primarily sold in Scandinavia, which is regarded as their home market. The product range consists of a large number of traditional vegetarian products, canned or frozen as well as grocery products. The products are primarily purchased by a group of very loyal vegetarians. Part of the company strategy is to offer products aimed at a wider audience, including modern health-focused families. At the time of the project start, sales of organic products was increasing and the Danish '6 a day' campaign aiming at increasing the daily intake of fruits and vegetables had just been launched. The company's aim was to expand their position in the market for processed products. The existing products within this category were falafel and vegetable burgers, well-established and successful products. They experienced a short adventure producing ready meals for a large British retailer but the company was not competitive on price because parts of the products were produced by hand. The focus on products for families with children was new for the company. They wanted to evolve from being 'a vegetarian food company' to being a 'supplier of vegetable-based food products to Danish families'.

## 16.3    The means-end chain (MEC) theory and practice

We used the means-end chain (MEC) theory for the generation of market information for the NPD process. This theory is widely used in consumer research and for development of advertising strategy (Gutman, 1982; Reynolds and Whitlark, 1995). The means-end chain theory is based on the assumption that consumers demand products because of the expected positive consequences of using the products (Gutman, 1982; Walker and Olson, 1991; Grunert *et al.*, 2001). Products are traditionally described in terms of attributes, even though the consequences of using them are just as important for consumers.

A means-end chain illustrates the links between product attributes, consequences and values, where the *means* is the *product* and the *end* is the desired *value state*. The purpose of the MEC theory is to explain how product preference and choice is related to the achievement of central life values (Gutman, 1982). Figure 16.1 shows a means-end chain for apples. It illustrates that we buy apples not only for the sake of the apples but because they give us energy so that we can stay active, which helps us to a moment of quality in our lives.

The most important practical application of the means-end chain approach has been for the development of advertising strategies. For this purpose, the 'Means-Ends Conceptualisation of the Components of Advertising Strategy' (MECCAS) model has been developed. It builds on product positioning according to relevance to the consumer (Reynolds and Craddock, 1988; Reynolds and Whitlark, 1995). According to Gutman (1982) and Olson and Reynolds (1983) the means-end chain approach is also applicable in product development (Gutman, 1982; Olson and Reynolds, 1983) although it has only been used sparsely in this area (Grunert and Valli, 2001; Valli *et al.*, 1999). Information about consumers' high-priority means-end chains for a product category gives companies the necessary knowledge to develop products that offer consumers

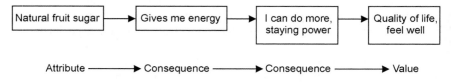

**Fig. 16.1**    A means-end chain for apples.

the desired consequences and which help them obtain central life values. A means-end chain approach to product development is therefore expected to support market-oriented product development defined as the gathering, spreading and use of market information in the NPD process.

## 16.4   Study design

A study design with two intervention points was set up in order to investigate the effect of the spreading and use of means-end chain information in the development process. In two phases of the development process MEC data was collected and made available for the development team. Figure 16.2 shows an overview of the PD Process and the two intervention points. We wanted to examine the difference in viewing MEC data as both explorative and confirmative market information. In the early stages of a development process, i.e. in the idea generation/screening phase, explorative data supports the likelihood of successful NPD (Hart *et al.*, 1999). In the later stages of the development process, market information should be more confirmative, helping the NPD team compare and select between prototypes. MEC data was made available to the inter-functional NPD team in the company, consisting of representatives from both the marketing and the technical functions.

MEC data was collected by conducting laddering interviews, which is the most common way of collecting MEC data (Grunert and Grunert, 1995; Reynolds and Gutman, 1988). Respondents for both studies were recruited in Copenhagen, Denmark. A total number of 90 parents were interviewed using the laddering method. The recruitment criteria were main responsibility for shopping and cooking and having children aged between 3 and 11 years old. Data were collected by a Danish research agency using an interview design developed by Søndergaard (2003). Each laddering interview lasted approximately 30 minutes. In both rounds of laddering interviews, respondents were asked to rank a number of products according to which they would prefer to buy for their families. The product order was randomised. After the ranking,

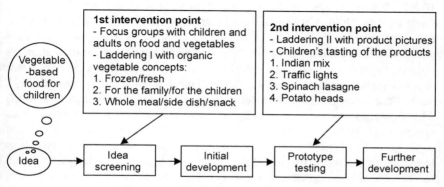

**Fig. 16.2**   The MEC intervention points.

respondents were asked to motivate their ranking and to identify product attributes. The interviewer started with the respondent's first priority asking, for example: 'Why is it important that "Product 1" is made from vegetables?' This started off the laddering interview and the interviewer asked questions along this line until the respondent reached a value level or did not know what to answer. The software program Ladder Map was used to develop hierarchical value maps (HVM) from the coded interviews. The lines in the HVMs differ according to thickness so that the thicker the lines, the more this link was mentioned by the respondents. Some lines are also dotted, representing a negative link, meaning that an attribute may lead to a situation where the positive consequence cannot be fulfilled, and the same applies to the link between consequences and values.

Preference tests with children were performed after the second round of laddering interviews. A total of 92 children between 6 and 11 years old participated. For the children's test, we chose the four products ranked highest by parents in the laddering interviews. The setup included two tests: a pre-consumption test where children stated their desire to eat the products from merely looking at them, and a consumption test, where children tasted the products. The products were evaluated in randomised order on a 5-point hedonic facial scale from (1) dislike very much to (5) like very much. Testing took place at a Danish primary school during lunchtime to create a near natural consumption situation.

## 16.5    Generation of new ideas in the early stages of the New Product Development process

The first round of MEC data collection was carried out during the early stages of the development process and gave the development team initial information to work with. The first intervention point was organised as an explorative study.

### 16.5.1    Focus groups for the identification of food categories

Since food for children was a totally new area for the company, which therefore had little prior knowledge of the category, focus group interviews were conducted with both children and adults to gain general knowledge on their attitudes towards vegetables. Two focus group interviews with parents were conducted: one with parents of children aged 2–5 years and another with children aged 6–11 years. Four focus groups with children were carried out: two with boys and two with girls aged 6–8 years and 9–11 years. The main conclusions from the focus groups were as follows: parents do not like ready meals, since they are not expected to be of good quality. Parents prefer to cook meals themselves, or at least part of the meal. Parents prefer buying fresh vegetables to prepared and frozen vegetables. Vegetables are eaten at lunch, dinner and between meals as a snack. Children are reluctant to eat new and mixed foods. Children want to be able to select what vegetables to eat themselves. It is a challenge for parents to make children eat a broad selection of foods and vegetables. These results suggest that there are great possibilities but also a number of challenges in developing frozen vegetable-based food for children.

### 16.5.2    Laddering interviews with general food categories

Drawing on these initial focus group interviews with children and parents, six meal categories were identified for the first round of MEC data collection. The categories were each described on a piece of paper and the description was placed on the table in front of the respondent:

1. *Frozen side dish for the family*. The product is based on organic vegetables (e.g., vegetable mix) and can be used as a side dish to the evening meal.
2. *Fresh side dish for the family*. Organic vegetables used as side dish to the evening meal.
3. *Frozen side dish for children*. The product is based on organic vegetables (e.g., vegetable mix) and can be used as a side dish to the evening meal.
4. *Frozen ready-made meal for children*. The product is based on organic vegetables and can be used as an evening meal.
5. *Frozen snack meal for children*. The product is based on frozen organic vegetables and can be used as a snack meal.
6. *Fresh snack meal for children*. The product is based on fresh organic vegetables and can be used as a snack meal.

Respondents were first asked to rank and then generate ladders for the six food categories.

### 16.5.3    Hierarchical value maps

Respondents in the first round of data collection ranked the category 'Frozen side dish for the family' highest. See Fig. 16.3 for the hierarchical value map.

As can be seen from the HVM, parents perceive this type of product in a positive manner due to the fact that it is a side dish for the *whole* family and not just for the children. It is easy to prepare and can be served for the whole family, instead of having to prepare separate food for the children. This is important for the manner in which parents wish to bring up their children regarding food and meals. A side dish also gives the parents freedom of choice on what the side dish accompanies.

The frozen ready-made dish for children is among the less preferred meal categories (see Fig. 16.4). It is perceived positive that the product is organic and it makes cooking easy, that the product is frozen and can be prepared quickly. However, the positive picture ends here and the rest of the perceptions are negative. Since the product is frozen it is perceived as unhealthy and of less quality. Being a ready-made product means that its exact content and preparation is unknown and there is no freedom in choice of ingredients for

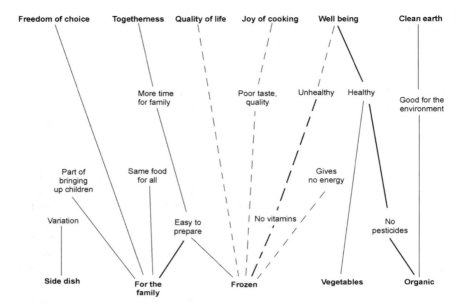

**Fig. 16.3**  Hierarchical value map for 'Frozen side dish for the family'.

the meal. Apparently, parents do not seem to be interested in ready-made products addressed to children.

## 16.6    The development of new product concepts by the New Product Development team

The scenario which can be described from the first round of MEC data collection is interesting in itself but in relation to the project goal only really interesting when applied for the development of new food product concepts. The six hierarchical value maps were introduced and explained to the development team, who worked with the results during a six-month period. Before conducting the laddering interviews we introduced the MEC theory to the development team and made sure they understood the basic approach. When the HVMs were available we had a meeting where the results were presented and discussed in relation to the project. The NPD manager developed a table for the use of the MEC data at team meetings, showing links between overall product types, new product ideas and first laddering results (see Table 16.1). This table illustrates that the results from the first round of hierarchical value maps influenced the product ideas, e.g. the Indian mix was a family portion and contained a serving for the whole family. Several of the products could furthermore be served either as a whole meal, side dish or snack in between meals. The table was developed

Frozen ready-made dish for children

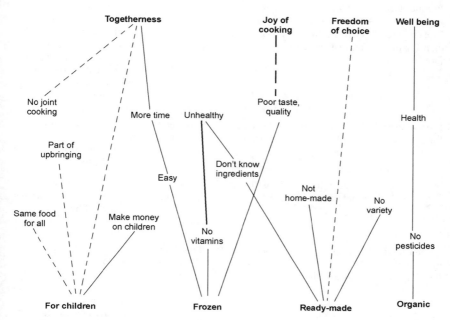

**Fig. 16.4**    Hierarchical value map for 'Frozen ready-made dish for children'.

with the aim of gaining overview and connecting the results from the hierarchical value maps to the ongoing development work.

Table 16.1 clearly shows how the laddering results were actively used in considerations on new product ideas. The development team used the table during meetings and to keep the development process on the track.

## 16.7    Testing product concepts in the later phases of the New Product Development process

The development team in the company came up with seven different product concepts based on the first round of laddering. All products were included in the second round of laddering. Four of the products are described in Table 16.2.

### 16.7.1    Laddering interviews with pictures of prototypes

The second round of MEC data collection at the second intervention point was designed as a confirmative test of the developed product concepts. The MEC data was again achieved by conducting laddering interviews. Respondents were introduced to the general theme of the interview and asked to rank three to four

**Table 16.1**   Table describing new product ideas by the NPD team

| Product type | Name | Ingredients | Packaging | Notes/ladders |
|---|---|---|---|---|
| Frozen side dish for the family | Organic Indian mix (wok dish) | Peas, carrots, corn | Family portion | + the whole family → easy → togetherness<br>− frozen → no vitamins → unhealthy (should be noted on the packaging) |
| Frozen snack for children | Traffic lights (fried croquettes) | Peas, carrots (spinach) | Croquettes on a stick | + side dish/healthy snack<br>− frozen → no vitamins → unhealthy |
| | Vegetable container | Peas (corn) | Individually frozen in pez-like container | + healthy snack<br>− frozen → no vitamins → unhealthy |
| Frozen ready-made meal for children | Bimmers broccoli (pasta dish) | Broccoli and milk sauce | Tray with children's motif | + pasta dish: (girls 6–9 years)<br>− Parents are generally negative towards ready-made dishes. Does this product have a chance with parents/children? Should be tested.<br>+ can be used as side dish → freedom of choice |
| | Spinach lasagne (Pop eye) | Spinach and boiled lasagne with milk sauce | Tray with children's motif | + lasagne is a favourite dish among children (6–11 years)<br>± spinach taste for children |
| | Space pasta (pasta dish with pasta figures) | Vegetables (can be varied) with mild tomato sauce | Tray with children's motif | + can be used as side dish → freedom of choice |
| Frozen side dish for children | Potato heads | Potatoes in slices (with batter) | Bag/tray | + frozen → easy → togetherness<br>+ side dish → freedom of choice<br>(can also be used as side dish by parents) |

**Table 16.2**    The product concepts developed by the NPD team

| Product | Vegetable ingredients | Other ingredients |
|---|---|---|
| Potato heads | Potato (87%) | Corn flour and wheat flour |
| Indian mix | Green pea (24%)<br>Sweet corn (24%)<br>Carrot (20%) | Rice |
| Traffic lights | Red bell pepper (31%)<br>Carrot (34%)<br>Green pea (33%) | Potato, onion and corn starch |
| Spinach lasagne | Spinach (22%) | Pasta, milk, spices, cheese and oil |

products within two of three meal categories; whole dish, side dish or snack meal. Parents ranked Potato heads, Indian mix, Traffic lights and Spinach lasagne highest in the laddering interviews. These four products were subsequently evaluated by children in the children's preference test.

### 16.7.2    Hierarchical maps
The ingredients in the Traffic lights (Fig. 16.5) are perceived as wholesome, full of vitamins with vitamins leading to overall health. The toy element is perceived as fun for the children, it gives them enjoyment and leads to quality of life. The attractive colours/pictures are expected to make all family members want to eat and share the meal together, which is a positive benefit for parents. The negative attributes are the mixed ingredients that limit liberty of choice and that the product is deep-fried which concerns parents due to risks of overweight.

### 16.7.3    Children's preferences for the products
In the first test, where the children only looked at the products, they preferred the appearance of the Potato heads better than the other three products. The children were familiar with potatoes and felt they knew the product. They said almost the same for the Indian mix. The children were more ambivalent towards the Traffic lights. Two large groups were at the extreme of the scale for this product meaning that the children either liked or disliked the product. The children felt least like trying the Spinach lasagne.

The consumption test showed that tasting the products did not change the children's overall ranking of the products very much. Children liked the taste of the Traffic lights as much as the Indian mix, although still not as much as the Potato heads. The results are shown in Table 16.3.

It is clear from the results that children prefer familiar or recognisable foods and if a new product does not lie within this category, children may be unwilling to taste the product. However, results corroborate that repeated exposure may modify children's preferences and food acceptance (Liem et al., 2004).

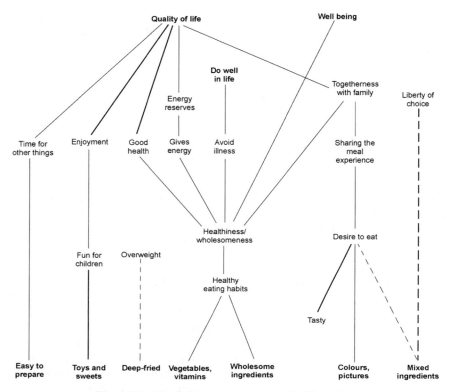

**Fig. 16.5**   Hierarchical value map for Traffic lights.

**Table 16.3**   Evaluation and ranking by children and parents

| Product | Mean evaluation | | Ranking (pre-consumption) | |
|---|---|---|---|---|
| | Pre-consumption | Consumption | Children | Parents |
| Potato heads | 4.7 | 4.7 | 1 | 1 |
| Indian mix | 3.7 | 3.7 | 2 | 2 |
| Traffic lights | 3.1 | 3.5 | 3 | 3 |
| Spinach lasagne | 1.9 | 2.2 | 4 | 1 |

Mean separation within columns by Duncan's test at $p = 0.05$.

## 16.8    Selection of products for further development

After considerations about economic feasibility and strategic fit, the company decided not to launch Potato heads and Indian mix. Similar products already existed on the market and these products are fairly easy for the families to prepare themselves. The idea behind Traffic lights, i.e. mashed potatoes and vegetables formed into croquettes and bites (mini-croquettes), led the company to launch a number of different products within this category. Since parents were reluctant towards products aimed only at children the company chose to focus on products for the whole family. See Fig. 16.6 for one of the new croquette products.

## 16.9    Problems met and lessons learnt

The MEC intervention clearly led the NPD team to use the available information on consumer motivation as inspiration for the development of product concepts. Furthermore, the instrument gave the team a solid basis for discussion during project meetings and helped maintain clear and fixed development goals. The effects of the MEC interventions on the NPD process are shown in Fig. 16.7.

The team members expressed a high degree of satisfaction with information from the first round of MEC data collection. Regarding the second round of MEC data collection, the NPD team would have preferred respondents to be able to taste the prototypes. This indicates that when using a means-end approach for product development in the food industry, it may be advantageous to include consumer tasting. It was clear from interviews with team members that the results from the first round of MEC data collection were used to a much larger extent for the core product development activities than the results from the second round. The team indicated that the hierarchical value maps from the

**Fig. 16.6**   New croquette with spinach and cheddar.

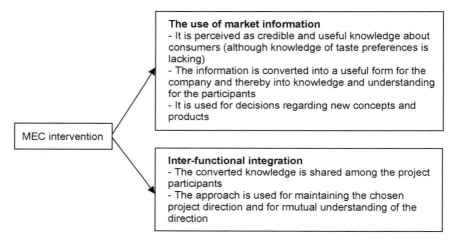

**Fig. 16.7**    Empirical results of the MEC interventions (Søndergaard, 2005).

second round of data collection (testing the developed concepts) supplied information at a different level of abstraction and that the information gave little further inspiration to the development process but that it was very useful for the design of a communication and advertising strategy. This result is consistent with the expectations regarding the MEC-approach in product development, as it is expected to have highest utility in the early phases of the development process where ideas are generated and concepts developed at a more abstract level. The application of MEC-data in the commercialisation phase for the conceptualisation of market communication is well established within the literature on the MECCAS approach. Using a MEC-approach as the basis both for the development of products and for the development of market communication would give companies an opportunity to send an integrated and coherent signal to the market and their customers.

The means-end chain approach shows clear advantages for market-oriented product development. Product developers perceive the MEC data presented in hierarchical value maps as easily accessible information about consumer perceptions regarding product attributes and perceived consequences. They understand that negative consequences identified by consumers should be avoided while the positive should be emphasised as suggested in the MEC literature (Reynolds and Whitlark, 1995). The information in the hierarchical value maps can trigger discussions about the goals of the development process and can help keep the project on track. Without the element of 'taste', the MEC-approach as a concept test is primarily useful for indications of trial purchase. For repeat purchase it is necessary to include sensory preferences in the MEC design model. It would prove useful to develop a design for data collection in the food industry taking into account its special demands regarding the development process as both appeal and sensory preferences are paramount.

## 16.10    Further information

This chapter draws on literature from a number of different research areas. The theoretical approach is set within the area of market-oriented New Product Development (Hurley and Hult, 1998).

The means-end chain approach is drawn from the literature which also specifies how to analyse the collected data (Reynolds and Gutman, 1988). Another example of a MEC-approach applied to the development of food can be seen from Grunert and Valli (2001). See also Grunert (2005) for a discussion of consumer behaviour with regard to food innovations.

The collaboration project formed part of a PhD thesis and other results are reported in Søndergaard (2003) and Søndergaard and Harmsen (2007). For detailed results on testing the products with children see Søndergaard and Edelenbos (2007). The collaboration project is part of the Danish Vertical Network for Fruits and Vegetables supported by the Danish Directorate for Agriculture and Development, Dæhnfeldt, Nutana, AGA and Biodania.

## 16.11    References

ADAMS, M. E., DAY, G. S. and DOUGHERTY, D. (1998) Enhancing new product development performance: an organizational learning perspective. *Journal of Product Innovation Management,* 15, 403–422.

GRUNERT, K. G. (2005) Consumer behaviour with regard to food innovations: Quality perception and decision-making. In Jongen, W. M. F. and Meulenberg, M. T. G. (eds) *Innovation in Agri-food Systems*, 1st edn. Wageningen, Wageningen Academic Publishers.

GRUNERT, K. G. and GRUNERT, S. C. (1995) Measuring subjective meaning structures by the laddering method: Theoretical considerations and methodological problems. *International Journal of Research in Marketing,* 12, 209–225.

GRUNERT, K. G. and VALLI, C. (2001) Designer-made meat and dairy products: Consumer-led product development. *Livestock Production Science,* 72, 83–98.

GRUNERT, K. G., BECKMANN, S. C. and SØRENSEN, E. (2001) Means-end chains and laddering: An inventory of problems and an agenda for research. In Reynolds, T. C. and Olson, J. C. (eds) *Understanding Consumer Decision-making: The Means-end Approach to Marketing and Advertising Strategy.* Mahwah, NJ, Lawrence Erlbaum Associates.

GUTMAN, J. (1982) A means-end chain model based on consumer categorization processes. *Journal of Marketing,* 46, 60–72.

HAN, J. K., KIM, N. and SRIVASTAVA, R. K. (1998) Market orientation and organizational performance: Is innovation a missing link? *Journal of Marketing,* 62, 30–45.

HART, S., TZOKAS, N. and SAREN, M. (1999) The effectiveness of market information in enhancing new product success rates. *European Journal of Innovation Management,* 2, 20–35.

HURLEY, R. F. and HULT, G. T. (1998) Innovation, market orientation, and organizational learning: An integration and empirical examination. *Journal of Marketing,* 62, 42–54.

KOHLI, A. K. and JAWORSKI, B. J. (1990) Market orientation: The construct, research propositions, and managerial implications. *Journal of Marketing*, 54, 1–18.

LIEM, D. G., MARS, M. and DE GRAAF, C. (2004) Consistency of sensory testing with 4- and 5-year-old children. *Food Quality & Preference*, 15, 541–548.

LUKAS, B. A. and FERRELL, O. C. (2000) The effect of market orientation on product innovation. *Journal of the Academy of Marketing Science*, 28, 239–247.

MOORMAN, C. (1995) Organizational market information processes: Cultural antecedents and new product outcomes. *Journal of Marketing Research*, 32, 318–335.

NARVER, J. C. and SLATER, S. F. (1990) The effect of a market orientation on business profitability. *Journal of Marketing*, 54, 20–35.

OLSON, J. C. and REYNOLDS, T. J. (1983) Understanding consumers' cognitive structures: Implications for advertising strategy. In Percy, L. and Woodside, A. G. (eds) *Advertising and Consumer Psychology*. Lexington, MA, Lexington Books.

OTTUM, B. D. and MOORE, W. L. (1997) The role of market information in new product success/failure. *Journal of Product Innovation Management*, 14, 258–273.

REYNOLDS, T. J. and CRADDOCK, A. B. (1988) The application of the MECCAS model to the assessment of advertising strategy. *Journal of Advertising Research*, 29, 43–54.

REYNOLDS, T. J. and GUTMAN, J. (1988) Laddering theory, methods, analysis, and interpretation. *Journal of Advertising Research*, 18, 11–31.

REYNOLDS, T. J. and WHITLARK, D. B. (1995) Applying laddering data to communications strategy and advertising practice. *Journal of Advertising Research*, 35, 9–17.

SØNDERGAARD, H. A. (2003) *Markedsorienteret produktudvikling med en means-end chain tilgang,* Århus, Handelshøjskolen i Århus.

SØNDERGAARD, H. A. (2005) Market-oriented new product development. How can a means-end chain approach affect the process? *European Journal of Innovation Management*, 8, 79–90.

SØNDERGAARD, H. A. and EDELENBOS, M. (2007) Vegetable-based food for children what parents prefer and children like. *Food Quality and Preference*, 18, 949–962.

SØNDERGAARD, H. A. and HARMSEN, H. (2007) Using market information in product development. *Journal of Consumer Marketing*, 24, 194–204.

VALLI, C., LOADER, R. J. and TRAILL, W. B. (1999) Pan-European food market segmentation: An application to the yoghurt market in the EU. In Kaynak, E. (ed.) *Cross-national and Cross-cultural Issues in Food Marketing*. Haworth Press.

WALKER, B. A. and OLSON, J. C. (1991) Means-end chains: connecting products with self. *Journal of Business Research*, 22, 111–119.

# 17

# Customer-centric product development – supermarket private label brands

**Andrea Currie, Australia**

*This chapter demonstrates the significance of knowledge generated from the purchasing customer as well as the consumer and how it can be targeted, obtained and incorporated into the continuing Product Development Process. There are two important ideas in this title. Firstly the term customer – the customer is the person (primary customer) who buys the product as compared with the consumer (secondary customer) who eats the product. In supermarket shopping, there is usually one person in a household who buys the food, although the other people in the household influence them. Customers have a broader spectrum of product attributes they are studying: price of course, special or couponed, size of package, life of the product in the kitchen cupboard or the refrigerator, ease in cooking. With some products such as takeaways and snack foods, the consumer often buys the product.*

*The other concept is that product development begins and ends with the customer. It is important to have both the customer and the consumer in the initial development of the product ideas and the product concept, but more important that there is input throughout the project. Both need to test the product prototypes and look again at the product concept; test the product at the end of the process development and often during the process development; test the production product (usually in a test market); help develop the product concept for promotion and advertising; test the packaging design and the advertising; and after the launch audit the product and the marketing. This weaving of the technical research and the customer/consumer research in total product development is not always achieved because of resources and time, but the aim should be to do this.*

*Customers buying food want to keep the decision process as simple and quick as possible, except when buying expensive and unusual foods for special occasions. But it is very important for the food product developer to realise that there is detailed and critical thought at certain points in time, set off by price rises, advertising, nutritional advice, illness or fashion. The product developer needs to keep up-to-date with changes in the society, culture, economic circumstances, shopping habits, health, nutrition and dieting, and of course the advertising and PR messages, and the new products in the marketplace.*

*There are problems in keeping together the changing consumer behaviour and the technological innovation. Minor changes in product improvement and packaging cause few problems, but it is timing of the major changes, in either a technological or a marketing innovation, that are hard to plan. Technological innovation often takes a long time and its timing for the market is often delayed because of raw material or processing problems, and this delay causes major problems to the market planner. Similarly the market innovation can cause production facilities to be too small or even out-of-date.*

*Questions that always arise – should there be a test market or not? How big and how long should it be? For many new products, a simple roll-out through the market is suitable, especially where there are only minor product and marketing changes. But where the product is really new or the marketing is modernised, then the test market becomes important. Are there risks in launching the product with the present knowledge or are there dangers of losing the advantage of 'first in the market'?*

*After the launch, there is a need to audit not only sales, but also consumer and customer attitudes to the product in the market place and their buying behaviour. Certainly at this point, it is the customer who is the most important factor in the product development – who is buying, are they repeat buying, how much are they buying?*

*The chapter particularly relates to pages 102–106, 195–252 in* Food Product Development *by Earle, Earle and Anderson.*

## 17.1    Introduction

Truly successful product development excels at creating products that customers *really* want to buy – sometimes when they didn't even know they wanted it! At first glance, customers can be a tough bunch – hard to know, hard to please, driven by multiple, combined and (sometimes) conflicting desires, at both conscious and sub-conscious level. But 'customer-centric' product development (see Fig. 17.1) is a researched, planned and structured cycle – there is a place for intuition and brilliant ideas, but only when you're sure you really know your customer!

In this chapter, I'll illustrate customer-centric product development with examples from a newly launched supermarket housebrand strategy. To set the scene, the new housebrand strategy had three quality tiers, something common to well-developed housebrand programmes internationally, but not yet

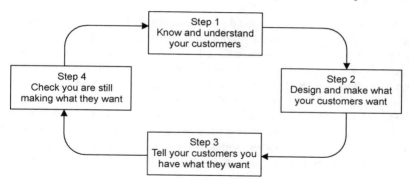

**Fig. 17.1**   The customer-centric process.

successfully executed in Australia. Up to the point of launch, all major supermarkets had housebrand programmes with a focus on value – 'generic' quality – literally the best product available for the price the business was prepared to pay for it. One of the key hurdles that we needed to overcome was that, owing to the value focus that dominated the supermarket housebrand offer for a number of years, a lack of trust in the quality of these products had become entrenched in our customers' attitudes towards housebrand products. It was a gamble when they purchased a housebrand product – sometimes it was terrific, the same quality as the market leading national brand – and indeed, could have been the national branded product due to the practice of some suppliers packing their national brand product as a housebrand to improve efficiency and drive down costs. At other times, often with high involvement categories or those having a national brand product with a prevailing market share, it was a struggle to find a manufacturer willing or able to make a comparable product at the right cost, with a resultant negative impact on quality.

While we knew there was loyalty to individual products in our housebrand range, there was very little to the brand as a whole. It was time for a complete refurbishment of our approach to housebrand.

## 17.2   Step 1 – Know what they want

### 17.2.1 Defining the customer
The key to this step is to define the customer and what drives their purchasing decisions. Marketers typically identify customers as:

- *Primary* – those making the purchase decision, usually those with or controlling the money.
- *Secondary* – those who have not purchased the product, but will consume or use it – in addition to the primary customer, or in some cases, instead of them.

Research that we had previously conducted showed that our supermarket business drew customers from virtually the entire population spectrum. This means that our product offer needed to cater for a diverse range of purchase needs and desires. Data mined from a customer loyalty card introduced across our corporation a decade previously allowed us to segment our supermarket customers in to five main demographics (with approximately 16 sub-categories), based on their shopping locations and purchase behaviour. As expected, the classic family unit was still one of the largest of our five major customer demographics, and I will use them as the primary example in this section. In the family unit, the primary customer is the main grocery buyer, and in this day and age, it can be either the mother or the father or (confusingly) a combination of the two. The secondary customers are the partner and children of the main grocery buyer.

The family unit is also a major customer source for a well-executed house-brand strategy. They are looking for ways to make their grocery money go further, but are not necessarily willing to compromise much with the quality of the product.

So how did we go about determining what our primary and secondary customers' key purchase drivers were?

*Retail information*

As mentioned earlier, data obtained from loyalty card use has proved invaluable in determining what products customers in specific demographics and geographical regions are buying, when they are buying, how often and with what other products. Although not all of our customers have or use their loyalty card, we commissioned a validation study to confirm that the data collected represents our customer base. This information, combined with figures from an external retail sales analyst and publicly available studies, underpinned the launch of products into our top tier brand (*Coles Finest*$^{TM}$). This brand is oriented towards special occasions, restaurant-quality food, gift giving and indulgence – making its launch at Christmas time a natural choice. The retail information we had to hand showed that we were losing customers to specialty stores at Christmas for purchase of key premium food items – turkey, ham, fruit mince pies, Christmas cakes and puddings. The items we provided in these areas were high unit price points – $AU60 for a leg of ham, in a market where a price of more than $AU10 for a unit of product is considered expensive – but the products, supported by in-store sampling, had great acceptance and an excellent sell through.

Sales analysis and trending of product categories is a fundamental part of determining opportunities for housebrand product development and is embedded in our development process. Data from our own sales is combined with that obtained from an external retail sales analyst, to formulate housebrand develop-ment on both a category and product level at the start of every project. The externally sourced sales information is used to gain a complete picture of Australian supermarket sales and can also provide insight into overseas super-market retail trends. This information was used to demonstrate the early success of the launch of the branding strategy, indicating substantially higher recognition and repurchase intent against major competitors' housebrands.

*Commissioned research*

We use commissioned research at this part of the customer-centric cycle to give us vision into the broad-based decision drivers that influence our primary and secondary customers' purchase behaviour. It was employed when the new housebrand strategy was being developed, to confirm that a three-tier strategy was viable, that the quality and price positioning for each tier was accurate and that the brand names and design elements for each tier were delivering permission to buy and creating the desired empathy with the brand. We also use commissioned research in a more focussed way in Step 2 for key, high involvement categories when we want to establish that we have created products that cater for specific customer needs.

*Benchmarking*

While in the planning stages, a key aspect of our new housebrand strategy launch was to observe execution of successful housebrand programmes overseas, particularly in the UK and Europe, as retailers there are acknowledged best practice in housebrand development. Key learnings were the importance of having clear quality points of difference between tiers, ensuring the quality levels are consistently maintained across product categories and that the graphics for each tier create clear recognition for your brand, and link different categories within the brand. In addition, approximately six months after the launch we benchmarked our mid-tier brand against a major supermarket competitor who had launch a second tier (equivalent to our mid tier) housebrand range and were pleased to find that our brand had a superior recognition and acceptance from our customers, underlining that we were on the right track with the day-to-day execution of our strategy.

Returning to our key primary and secondary customers, the family unit. Through the above mechanisms, we determined when shopping, our primary customers need to make purchase decisions on behalf of their family and can be motivated by a number or combination of influences, some of which are discussed below.

*Permission to buy*

Research conducted at the time our new housebrand strategy was being constructed showed that there was a 'cringe' factor associated with purchasing of housebrand products, linked to their variable quality. While the new housebrand approach clearly underpinned defined and consistent quality standards for each tier, we still needed to communicate this consistency to our customers, to help them feel comfortable to put these products in their shopping trolleys and serve them to family and friends. In-store demonstrations were planned at the launch of major products, but the products' branding, graphics and styling were pinpointed as key ways to establish rapport and deliver the message of credibility to our customers, every time they were considered on the supermarket or pantry shelf. To this end, the brand name used for our mid-tier products had an emotional aspect (*You'll Love Coles*™), were personalised using photographs and endorsement statements from advocates on front of pack, and employed eye-catching styling, colours and graphics typical of market-leading products. As these products looked like national brand products, customers were more relaxed about being seen putting them into their shopping trolleys.

### Are they accompanied by the family?

Secondary customers may be influenced by advertising and other promotional activities, which the main grocery buyer can be more immune to. Newly launched products can be accompanied by heavy advertising and in-store promotions, all designed to attract attention and make products more desirable. On the surface, this did not seem like good news for our housebrand strategy for, while our products were equivalent in quality and packaging to the market leaders, we did not have a promotional budget to match. But this purchase motivator added extra incentive for us to successfully establish a rapport between our customers and our mid-tier product offer and, in particular, create an aspect of fun for those products aimed primarily at children. An example of this is with our canned baked beans and spaghetti in tomato sauce – the graphic depicts an active eight-year-old boy playing soccer with an enlarged baked bean substituted for the ball, while the spaghetti shows a six-year-old girl using a strand of spaghetti for a skipping rope.

### Time pressure

We were aware of the statistics showing that increasingly, both parents of the family unit were in the workforce; if shopping for meal solutions in our stores (an activity regarded as necessary but not usually fun) was easier, particularly through the use of our housebrand products, this would encourage loyalty to products and the brands. Convenience was factored into our strategy from the beginning, and products such as bagged salad kits, marinated roasts and premium filled pasta were early products on offer. Having distinctive pack designs with common design themes across product ranges and the brand as a whole was also intended to promote customer recognition – many customers shop using pack designs, features, colours or shapes as a cue to product selection – and make it easier for customers in a hurry to find and select our products.

### Nutrition

Nutrition is a powerful purchase influencer in our primary customers' lives. Government and community pressure is requiring mainstream food products and major fast food businesses to deliver balanced nutrition profiles, with 'treat' foods banished as occasional indulgences, substituted with a more suitable product or reformulated to a more acceptable nutrition status. In discussion of Step 2 of the cycle (below), I will discuss how nutrition profiling was used when a treat product, microwave popcorn, was converted from our previous housebrand range into the new.

### Indulgence

Paradoxically, another driver is the desire to indulge, either the primary customer themselves or family members. Celebrations, dinner parties and other special eating occasions are related to this desire; not just to indulge the family, but as a chance to impress. I have covered above how we assisted our customers through the stress of the festive season (and our research underlined that it is a very tense time for some) by providing a range of Christmas meal components prepared from high quality ingredients that absolutely delivered customers' expectations.

### Cultures and cuisine styles

Increased multiculturalism and international travel have introduced a wide variety of different cultures' cuisine styles to most developed countries, and a corresponding number of product development prospects. Authenticity is regarded as important, but often needs to be tailored to Western tastes. We recently encountered this dilemma when reviewing our existing range of Mexican products in preparation for their conversion to the new brand. It is probably fair to say that authentic Mexican cuisine is not well understood by the majority of customers who purchase the various products currently available in Australian supermarkets. Although our Home Economist had prepared a range of recipes that accurately reflected the style, these were a long way from products our customers were familiar with, necessitating a compromise in the outcome expected for the finished products.

### Social responsibility

Social responsibility and environmental impacts are emerging concerns that we know, through our analysis of loyalty data, have relevance across significant subsections of our customer base. Organically derived products have been a feature of our housebrand range for some time; to improve their prominence and accessibility, we have recently taken them out of our 'Health Foods' aisle and are merchandising them next to their conventionally grown counterparts, with all products enjoying corresponding lift in sales. An additional example is our Coles Finest<sup>TM</sup> Coffee Beans; while delivering a superior flavour for all three variants, we could add to their exclusiveness by having the beans sourced through the UK-based FAIRTRADE programme.

At the end of this phase of the customer-centric cycle, you should have constructed and recorded a comprehensive picture of your primary and secondary customers, their social circumstances, demographics, their key lifestyle motivators and choices and be clear about how current and future products are going to assist them. You are now ready to move into the creation of products that meet their needs.

## 17.3   Step 2 – Design what they want

To commence this section I will discuss the conversion of an existing housebrand product, microwave-prepared popcorn, from our previous housebrand into our new strategy.

### 17.3.1    Customer insights

Our target primary and secondary customers were from our classic family unit demographic. Customers understand the product is treat-oriented, but believe it is a more nutritionally sound alternative to other savoury snacks such as potato crisps and seasoned extruded snack foods. An analysis of customer contacts and complaints regarding the existing product indicates they think the product is too greasy, and has a fatty taste.

### 17.3.2    Competitor benchmarking

All of our product development projects start with a market review that appraises the attributes of existing national brand and housebrand products in the category. Using this, combined with sales trend analysis, customer insights and internal or external research (either commissioned or obtained through existing sources), we build a clear picture of the sort of product required to successfully fulfil customers' needs. In this case, the product range was to be added to our mid-tier, *You'll Love Coles*™ brand; the quality benchmark for this tier was to be as preferred as the market leading product in the category. As the flavour and aroma of this product were intrinsic to its acceptance and success, the buttery flavour profile of popcorn available at cinemas was also considered in addition to the existing microwave-prepared products on sale through our own and other major supermarket chains. As well as assessing flavour styles, we reviewed the ingredient lists and nutrition information of competitor products, paying particular attention to the national brand product. We determined that this product in particular had lower total and saturated fat and lower sodium levels compared to our current product, and also used a polyunsaturated vegetable oil, in contrast to ours, which was based on coconut oil. It was surmised that the fatty taste highlighted by customers in their feedback was related to the taste profile of the coconut oil; coconut oil has the advantage of being solid at room temperature and thus, minimises oil patches on the packaging, but its lower melting point does negatively affect mouth feel, and the flavour of the oil tends to interfere with the added butter flavour.

### 17.3.3    The product brief

All of our product development projects prepare a product brief, which is provided to prospective suppliers in order for them to prepare and submit samples of product they believe meet the brief and deliver the outcomes we are seeking. In this case, as with many housebrand lines, the product is 'me-too' in style and all the key attributes being sought were for the product to match its competitors. In addition to specifying required pack sizes, weights and multiples, required product attributes were:

- Not more than 5% of corn kernels to remain unpopped after nominated cooking time; inherent to this was that all suppliers under consideration had to have access to the relatively sophisticated heat transfer packaging that is intrinsic to successful popping.

- Flavour to be rich and buttery, with excellent aroma, reminiscent of product available at the major cinema outlets.
- Not more than 25% total fat.
- Not more than 10% saturated fat.
- Not more than 1000 mg of sodium per 100 g of product.

After several rounds of resubmissions to fine-tune the product range, the lines were finalised, suppliers were awarded contracts and the product moved into the next phases of artwork development, specification finalisation and launch.

### 17.3.4    Testing the success of the brief

Some aspects of the brief – quantities of key ingredients, nutrition improvements and claims – you can confirm through formulation breakdown and nutrition analysis. In the example of the popcorn above, the amount of unpopped kernels was assessed for each submission and potential suppliers were asked to demonstrate they had met nutritional limits through provision of calculated or analytical nutrition data. Some key success measures, such as flavour and overall quality acceptability, are less easy to quantify, but are, clear 'customer-wanted' attributes of the product.

Sensory and market research are clearly indicated in order to determine whether you have achieved these important attributes. Sensory research, in particular, is covered in considerable detail in other chapters of this book, so what I will share here is how and when we usually employ these two measures through our housebrand development programme.

In the first instance, the sensory profile of submitted product samples is assessed using a small group of semi-expert team members – representatives of the Merchandise, Housebrand Product Development, Marketing and Technical teams responsible for the category that the product will be ranged in. The sensory profile is adjusted according to the consensus of this group, and for many categories, the product proceeds to launch after being signed off by this group. While this approach may lack the robustness usually preferred for statistically based sensory analysis, it has the clear advantage of requiring less time and resource to complete.

Where consensus is not achieved, where the product is to be launched into a high profile category and/or in preparation for commissioned market research, a statistically based taste panel is conducted by our Technical team, using team members from within the building. Typically these panels would include a predominance of Technical team members who have had sensory training and have well developed palates. The usual method employed is a nine-point hedonic scale, with the proposed product required to statistically equal the preference score of the benchmark in order to progress. Comments about each product are collected to give direction for any improvements necessary.

Where the added flavour of the product is intrinsic to its sensory success, the companies providing our flavour have provided significant sensory input and support. Our soft drink range was developed in conjunction with a flavour house, to ensure consistency of flavour across the range, as the volume of product would require at least four bottling plants over two companies to deliver. The flavour companies concerned undertook significant levels of internal sensory analysis to ensure the requested flavour profiles were delivered.

Where the product is intended for consumption by a specific target customer not well represented within team members in the building (examples: children, vegetarian, home brewers) 'take home' volunteers are sought and the products sent home for assessment with detailed questionnaires. As the assessment is taking place outside of controlled conditions, bias is the concern with this method. On occasions, we have found hedonic scales problematic for younger children, even using faces with expressions varying from 'frowning, yucky' through to 'smiling, yummy' and we have sometimes obtained results at extreme ends of the scale, which may or may not bear a resemblance to the child's actual opinion of the product!

Commissioned research does have a role to play, particularly where the product is 'high involvement' (examples: the category is dominated by a major brand with significant market share, the product is of high unit value or there is a significant emotional decision for the customer to make when selecting product, such as indulgent products and those intended for children). Recent launches in the soft drink and block chocolate categories were both assessed through commissioned research, which included assessment of the parameters outlined in Table 17.1. It was pleasing to note that in both cases, our relatively

**Table 17.1**  *You'll Love Coles*™ brand – some commissioned research parameters

| | |
|---|---|
| Participant selection (matches predetermined customer profile) | • Main grocery buyer<br>• Regular shopper in Coles stores<br>• Regular users of the product<br>• Not rejecters of housebrands<br><br>(Optional)<br>• Regular users of the benchmark product |
| Questionnaire construction | • Sensory characteristics (appearance, flavour, texture)<br>• Focus on detail to define whether key sensory attributes have been delivered and to give direction on improvements if required<br>• Product claims/key selling features – obvious, appropriate, compelling, credible?<br>• Overall liking<br>• Package design and graphics – distinctive, congruent with product features, visibility of product claims, functionality<br>• Purchase price and purchase intent – blind and with brand revealed |
| Outcomes | • Recommendation to proceed to launch<br>• Recommendations to improve product – sensory, selling features, overall liking, packaging, purchase price |

unstructured internal sensory analysis was supported by positive outcomes for the products submitted to the commissioned research.

We also use commissioned research as a 'would you buy it' measure for all of our *Coles Finest*™ products. These products have been fine-tuned by the same cross-functional category group as products launched in our other housebrands, but pass through an additional assessment by a panel of senior Merchandise, Housebrand and Marketing team members (including our Home Economist) with the food skills appropriate to gauge the suitability of individual products for this range.

## 17.4   Step 3 – Tell them you have what they want

Although the focus of this book is essentially on product development, it is appropriate at this point to touch on how the advertising and promotion of the brand new product affects customers' purchase decisions. There is little point in creating a great product if customers don't know it exists, or if they do, how it works for them.

### 17.4.1   The communication plan

We use communication plans as the basis of cohesive marketing campaigns for our products. This allows us to identify the product's primary customers and key benefits to those customers, distilled down into the few key messages it takes to influence customer purchase behaviour. The success of the marketing campaign hinges on identifying and sticking to these messages – staying 'on-message' – while customer engagement often relates to the inventiveness with which this is achieved.

There are four aspects to the communication plans we employ:

1. Who are our target primary and secondary customers? This is just confirmation of what has already been identified at the start of the product development project.
2. What do they want or need to buy? How does this product fulfil their needs?
3. What do they need to know about this product to decide to buy it? How should they find out about it?
4. What factors *apart* from the product's attributes will influence their decision to purchase? What will give the product credibility, and your customers permission to purchase?

Of course, the first two aspects have already been decided and acted upon to build the product to this point – they are the platform upon which the marketing campaign will be built. It is the last two aspects that are the basis of the marketing and communication campaign, and involve decisions on:

• In-store signage – this is the major vehicle we use to promote our housebrand products and includes everything from large window posters, through double-sided sleeves placed over our loss prevention sensors, floor decals, head-height aisle banners, shelf fins, wobblers and tickets highlighting the location of specific product ranges, to banner advertising on the computer screens at each checkout. Our intent is that the customer is exposed to aspects of our housebrand range from the time they enter the store to the time they leave.

- Advertising – this can include catalogue, print media, billboard, radio, television, cinema, SMS, Internet or combinations of these. Being a major supermarket, catalogues are our primary communication method and are used to promote range launches, specifically discounted housebrand products (to encourage trial) and mass (full- or double-page spreads) of products in a particular tier. We have an in-store magazine which we again use to feature product launches and which can carry advertisements for housebrand products, placement of products in editorial and use of products in specially developed recipes. The magazine was a key communication vehicle for our *Coles Finest*[TM] launch with the products featured strongly in the Christmas edition.

- Creation of the advertising and promotional campaign – what will resonate best with target customers? Comedy, drama, popular culture, celebrity or expert endorsement? Our *You'll Love Coles*[TM] television campaign focussed on depicting two key messages – that the quality of the product would be evident the very first time you tried it, and that customers enjoyed the products, reinforcing the emotional connection we wished to establish between the brand and our customers.

- Are there key special interest, support or professional groups that influence customer decision making and enhance sincerity? These can be useful for very targeted products, such as children's ranges or 'free from' offers.

- What events can be created or sponsored to further deliver the message and put product into customers' hands? When we launched our *You'll Love Coles*[TM] bottled water, we arranged for the product to be given out as part of an Australian capital cities' major women's fun run. This was a great opportunity to put the product in front of a large selection of one of the target customer groups for the product – women grocery buyers who were interested in keeping fit. A related opportunity which we have used for our *Coles Finest*[TM] products is placement in credible editorial coverage for specific events – again, at Christmas time, in respected cuisine print media. As journalistic standards need to be maintained, we don't expect that the product will always be featured, but our intent with our top-tier brand is to create products that have such superior sensory characteristics that they *should* be included.

## 17.5    Step 4 – Keep checking that you are still providing what they want

Well – we are over the last hurdle and the product is now launched to the market. The advertising and promotions campaign is going well, and initial sales look very promising. Time for us to relax? No!

The means to maintaining sales momentum for the product range is to ensure it continues to deliver what our customers want it to. Metrics we monitor post launch include:

- Sales – put simply, upwards or steady means the product is doing the job; downwards means its relevance has been superseded or hadn't been properly delivered or communicated. Downward trends (particularly in the three months after launch) mean that we will take a good look at the product, customer feedback about it and the communication plan and revise based on the outcome.
- Customer contacts – we always expect to receive more negative feedback than positive from our customers – it is human nature to complain more than congratulate. In our customer feedback, we measure:
  - Complaints versus sales – most supermarket items are measured on a CPMU (complaints per million units sold) basis; an upward trend signals trouble and should be investigated for specific trends
  - Trend analysis – we regularly characterise customer contacts by type, and monitor which way the key ones are trending. We focus on removing the most common ones, or have them trending downwards.
  - Serious incidents – there are some things that require immediate investigation and action – metal, glass or hard plastic contamination, alleged food poisoning or product tampering. We have mechanisms established which allow prompt escalation, to ensure they receive the proper attention.
- Compliance – as a matter of course, we ensure that our products' key selling attributes continue to be delivered – if claims are based on nutrition or composition, then we make certain, through regular analysis, that this continues to be provided.
- Competitor benchmarking – are our products maintaining a point of difference against other offers in the marketplace? Have there been new entrants since launch? We check this regularly to ensure our products still have significance.
- Customer insight – what is happening in our customers' world? It is the nature of life that nothing stays the same, and we keep tabs on the evolution of new insights and imperatives that will influence our customers' shopping behaviour.

If you think this is starting to sound like the start of the chapter, then you are right – because product development *is* a cycle. From the time a product is launched, the focus should be on maintaining its importance in the customer's ever-changing environment. This may require several 'renovations', but at some point it is inevitable that you will need to recommence the development cycle, and this is most easily achieved by keeping a constant vigil on the imperatives that dictate your customers' behaviour.

## 17.6  Left field ideas, individual opinions, sudden opportunities – how they fit

In the fast paced world of supermarket retailing, it is inevitable that, at some point, unforseen circumstances arise – a rapid trend will develop in the market (consider the sudden acceptance of climate change in recent times), legislation changes, senior management intervenes with a 'brilliant idea' or a major competitor exits the existing or a related market. Our organisation's success depends on our ability to respond rapidly. In these circumstances, should we abandon carefully laid plans, if they no longer seem relevant under a more pressing situation?

Of course not – strength in these circumstances is our fundamental understanding of our customers, putting us in a great position to assess how they might react to this situation. We focus on how this new opportunity can meet their needs.

Product development is an exciting and creative process – something that has inspired me for more than 20 years in the food industry. I learnt early on that innovation and great ideas are not enough – they need to be important to your target customers, otherwise you are the only one who thinks it is a good idea! Your fundamental role as a product developer is to be an advocate for your customers, always asking, 'What's in it for them?'.

# Part VI

# Product development in practice

# 18

# Product design, process development and manufacturing – a road map for the technologist

**Ed Neff, Australia**

*This chapter examines the important steps of product and process design together with industrialisation for product launch. It examines the Product Development Process (PD Process) and the role of the technical department from a food technologist's or food scientist's perspective, looking at the importance of management procedures, communications, teamwork and close cooperation between technical and marketing personnel. The chapter reflects the extensive experience of the author in several multinational companies in Australia and the case studies described in the book.*

*Apart from the prime objective of launching a successful product in the marketplace, the company's Chief Executive and the Marketing Manager would list the following as key success outcomes in a product development project:*

- *on time delivery*
- *initial volumes distributed and sales reaching predictions*
- *cost and profit targets met*
- *launch product meets design and quality standards.*

*Of course, these are fairly fundamental objectives. To achieve them consistently requires discipline, teamwork and excellent project management.*

*This chapter particularly relates to Chapters 6 and 8 in* Food Product Development *by Earle, Earle and Anderson.*

'The ultimate determinant of success of the Product Development Process is the consumer.'

## 18.1   Introduction

We will examine the PD Process in three stages: see Fig. 18.1:

- Feasibility stage. This is the important first stage when the product concept is developed into a development brief. It can be referred to as the product strategy development phase (Earle *et al.*, 2001).
- Development stage. The real work horse of the PD Process when the development brief and product design specifications are developed into a viable product.
- Industrialisation stage. The product must now be manufactured commercially and delivered for launch to the market.

The key steps in each stage will be considered with discussion focussed on managing the PD Process towards a successful outcome. Lessons arising from case studies in earlier chapters add illustration on the roadmap for success.

## 18.2   Managing the Product Development Process

The three stages of PD are presented in Fig. 18.1. The key activities are shown for each stage, together with the important milestones for product evaluation and project approval. This PD Process must be integrated with the consumer research activity, which may be coordinated from within the marketing function. (Goldman and O'Neil, Chapter 14). The key steps will be applicable for any PD project, whether this is a simple line extension or a more challenging new product not previously manufactured by the company. The food technologist must continually monitor progress and ensure that the key steps are observed and satisfied.

It is interesting to reflect on the PD Process in a small company versus a larger or multinational organisation. For various reasons a smaller company may attempt the PD Process more by common sense, or entrepreneurship, with limited technical expertise and poor adherence to the principles of good PD. Sorensen (Chapter 12) provides three sobering case studies from small enterprises where deficiencies in the PD Process leading to failure are summarised as:

- A lack of understanding of the properties of the raw material used, and an inability to control or monitor these materials.
- A reluctance to invest in resources not seen to contribute directly to the bottom line, and an inability to control the production environment.
- Undertaking projects with requirements that exceed the skills of the development and production teams.

In Chapter 13, Nghee shares with us the challenges and frustrations met by an entrepreneur in launching a new product. An entrepreneur will take more risks and is prepared to cut corners but all the key steps are still relevant. Edwardson and Best (Chapter 5) show us how the process can be applied to an agri-food

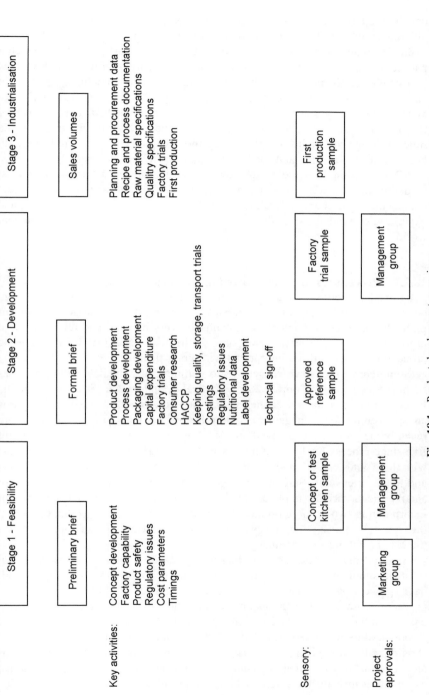

**Fig. 18.1** Product development overview.

development project in a developing country only if there is a single controlling aid agency or at least strong coordination between different aid agencies.

In a commercial environment, there will always be time pressures to reach a particular stage sooner or to launch the product in a particular time window. This is a challenge for the food technologist and short cuts sometimes need to be taken or compromises agreed to, but there should always be critical checks. When do you say yes and when do you insist on a more realistic reassessment of the programme? In Chapter 7, Wilkinson and Palamountain disagreed on the time to design and build the plant; they took the quicker but more risky path because of lack of time and money – fortunately it was successful!

PD is exciting and the recipe for success lies in good teamwork, honest communications, adherence to some essential disciplines and good project management. Of course, there are other attributes for success such as technical expertise, creativity and a dash of experience. Welcome to the commercial world.

### 18.2.1    Procedures and the marketing/technical interface

PD procedures will exist and these will vary in complexity and detail depending on the size and culture of the company or organisation. A large multisite or multinational company may have established very detailed and relatively rigid procedures. A smaller private company, on the other hand, may have only basic procedures which will be interpreted with great flexibility.

Some companies have subscribed to commercially available PD Processes such as Stage-Gate® (Cooper, 2005). Typically, Stage-Gate® is a five-stage PD Process from idea stage to launch with a gate between each stage. Each gate is an opportunity for management review leading to a go/no-go decision. More recently the process has evolved into NexGen Stage-Gate® variations to the full five-stage process for less complex projects. For example, Stage-Gate® XPress with three gates can be used for projects of moderate risk such as improvements and range extensions. Stage-Gate® Lite has only two gates and can be used for small projects such as simple customer requests (Cooper, 2006). The principle of incorporating two or more review gates in the PD Process is encouraged even if the rigidity and discipline of a system such as Stage-Gate® is not appropriate for the size or culture of a company.

A sound PD Process, as described in Fig. 18.1, is essential for success but must be supplemented with some lessons and commonsense: a company will have procedures to manage the various stages of the PD Process; these procedures are essential to ensure that the knowledge is available for critical analysis at the check points. Rigid interpretation of these procedures can become a constraint to progress, particularly where timelines are being challenged. How should the technologist interpret these instructions? It is tempting to use them as an excuse as to why something cannot be achieved. On the other hand, it is just as tempting to bend the rules, make a compromise, and remain on track. The challenge is to know when to be flexible, find a way forward, but still meet the key requirements of the PD Process. The use of checklists, and the concept of a technical sign-off, as discussed later, helps to maintain a balance. Discussion of an issue with a more senior or experienced colleague can be helpful. Timelines are often challenged by the need to wait for the next new product meeting or management group meeting in order to gain approval. It may be possible to obtain a preliminary review of the project to allow some critical action, such as ordering an imported ingredient, to be actioned. Procedures must be observed but there is scope for flexibility in their application.

The PD Process is a multi-functional activity, which will occupy the technical function (food technologist, technical manager, R&D manager) and the marketing function (product manager, product assistant, marketing manager) as its key participants. Senior executives must also be involved in order to review progress, provide direction and approval to proceed to the next stage. As there will be many projects in a company's portfolio, each project must compete for resources based on its viability and priority. This can cause tensions and friction between departments.

In Chapter 2, Inwood discusses the role of the NPD manager and stresses that he first and foremost has to be a businessman.

> 'The NPD manager's role is to keep the NPD programme on the company agenda, its resources appropriate to its agreed objectives, the strategy alive over the relative timeframes, and ensure the opportunities for investment in NPD are properly represented when business strategy is considered.'

The representatives of technical and marketing will be in day to day contact in steering and managing progress. Who is the leader? This is a question often posed by technologists. Unless it is a highly technical driven project, then it is wise to consider the technical function to be a service to marketing and hence for marketing to be the leader. This can be justified on the basis that marketing are representing the customer and the consumer, and in medium to large companies are most likely to be responsible for the consumer research and product testing. Goldman and O'Neil (Chapter 14) provide us with a good insight into the evolution of the sensory research consultant into a trusted product and consumer advisor to today's company. Currie (Chapter 17) stresses the importance of the customer in the development of housebrand products for a supermarket chain

The notion of marketing leadership can be a little hard to accept for a mature experienced technologist when working with the newly appointed product assistant considered to be still green behind the ears.

Where there is a product and process innovation, management is usually by the technical function as the experimental work is long term and difficult for marketing to fit into their short-term market planning. But there is a need to keep them informed as the project progresses so that they understand the innovation and start to fit it into their planning as it nears completion. As Russell (Chapter 11) shows in the project on continuous ice cream processing, the basic research took a long time, but at the end marketing did not immediately understand that this creamy, low fat ice cream was a strong candidate for the new trend to have a nutritionally acceptable ice cream.

### 18.2.2    Cross-functional team and team dynamics

Success in PD will require teamwork, trust and a close working relationship between marketing and technical. As a technologist, you should work hard to develop and maintain this trust and positive relationship. Take time to discuss the project with the marketing person and understand the marketing strategy. If possible, attend some of the consumer research panels. Invite marketing to the laboratory, pilot plant and to the factory. Avoid the development of an 'us and them' attitude, maintain honest dialogue and enhance the PD partnership.

The formation of a larger cross-functional team, comprising representatives from key functions involved in developing, commercialising and launching the product, may be very beneficial in situations such as:

- high profile, strategically important projects
- new to company product or new technology
- very demanding timelines.

Team leadership and team dynamics become a critical success factor but the pay-off can be considerable when all functions are on board with the project. Usually, a senior executive will lead or be a member of the team and this helps to greatly accelerate decision making and approvals. The team leader must have formal authority to make decisions. PD using cross-functional teams is encouraged in the Stage-Gate$^{®}$ process (Cooper, 2005).

The size of the project team will vary with the complexity of the project and may vary as it progresses through the development stages. New team members will join as additional skills and knowledge are required. A successful team will maintain a positive and cooperative culture both across and within functions. As technical team leader you will have responsibility to develop and maintain cooperation amongst all members of this function, including technologists, engineers, quality scientists and production personnel.

In Chapter 10, Perreau describes the challenge presented when the team leader and originator of the concept departed for another job in the company.

'We were fortunate that by this stage we had built up a robust team of competent people who knew the plant well and furthermore, had access to further human resources if they should need them.'

It is also interesting to note that their technical team was enhanced by an external consultant. Especially in the smaller companies with just a few people, there is a need to have external consultants brought into the team to provide the necessary knowledge and skills (Wilkinson and Palamountain, Chapter 7). But it is often difficult for the company to identify the knowledge needed and the people who may have it. Larger companies seem to be contracting out various areas of the PD Process, so that they, too, have the problems of selecting and coordinating external consultants into the team.

### 18.2.3    Establishing priorities

Too many projects are designated high priority. Regular reviews between marketing and technical will help to overcome this problem but a sobering determinant of priority is provided by setting a realistic timeline to launch as aligned to a feasible launch window. Many products are seasonal and can only be launched in designated months. Priorities must also be reviewed as part of the overall company portfolio management, which is discussed below.

### 18.2.4    Co-manufacturing

Co-manufacturing or contract manufacturing is becoming more and more prevalent as companies rationalise their manufacturing facilities and resort to outsourcing. It is important to observe all the key steps in the PD Process and to treat the co-manufacturer as an extension of one's own facility. It is tempting to assume that the co-manufacturer has all the necessary competence and expertise and to delegate care and responsibility. This can be a recipe for disaster. Many companies subject potential co-manufacturers to a comprehensive and demanding quality audit, which must be achieved before negotiations can proceed.

The level of co-manufacturing can range from relatively simple to more complex:

- label or packaging promotion, with or without add on such as leaflet or gift
- product supplied to co-packer for packaging
- existing product outsourced to co-manufacturer
- new to company product.

Here again, it is tempting to allow marketing to handle and coordinate simple co-packing projects, such as special pack promotions, directly with the co-packer. This also can be hazardous and it is recommended that all projects, even simple co-packs, are handled within the PD Process.

In Chapter 13, Nghee describes the development and launch of an innovative coconut product by an entrepreneur. Commercialisation required a suitable co-manufacturer as partner. In this case, she was an experienced food technologist and had developed similar processes, so she was able to specify the process and the product quality. Sorensen (Chapter 12) describes a project where part of the sauce making was in the company's own factory and the packaging was done by a contract manufacturer. This was a disaster because of contamination during storage and transport.

## 18.3    Company strategy and product development management

It is important for the technical department and the PD team to be aligned with company strategy. In this era of frequent company restructuring, mergers and acquisitions, the strategic direction and priority for PD can be a changing canvas. Priorities for a PD strategy must be derived from the business strategy.

Inwood (Chapter 2) discusses the importance of the strategic plan relative to multinational food businesses:

- The NPD programme should be aligned with and part of the strategic objectives of the company.
- The product development programme needs 'top management commitment' to be truly successful.

Examples of strategic priorities include:

- Growth by acquisition. This will involve the technical team in the acquisition process, review of products and technologies and subsequent integration of acquired businesses.
- Innovation and new products. An exciting and challenging strategy which calls for technological excellence.
- Product renovation and range extensions.
- Cost reductions. This will always be on the agenda but there may be a strategic direction calling for limited capital expenditure and improved financial return from the existing business.
- Rationalisation of manufacturing plants.

### 18.3.1    Project briefs

In planning, the company needs to develop a project brief and give a project approval to this brief. The most successful and effective brief for a development project will be developed by teamwork, cooperation between marketing and technical, and will be improved sequentially through the development process. Too often, technologists expect the marketing department to deliver a detailed accurate brief at the commencement of a project and then become armchair critics of marketing's technical ineptitude; not a good start to a project.

Technologists must strive to understand the consumer need and the marketing rationale and then marketers must cooperate in developing a technical brief or product design specification. In return, marketing should seek to understand the technical parameters associated with the desired product concept.

### 18.3.2    Project approvals

Figure 18.1 presents three key stages for project approval:

- A project should be approved before time and resources are devoted to a feasibility evaluation. However, where a positive relationship exists between marketing and technical, and for simple product improvement, it may be adequate to obtain the approval of the senior marketing executive rather than the approval of a full management committee.
- Management committee approval should be obtained to proceed to Stage 2, full development of the product and process. Approval will be in association with a formal brief for the project and this is a formal go/no go decision, or gate, as described by the Stage-Gate® process (Cooper, 2005).

- Similarly, management committee approval must be obtained to proceed to the industrialisation phase. This will involve the approval of the marketing and business plan for the product, complete with sales volumes, costs, budgets and timings. This is, obviously, the most critical review as from this point on the project costs will escalate exponentially as commitments are made for materials, packaging, equipment and marketing expenses.

### 18.3.3   Portfolio management

The progress and status of projects must be tracked and reported on a regular basis. Many companies conduct regular, say monthly, new product meetings, involving senior management, to review progress and consider proposed actions. These meetings may well be the forum for project approvals as discussed in Section 18.3.2.

The agenda for such project reviews should be an overall new product portfolio status report. The report should highlight key results on each project such as:

- progress/achievements
- constraints/problems
- timing issues
- financial issues
- recommendations.

The report should be the result of deliberations by each project team, prior to the meeting, with recommendations available for discussion or approval by management at the new product meeting. Meetings should not degenerate into prolonged and detailed discussions of every project, which should have been addressed by the project team ahead of the meeting. Priorities can be discussed, new projects can be introduced and stagnant projects can be deleted or deferred. Good portfolio management is essential to maintain a healthy, balanced and motivated environment for PD.

## 18.4   Stage 1 – feasibility

The manner in which feasibility studies are undertaken will say a lot about an organisation's culture, and ultimately, success in PD. The objective is to develop and promote a fertile and cooperative interchange between marketing and technical, aimed at identifying new product opportunities and facilitating an expedient and effective evaluation process. The process may be sequential and evolutionary and the best new ideas will evolve from a team effort with contributions from all parties.

This first stage is the opportunity to develop an idea, or concept, or opportunity with collaborative effort, to undertake a feasibility evaluation and to determine if it should proceed to full development in Stage 2. Not all projects will require a feasibility evaluation. PD briefs for line extensions or product reformulation can bypass Stage 1 and proceed directly to Stage 2. It is always difficult to determine the division between feasibility and full blown Stage 2 development as the tendency will be to overlap into the development phase. The feasibility stage should be taken only to the point where indications are positive towards further development, allowing the preparation of a formal brief. If continued too long, they may reduce the available active research time.

### 18.4.1   Innovation and can do attitude

A can do attitude is essential for success and the generation of true innovation. It is all too easy for technologists to receive a new product proposal with the best of intentions but then to undertake an evaluation that overwhelms them with the problems, the barriers, the negative issues. The first step for a technologist is to fully understand the proposal. Take time to meet with marketing and discuss the consumer need, the opportunity, and the concept. Discuss the technical difficulties with an experienced person in the company or a consultant. Adopt a positive attitude by understanding the rationale.

Bogue and Sorenson (Chapter 15) advise us that a market-oriented approach to innovation will be the most successful. 'The integration of the consumer with the NPD process can best be achieved at the pre-development stages of concept ideation, concept screening and optimisation.' In Chapter 16, Søndergaard and Edelenbos describe consumer research in the early stages of product development of a vegetable-based product for children, in which the results of the consumer research were used by the company personnel in developing product concepts and prototypes.

True innovation can often be generated at this early stage by collaboration and blending of the marketing concept with technical know-how in product design specifications, leading perhaps to technical breakthrough. How satisfying it can be for a technologist to be able to contribute a technical innovation, usually very simple, which suddenly takes the project to new possibilities. In Chapter 6, Woodhams demonstrates well the importance of technical knowledge in developing product design specifications and quality standards at a very high level as integral to the product and concept.

It is also important to keep an open mind and an adventurous spirit in this early phase of product development and formulation. Be prepared to experiment and test those product combinations or process variations which seem unconventional or defy theory. This may be the first step to a really innovative development. (Ashworth) Bowie, in Chapter 8, invented her unique oat bake product in her kitchen and testing with her children and then persuaded a company to take it from a kitchen concept to commercialisation. She offers the lesson: 'never be afraid to try off-the-wall ideas'; Ooraikul in Chapter 9 and Russell in Chapter 11 provide examples of technological processing innovation which led to new products.

## 18.4.2   Key steps

*Concept development*

Take time to understand the concept. Research competitor products, both local and overseas, and use these as a benchmark of what can be achieved. Use these benchmarks to expand the concept and search for that elusive technological innovation. Be creative. This is an important time to develop new ideas. Develop the product, process and packaging parameters to a point where a bench scale or kitchen prototype can be demonstrated. This was a very important step in the development of (Ashworth) Bowie's oat bake product (Chapter 8). Remember that this is a feasibility study and it is not necessary to develop the final product. The kitchen and laboratory trials are developing alternative ideas.

*Factory capability*

It is necessary, at this early stage, to determine if existing production facilities are suitable and available to manufacture the new product ((Ashworth) Bowie, Chapter 8). Some new equipment or enhancement to existing equipment may be necessary and, if so, then an estimated capital cost must be provided. A totally new process may require substantial capital investment or if money is not available, old equipment can be formed into a processing line (Wilkinson and Palamountain, Chapter 7). For a small company, avoiding or minimising capital expenditure may be a prime objective and creative improvisation can be very rewarding. But technical expertise and adherence to the principles of the PD Process are essential or these compromises can lead to failure, as demonstrated by Sorensen in Chapter 12. Alternatively, the new product may be a candidate for co-manufacture through a third party.

A preliminary process evaluation will be a major part of the feasibility study. It is important to remember the objectives of a feasibility study. Maintain a positive attitude to a proposal rather than listing all the reasons why it cannot be done. Capital cost estimates are preliminary and should indicate an order of magnitude only, plus or minus 25 percent.

*Product safety*

The integrity, shelf stability and safety of a product are essential considerations of a feasibility study. If existing preservation techniques are employed then there will be well-established criteria available for product safety. On the other hand, if new processes or technologies are being developed then consideration of product safety will be a critical part of a feasibility study. For example, in developing a chilled food using multiple hurdle technology, the establishment of a safe shelf life may be the critical determinant of a viable new product, requiring challenge testing involving inoculation of the potential spoilage organism to validate the stability and safety of the formulation and process. Hurdle technology utilises the combination of several food preservation techniques, such as temperature, water activity, pH, acidity, preservative, to provide shelf life stability to a food. (Leistner, 1995). Sorensen (Chapter 12) provides a case study of failure in the packing of single serve tomato sauce sachets after a process change. This could have been avoided with a HACCP study and, perhaps, a challenge test after the process change. But it was a small company with inadequate funds and knowledge and did not recognise the possibilities of contamination. A strong recommendation for courses in food safety!

*Regulatory issues*

Product development technologists have become ever more fluent in the interpretation of food regulations. A new product concept must be assessed with regard to its acceptance within the legal framework. This may extend beyond a product formulation to the claims that marketing would like to accompany a product, such as, health claims or nutritional attributes. A feasibility study should include clear direction on what is permitted and the limitations imposed by food law. Disregard of the need for animal testing of pet food can cause serious marketing difficulties (Wilkinson and Palamountain, Chapter 7). With the increasing use of health claims in human foods, the need for recognised medical testing has become urgent (Woodhams, Chapter 6).

*Cost parameters*

Product costings can become an area of tension between technical and marketing. It is important that marketing identify the price range early in the product and process development. It is not unusual for marketing to set unrealistic cost targets which translate to a residual cost of zero for ingredients after accounting for all other costs. A feasibility study is the appropriate stage to establish realistic cost parameters. Take the time to explain product costings to marketing and explore various options to achieve viable cost parameters. If it is not possible to reduce product costs or increase selling price by marketing to different groups (see Wilkinson and Palamountain, Chapter 7), then it is better to face a no go decision now rather than at Stage 2 when considerable work has been devoted to the project.

*Timings*

This can also be an area of tension. Products sold via retail supermarkets are subject to fairly rigid seasonal launch windows and meeting the next launch window often becomes a marketing imperative with consequent compression of the development phase. Project management techniques, identifying the critical path, and critical steps, are a valuable help in planning (Side, 2002). For relatively simple development projects such as line extensions, the critical path may well be determined by the label and artwork development, which will fall into marketing's responsibility. This can come as a bit of a surprise and reality check to marketing after they have insisted on a fast-track timing. For more complex projects involving process changes or new equipment, it is essential to establish a realistic timetable at the outset. Where there is pressure to market without adequate storage tests, the product may deteriorate and be forever a failure. Sorensen (Chapter 12) describes the failure due to spoilage of a single serve sauce product which could have been avoided if the storage trials had been completed. A costly short circuiting of procedures

### 18.4.3   Reviewing the brief

The final step in a feasibility study is for marketing and technical staff to review the project and agree on a formal brief for Stage 2. The brief will accompany a report and recommendation to management regarding the project proceeding to Stage 2. This is a go/no go point or gate.

## 18.5   Stage 2 – development

Given the green light by management the 'work horse' phase commences. Project management skills, matched with technical expertise and consumer/ market knowledge will lead to success. Nghee (Chapter 13) describes the challenges this presents an entrepreneur, requiring the cooperation of others and expert advice. 'The would-be entrepreneur cannot do it alone.'

The key steps discussed below include comments for product, process and packaging development which are very much integrated to deliver the desired product.

### 18.5.1   Key steps

*Product development*

One of the first tasks is to produce prototype samples with continual evaluation and formulation improvement and consumer testing. In many cases this step can be performed on a small scale in a test kitchen or laboratory but depending on the proposed process it may involve pilot-scale development. Some of the lessons to be observed are:

- Be sure you have the basic technical and scientific knowledge (Sorensen, Chapter 12).
- Have objective testing of product qualities (Sorensen, Chapter 12).
- Maintain good records of all prototypes with notes of evaluation results.
- Hold samples of prototypes.
- Always include a control or reference formulation in a series trial.
- Do not be afraid to try adventurous formulation or process options even if these appear to defy technical expectations. Innovation will often arise from the unexpected.
- Maintain a good relationship with marketing in this repetitive phase of trialling, evaluation and retrialling. Remember always that the final arbiter is the consumer.

*Process development*

The challenge in PD is very often to utilise existing equipment and processes, thereby minimising capital expenditure and lead times. It is important to decide early in this phase whether an existing process will deliver the product objectives. Excessive compromises on process can cause or contribute to new product failures. It may be a wise business decision to invest in a new item of equipment or to rejig a process in order to deliver the right product to the consumer. The later product prototypes need to be checked to see that they reach the technical standards and also have consumer acceptance. The effects of the processing conditions on both of these need to be tested.

For products new to the company, the process development phase may represent a major component of the PD Process requiring progression from laboratory scale to pilot scale to commercial scale up. Both Perreau (Chapter 10) and Russell (Chapter 11) show us the importance of examining the key unit operations in the process and then building these into the complete process. This will be a time-consuming process requiring testing and validation at each stage. It is important that adequate time is allocated in the project timelines to cover this demanding development.

In all processing development, the importance of the quality and costs of the raw materials must be stressed (see Sorensen, Chapter 12). This phase of the development process is also well illustrated in Chapters 10 and 11.

*Packaging development*
Food packaging has become an ever more technical and exciting field, leaving no doubt that the package is an integral part of the product presentation. Maintain a comprehensive audit and review of competitive products. Talk to suppliers and seek their contribution on new developments. Packaging innovation can provide that breakthrough and competitive edge in the market place. The protection of the product by the package needs to be studied and assured ((Ashworth) Bowie, Chapter 8 and Wilkinson and Palamountain, Chapter 7).

*Capital expenditure*
The process development phase will identify capital expenditure requirements. Depending on its magnitude, it may represent a key item in the business proposal to management at the end of Stage 2. The following need to be approached for possible costs: equipment suppliers, plant engineers within and outside the company, and builders if new premises are planned. Approximate costs need to be considered at this stage.

*Factory trials*
For a technologist coordinating such a project, it is important to enlist the relevant factory personnel, the factory manager and the factory engineer, into the project team at an early stage. The factory must take ownership for a new process and must be allowed to drive that aspect of the development. The penalty for trying to fly alone will be a factory reluctant to cooperate and too ready to criticise and observe failures.

Trials at this stage may be required as part of a process scale up, to validate a process or to produce samples using factory equipment and facilities. This was a key step in the acceptance of a new low temperature extrusion process for ice cream as described by Russell (Chapter 11). Here again, it is important to have the factory accept responsibility and ownership of the new product. A briefing, or new product meeting, may be the best means of selling the new product to the factory. Even though such meetings can be a lively event, it is essential to involve factory operatives and to enlist their help and suggestions. Factory trials will also allow the production of more realistic samples as compared to earlier prototypes and it is these samples that, if possible, should be used for consumer research.

*Consumer research*
The consumer research at this stage of the PD Process will be a key determinant of whether to proceed. From a technical perspective it is important to provide consumer research samples that are truly representative of what will be produced commercially. Too often technologists will be criticised because the commercial samples do not measure up to preliminary samples provided for consumer research. Of course, the challenge can be presented in reverse: the commercial process should be developed so as to produce a result that at least matches development samples. The consumer tests increase in magnitude at this stage, and indeed with some products test marketing can be used.

*HACCP*
Every new product or process should undergo a HACCP evaluation (Wallace and Mortimore, 1998; Mortimore and Mayes, 2001). If the product is a range extension, for example, the HACCP study will be a simple extension from the existing range but it needs to take into account any new materials or process variations (Sorensen, Chapter 12). With a product new to the company, a comprehensive HACCP study must be undertaken with clear identification of control points. Product safety is at all times paramount and cannot be compromised. The technical sign-off which is discussed below requires completion of a HACCP study.

*Keeping quality, storage, transport trials*
For a product new to the company this part of the NPD process can represent the critical step in the critical path (Wilkinson and Palamountain, Chapter 7). It is, therefore, important to establish a keeping quality test protocol as early as possible. Some accelerated storage techniques can be employed but must be interpreted with care. For new varieties or range extensions, it is often possible to use data from the existing range to establish keeping quality specifications but one must be alert to the effect of new materials or process changes. Sorensen (Chapter 12) provides examples of the risks in ignoring or short circuiting this phase of the development process. Transport trials should be conducted on the total package, that is, product from factory trials plus proposed packaging format.

*Costings*
Product costing at this stage will be the basis of financial projections included in a recommendation to proceed to Stage 3 and product launch. Care must be taken to ensure that realistic and achievable cost parameters are employed. Reliable

costs must be established for new ingredients and packaging. Avoid nasty surprises at Stage 3, when commercial quantities are purchased for launch. Too often, estimated costs are used for costing based on the prices for similar materials currently being purchased, only to find later that the new material commands a considerable premium. This can destroy a new product's viability. Processing and distribution costs are often affected by production levels, so costings should be done at different outputs. Overheads can be the Achilles heel; be sure to know what part of the overheads the new product is to bear. Product costs become the responsibility of the factory and they should be involved in their establishment. Discrepancies in costings are often the subject of post-launch evaluations.

*Regulatory issues*

Unfortunately, development technologists today have to become increasingly competent as food regulation lawyers. Innovation in product development can become frustrated by food regulations. Nevertheless, it is required to establish conformance to regulations and this can extend beyond the information on labels to the area of claims, such as health claims. A good approach in resolving regulatory problems is to workshop the issues with colleagues and other food regulation practitioners. This can be very fruitful in finding solutions. If wishing to have health claims for the product, be sure to determine the testing requirements.

*Nutritional data*

The requirement for nutritional data on labels is becoming more demanding and, in many cases, mandatory. Preliminary data can be compiled by estimation from a variety of sources, including compositional tables and analysis of similar products in a range. However, data should be confirmed by analysis of factory trial samples and then reviewed post launch. Woodhams, in Chapter 6, describes the extraordinary care and effort required in developing and commercialising a nutriceutical product which, in effect, has to comply to pharmaceutical-type standards.

*Label development*

This is often coordinated by marketing but will require considerable technical input on such items as product name, ingredient list, nutritional data, country of origin, weight or volume, cooking instructions, recipe suggestions, and claims. Declarations with regard to allergens have become a very important issue and will require appropriate labelling. A formal approval system should be implemented to allow a label to be developed through the drafting stage and final sign-off. Label development often represents a critical path step and, therefore, will receive pressure for quick approval. The importance of careful scrutiny of all label items cannot be stressed enough. A rushed approval can lead to launch delays when an error is detected later or, at worst, a recall for a breach of regulations or consumer information.

### 18.5.2   Project management

Technologists need skills in project management requiring, amongst other things, good record keeping, project planning, problem solving and excellent communications. Planning and monitoring of timelines will be a constant task. In addition, a technologist will often be managing not just one project but a portfolio of several projects. Use of a project management tool such as Microsoft Project will prove invaluable, particularly for the preparation of Gantt charts (Wallace and Mortimore, 1998; Mortimore and Mayes, 2001).

### 18.5.3   Sensory evaluation, consumer research and reference samples

It is important to always be able to evaluate a product against an appropriate reference. Initial concept research may have been conducted on a kitchen scale concept sample during Stage 1 then, as development progressed to Stage 2, pilot scale and then factory trial samples are produced. Sensory evaluation will be conducted to ensure that the product retains the essential and desirable characteristics identified initially. The reference sample should, in this way, proceed to an approved factory trial sample. This becomes the reference and the target for future production. Sufficient quantities of the reference sample should be retained for subsequent evaluations and stored to establish that there are no changes in product qualities.

Goldman and O'Neil (Chapter 14) discuss the dangers of product degradation resulting from sequential reformulations for the prime objective of cost reduction. 'This situation can occur when the sensory researcher does not fulfil a trusted advisor role.'

### 18.5.4   Technical sign-off

Before a project can proceed to Stage 3 it must receive a technical sign-off by the senior technical manager. A check list can be prepared detailing the key technical parameters with particular emphasis on product safety, HACCP, legal conformance and cost standards. The senior technical manager must be satisfied that the product can be produced commercially and meet the appropriate standards. Adherence to such a sign-off procedure will prevent a project being pushed through to Stage 3 prematurely or lacking resolution on a key issue. The costs escalate once a project moves to Stage 3 and severe penalties can be incurred if a project is aborted at this later stage.

## 18.6   Stage 3 – industrialisation

Management have given the go ahead for a launch and the adrenalin can really start pumping now. Many more parties become involved: marketing will work more closely with sales, and all the manufacturing functions, including planning and procurement, are fully mobilised.

### 18.6.1    Who is responsible?

Project management responsibility remains with the project team. As previously discussed, certain projects will employ a cross-functional team, under a team leader and this team will retain overall coordination. Smaller projects may be handled by a marketing/technical liaison working as a team.

The technologist's role throughout this industrialisation phase is best described as technical custodian of the product, providing all relevant departments with the required technical data and ensuring the technical integrity of the new product. As the project moves into Stage 3, it is important, or imperative, that each operational function takes responsibility for its respective part of the new product introduction. So, whilst the project team maintains coordination and provides all reference points, the factory must take responsibility to manufacture the new product in accordance with the standards established. As mentioned in Stage 2, the factory will have been briefed early in the project and will have been encouraged to accept ownership of the new product well before it receives management approval to proceed to launch.

### 18.6.2    Key steps

*Planning and procurement data*

Marketing have the task of providing sales forecasts for the new product together with anticipated take-off schedules. These are usually generated in the latter stages of Stage 2 so as to allow planning of production and launch but they must be continuously checked and updated as information becomes available. One would hope that the variation in predictions becomes less as time goes on.

New ingredients and packaging can represent critical path items, particularly where materials are imported. Artwork for labels and packaging must be finalised and this often causes delays. Sales forecasts at this early stage of a launch are notoriously unreliable. From a manufacturing perspective, the factory must be committed to meet sales demand but, at the same time, care must be exercised not to procure excessive quantities of unique raw materials which may end up being written off if sales forecasts are not realised. A positive partnership with suppliers should prepare for fluctuations in demand, up or down. Allowing for smaller than normal order quantities can be an effective means of managing initial sales demand.

*Recipe and process documentation*

As technical custodian, the technologist must provide recipe, packaging and process documentation, including quality assurance and safety procedures, to allow the factory to manufacture. These documents should be dynamic in the sense that they will evolve and be updated as they progress through the journey to commercial production. Objective testing should be used whenever possible (Sorensen, Chapter 12).

*Raw material specifications*
All new materials, ingredients and packaging, must be supported with specifications. Typically, the technologist will liaise with suppliers and the quality department to establish these standards and review them as required. This is a real problem for the small manufacturer who buys in small quantities, often through agents, and cannot always obtain adequate specifications.

*Quality specifications*
In-process and finished product specifications must be established together with tolerances. For range extensions, this task is facilitated by reference to existing standards of other varieties in the range. For a totally new product, draft specifications must be established and may need a greater level of review and updating as it progresses through to commercial production. Critical parameters should be highlighted and preserved. Care must be taken to avoid any erosion of the product characteristics of importance to the consumer.

*Factory trials*
The objective of a factory trial is to produce an acceptable product under commercial scale manufacturing conditions prior to proceeding to full-scale production. This is preceded by careful planning and preparation. If the process involves new equipment, or a new line, then considerable testing and commissioning will be completed before attempting production trials.

This is well demonstrated in Chapter 10 (Perreau), where a new and innovative cutting process for a confectionery product, delivered a significant breakthrough but required extensive work and dedicated effort. The time required for such a phase should not be underestimated.

'Some tasks can take very much longer than initially predicted. One particular task on the design of the computer software for the full-sized plant was thought to take an afternoon. It actually took two man years! Be prepared!'

The commercialisation of a low temperature extruder for ice cream, described by Russell (Chapter 11), took several years in different countries, as his company was multinational.

Factory trials should be conducted and managed by the factory with active coordination, support and supervision by the product development technologist. Factory personnel must be fully briefed on the new product and this is best undertaken by a pre-production meeting where all aspects of the product and process are presented. A reference sample must be available which will most likely be from a Stage 2 factory trial or pilot-scale trial. Critical quality attributes, essential to consumer acceptance, should be highlighted to allow the factory to fully appreciate the importance of delivering a winning product. Consumer research may have identified an attribute such as colour or texture to be critical (see Bogue and Sorenson, Chapter 15). Product safety is non-negotiable and critical control points should be identified. Recipe and process documentation, together with quality standards should be prepared for a trial, even though this may be interim or preliminary data, which will be reviewed and updated subsequently. Invariably, the launch timetable will be on a very tight schedule and optimists will prevail. Reality is that unexpected problems will arise and factory trials may need to be repeated, often several times, until a satisfactory commercial production can be assured. Provision for repeat factory trials in the production schedule is recommended. But production trials should be planned so that experimentation is not needed, although fate may determine it otherwise. There is nearly always a factor that is forgotten – often minor but one that upsets the processing.

*First production*
The technologist's role, as technical custodian, will be fully realised at this exciting stage of first production. The factory now has responsibility to deliver the new product for launch according to agreed manufacturing specifications. The technologist must monitor production and variation of process variables and of raw materials and be alert to any unexpected problems or quality variations. Sorensen (Chapter 12) highlights a new product failure due to quality variations in commercial production. The role will be a roving commission, observing the first-off production, checking the process variables, visiting the quality control stations, evaluating samples for conformance at frequent intervals, liaising with the factory supervisors and responding to the technical enquiries as they arise. Marketing personnel are likely to be eager to witness this first production and the technologist can proudly act as tour guide.

### 18.6.3    Handover to factory
The technologist should remain technical custodian for the life of the product. However, a formal handover to the factory, subsequent to one or more satisfactory production runs, is a sound procedure. Samples from these early runs must be fully evaluated, compared to reference samples, and approved. Such approved first production samples become the new reference. Specifications and quality standards need to be reviewed and updated in the light of commercial experience. Process issues must be discussed with production and engineering and corrective actions formulated. A formal handover can then be implemented with the factory's consent.

## 18.7    Problems met and lessons learnt

*PD Strategy*
- PD Process understood.
- PD strategy aligned with company strategy.
- Priorities reviewed as part of the overall portfolio management.
- Project approved before time and resources are devoted to a feasibility evaluation.

*Development of product and process*
- Procedures observed but there is scope for flexibility in their application.
- Technologists understand the consumer need and the marketing rationale and then assist in developing a technical brief.
- Marketing understands the technical parameters associated with the desired product concept.
- The factory involved and included at an early stage in development.
- Management committee approval obtained to proceed to full development of the product and the process.
- Hard work applied to develop and maintain trust and positive relationships between marketing and technical.

*Industrialisation of product and process*
- Technical sign-off obtained to proceed to the industrialisation phase.
- Management committee approval obtained to proceed to the industrialisation phase.

*The technologist's role*
- Retain the role of technical custodian for the product.
- Maintain a positive can do attitude.

## 18.8   References

COOPER RG (2005), *Winning at New Products*, 3rd edn, Reading, MA: Basic Books.
COOPER RG (2006), Formula for success, *Marketing Management*, March/April, 15 (2), 18–22, American Marketing Association.
EARLE M, EARLE R and ANDERSON A (2001), *Food Product Development*, Cambridge: Woodhead.
LEISTNER L (1995), Principles and applications of hurdle technology. In Gould GW (ed.), *New Methods of Food Preservation*, London: Blackie Academic and Professional.
MORTIMORE S and MAYES T (2001), *Making the Most of HACCP: Learning from Other's Experience*, Cambridge: Woodhead.
SIDE C (2002), Food product development, Chapter 3 in Kramer F, *Managing the Product Development Process*, Oxford: Blackwell Publishing.
WALLACE CA and MORTIMORE S (1998), *HACCP: A Practical Approach*, 2nd edn, Maryland: Aspen Publishers Inc.

# 19

# Consumers, customers, marketplaces and markets in New Product Development – a roadmap through marketing and launching in product development

Gordon Fuller, Canada

*This chapter studies the place of marketers and marketing in the Product Development Process (PD Process). In particular, it is concerned with ensuring that knowledge of the targeted consumers and customers is available and used at all stages from the development of the product concept to the final study after launching, and that marketing is designed based on this knowledge. Marketing departments develop the final descriptions of the targeted consumers, customers and their needs and wants; identify the marketplaces and the distribution methods; design the advertising, pricing and selling methods; and develop the market plan for the launch and introduce it to the companies.*

*The companies can be developing and launching products to different markets – consumer, food service and industrial; and in different market areas and in different countries and regions. All of these will need specialised marketing methods, so the marketing study in the PD Process can be extensive. It is important that it is coordinated with the technical product development described in Chapter 18, so that the new product can be developed efficiently, effectively and successfully.*

*This chapter predicts the changes that can occur in the future in consumers and their needs and wants, the marketplaces, the technical developments, and government regulations. It suggests how market research, and also the methods of marketing new products, need to change. Close integration of marketing into the PD Process offers the maximum probability for harmonious and comprehensive development, and thence for product success.*

*This chapter particularly relates to Chapters 5 and 6 in* Food Product Development *by Earle, Earle and Anderson.*

'. . . all exact science is dominated by the idea of approximation. When a man tells you that he knows the exact truth about anything, you are safe in inferring that he is an inexact man.' Bertrand Russell

# 19.1    An introduction: what roadmap to successful new products?

No roadmap exactly describes a route to a goal: exhortations to follow '10 Steps to a Successful Product' or use 'Smith's Method to Do Successful Products' are, therefore, all approximations. Roadmaps do not describe the actual journey, only the general direction. The journey itself is subject to changes caused by weather, road conditions and human errors in map reading. The path to product development is beset by skill levels, personalities, and company structure. New Product Development (NPD) is not transactive, i.e., based on rigid procedures: NPD is transformative, subject to change and flexible according to events. NPD displays an entrepreneurial flair in the application of the disciplines of the 'soft' (consumer research, psychology, and marketing) and 'hard' (the physical) sciences.

### 19.1.1    Directions and destinations
Goals vary in product development. Broadening an existing line of soups with new flavours is a fairly mundane technical and marketing task. To improve an old tired product with a new, improved one is again a task without too much risk or difficulty (Fuller, 1994: 65).

Much riskier is the introduction of a new product for tactical marketing reasons. That is, a product is introduced to establish a marketplace position against a competitor's product with an established shelf position. This challenges the competitor's product and shelf position: the competitor can be expected to retaliate. Such an introduction represents enormous risk with heavy advertising and promotional costs.

Whose needs has the marketing department described? Who is targeted? How well has the product concept been defined to satisfy these needs? Where is the NPD team going? Product development demands knowledge of who the targeted customer or user is; what they like and want and how well the product satisfies their likes and wants. Also important, as a follow-up to the launch, is knowing where customers purchase the product and where and when they use the product. For example, liqueur-flavoured coffees are purchased in cities usually in upscale marketplaces but often used on weekends at *après-ski* occasions or at cottages. Such awareness of use focuses marketing and advertising strategies.

The buying public are fickle: their identities, behaviour, likes and dislikes change with the next popular market craze. Successful development depends throughout the entire development process on accurate, real-time knowledge of how markets are changing and how marketplaces are responding to these changes. The development team must be aware of:

- new ingredients with unique properties
- advanced technologies that permit the manufacture of products with novel attributes
- the latest nutritional and nutriceutical information to enhance a product's appeal
- customers and consumers and how they are being influenced in marketplaces.

Any one of these developments can start a market craze attracting the buying public. Awareness of market trends requires competent research of all markets and marketplaces to discover what is going on with the competition, retailers, customers and consumers. Research is required on how local and regional populations are changing by numbers, by ethnicity, by age, and by income, valuable information in selecting test market areas. Any segment of the public for whom a new product is designed must be characterised precisely so that the needs and wants of this segment are met with products satisfying those needs. Much more sophisticated market research data are required today than the per capita food consumption data or case movements of food product categories between warehouses in vogue at one time by market researchers. Such statistics tell nothing except that product is disappearing or being moved about. Sometimes product is deliberately moved about to create marketing misinformation.

Successful product development obviously starts with the destination, i.e. the customer or consumer: the customer/consumer interface dictates the path, not some ordered set of steps.

### 19.1.2   Signposts along the way: consumers, customers and marketplaces

Customers, consumers, markets and marketplaces must be understood clearly by the marketing department and studied carefully for the signposts that they are. Customers and consumers are different entities, a distinction marketing personnel must recognise (Fuller, 2005: 10–11; see also Gibson, 1981). Currie (Chapter 17) correctly recognised this distinction using the terms primary and secondary customers to distinguish customer and consumer. She used the distinction in the successful development and retailing of in-house brands.

The marketing department must clearly identify for whom their new product is intended, i.e. its targeted segment of the public. They create, through their research, a product concept for the technical team to make real. They then position their advertising, promotional material and media vehicles to communicate the message about this new product and how it satisfies the needs of the targeted public.

A customer is one who buys and therefore acts as a gatekeeper for consumers. For example, a mother buying for her family, a purchasing agent buying for a canteen, an institution, or for a food manufacturing company, and a chef buying for a restaurant are customers and gatekeepers. The mother, the purchasing agent and the chef control what is purchased; they buy to standards and specifications required by their consumers. The mother buys according to her family's likes, dislikes, allergies and her budget: the chef purchases according to the style of cuisine customers demand. Gatekeepers cannot always follow these standards. Factors influencing their decisions are:

- pricing to fit within a budget
- availability of products in the marketplaces
- standard of quality of available products in the marketplaces
- and reliability of suppliers. That is, brands they are comfortable with.

The latter two criteria, consistency of quality and reliability of supply, are important considerations for industrial purchasers.

Customers are also consumers. They consume the food purchased as is attested by the vast array of finger-food items available for eating on-the-run. But a consumer does not always eat the food. The industrial consumer is the fabricator of another product in which the product purchased is used. This altered product may then be sold to another customer for use by that new consumer, who uses the altered food item in yet another food product and so on in a complex chain. Thus we see a clear distinction between consumers; there are industrial consumers and non-industrial consumers and hence industrial customers and consumers. Market research must distinguish between the distinct needs of non-industrial customers and consumers and between those of industrial customers and consumers. When the needs of these different groups are defined clearly by marketing research then the development team can design a product to fit those very different needs.

Marketing personnel recognise customers and consumers as different targets. Customers forage the marketplaces. They must be communicated with. Customers, as they forage, are deluged with advertising and promotional blandishments to buy at a price and quality that they are willing to pay. The marketer's message must communicate and sell their product where the forager is. Customers, however, are targets of competing companies with similar products.

Industrial customers forage in different marketplaces using different foraging techniques: they forage in trade magazines and at trade fairs where ingredients and industrial goods are displayed and demonstrated in finished products. Industrial customers are served by a specialised sales force, technical sales representatives.

There is an interesting distinction between industrial consumers and non-industrial consumers. The industrial consumer is more likely to be the forager than is the non-industrial consumer. Purchasing agents are industrial customers but buy ingredients and raw materials from sources foraged by, and according to the standards prescribed by, industrial consumers. When these specifications cannot be met by suppliers except as a high cost made-to-order speciality item, then purchasing agents decide between suppliers based on the closest-to-demanded specifications, the most reliable and consistent quality standards, a dependable delivery schedule, and the approval of their internal consumers. All must be done within established budgetary guidelines.

Consumers are the targets of advertising campaigns which extol different values, e.g. usage, taste and flavour superiority, or crispness, smoothness, etc. They are looking for an 'experience'. They are often not concerned with budgetary restrictions, for example young children shopping in tow with their mothers. Children make their wants (specifications) known to their mothers, the gatekeepers. Similarly, culinologists (a new breed of food technologist) make their requirements for ingredients and raw materials known to their company's purchasing agents.

Two case studies, Bogue and Sorenson (Chapter 15) and Søndergaard and Edelenbos (Chapter 16), describe consumer research techniques. In both case studies the authors have been successful in coming to conclusions about product concepts. We do not know whether successful products came of these studies, whether a successful market was developed or what competitive forces were at play in the marketplaces that might have influenced customers and consumers and forced changes to the original concepts. There was no separation of customer (the gatekeeper) and the consumer (user) but Søndergaard and Edelenbos did interview parents (gatekeepers or customers) and children (consumers). We do not see the NPD process in its entirety. We are treated only to successful development of the concept and not to a successful launch with repeat purchases of a new product.

The terms 'markets' and 'marketplaces' need clarification. A marketplace has a distinct geographic location (but compare e-tailing). It is a corner grocery store, a supermarket (which may have many different mini-marketplaces within it), a farmer's roadside stand or a trade fair; it is a tangible, locatable thing. A market is not tangible; it is conceptual as in 'there is a market for organic produce' or 'the quick-serve food market is growing' or 'fair trade coffee is an environmentally safe and profitable market'. A market represents an opportunity for innovative products to be designed, crafted and sold to the correctly targeted customers wanting such a product. The products are purchased in an appropriate marketplace. Often it is the marketing department of a company that, through its research of the buying public's habits, discovers a hidden need or desire. Marketing personnel did not create the need, they uncovered something (a market) that the public realised they had always been looking for. A marketplace is the place customers go to buy and satisfy that need.

The creation of products or services for niche markets and larger publics will be sold in the marketplaces where the targeted customers are to be found. The customers will then supply these to consumers. Retailers, therefore, are a unique form of customer/consumer interface. They buy, display or prepare for display (a form of using) and purvey to other customers.

Sales and marketing are two different functions within the company. Sales departments carry out the tactical programmes established by marketing departments. They sell to purchasing agents, meet retailers, review effectiveness of the advertising and promotional materials and campaigns produced by their marketing departments. Sales personnel are in the marketplaces: they provide feed-back about their customer contacts and about retailer expectations.

Technical sales representatives meet with industrial consumers at trade fairs or at demonstrations in potential customers' factories.

### 19.1.3   Changes and obstacles

PD Processes are conducted in a dynamic and energetic environment that is constantly and erratically changing. Competing products challenge the interests of customers and consumers. The public follow fads in foods, in personal health regimes; they self-medicate with over-nutrified products or take up the latest diet-of-the-day pushed by some movie star or doctor. Fad diets provide opportunities for short-lived new products satisfying the fad but these will be replaced by the next popular diet. Knowledge of health, nutrition and toxicology changes suddenly and dramatically with new discoveries in science – coffee has been both good and bad for drinkers twice over in my lifetime, as have eggs and egg products, and eggs are now touted as beneficial in weight reduction. New knowledge and technology make available new foods, new ingredients and combined with the rediscovery of new 'old foods', cf. quinoa, amaranth and spelt, provides opportunities for products with unique properties.

Agricultural catastrophes or bountiful harvests drastically change the availability and costs of raw materials or ingredients. Governments can espouse agricultural or economic policies greatly affecting availability or pricing within the food industry: this Edwardson and Best (Chapter 5) found out with the collapse of their cassava project. Governments on the advice of experts and activists revise food legislation, their regulations and food standards often abruptly. Such disruptions throw established products and new product creation into disarray. Development and introduction of a new product in the real world does not take place in a vacuum.

Food companies competing for shelf space within the marketplaces change through mergers, acquisitions (see Ooraikul, Chapter 9), or failures. With these changes, the marketing objectives and strategies of all food companies in the marketplaces change. Directions change.

### 19.1.4   Errors inherent in 'roadmaps to success'

Obstacles on the 'roadmaps' to a successful product are never foreseeable: similarly a lock-step path to product development (the roadmap approach – turn here, turn there, go so far before turning) can not foresee construction road blocks, congested traffic areas, and detours encountered en route to a destination. There are human errors. With different targets, different means of communication and different market research technologies, consumer and industrial products cannot follow identical developmental procedures nor fixed-step methodologies. Development of successful short shelf life products, shelf-stable products and frozen prepared products are vastly different in their execution, marketing and distribution. They cannot follow similar steps for new products.

Examples of fixed steps to reach a successful product can be found in Mattson (1970), Skarra (1998) and Mansky (2006). Following them tells only what ought to be done to get to a destination. The steps cannot be applied to all product types without major modification. For example, steps in the development, marketing and launch of products with short shelf lives such as dairy drinks, baked goods, or fresh salads of mixed greens, etc., are very different for each of the products mentioned yet all are short shelf life products. Likewise, the development marketing and launch processes between these short shelf life products and products such as frozen prepared meals, REPFEDs (refrigerated processed foods of extended durability) and shelf-stable products are vastly different.

Retailers face environmental and handling concerns for short shelf life new products: these are also concerns for the developers. Where glass and aluminium containers must be returned for refunds and perishable product disposed of in a sanitary and environmentally safe manner, there will be extra costs that are applied to the product's launch. Many short shelf life products require highly efficient and costly returned-goods collection systems.

## 19.2   'Success' or 'failure': understanding the purpose of marketing and launching

Who dictates what is, or what conditions define, a successful new product launch with repeat purchases? Whether or not a new product launch has been a success is a question that has always confused me. From vantage points gained as an academic lecturing on New Product Development, a vice-president of technical development for an international food company, and a technical food consultant to many companies undertaking NPD, I have seen different interpretations of 'success' and 'failure' in new product launches. What I saw as a successful launch, a promising future or a purposeful positioning for a particular product, the company introducing the product took as a failure. Vice versa, obvious failures to me, were flogged persistently.

### 19.2.1   Success or failure?

A canned plum pudding made very modest sales as a seasonal item for one of our companies. It was easy to make on the company's equipment: it didn't interfere with other production schedules. Indeed, it kept production and skilled staff active in an off-season period. It was modestly profitable. It didn't use up great advertising monies and kept the company's logo in the customers' eyes over the festive season. In short, it successfully served several purposes and, to me, had no drawbacks. Management culled it solely because they had too many SKUs (stock keeping units) and felt it took too much of their sales force's time to push.

The financial department, an integral part of both the management and product development teams, are often a deciding factor in whether a product is considered a failure or not. If they determine during test market that profits are not being or will not be generated fast enough to justify continuing, they decide that the product is a failure and drop it. It is the short term return on investment that determines success or failure in the financial department's eyes. Many companies demand quick profits in as short a time as possible. Therefore new products are pulled, much to the annoyance of retailers who made space for them and of customers whose needs were satisfied and desired these products. With a longer-term view, a very successful product may have flourished.

Tweaking a product (i.e., constantly making little improvements in it) while it is still in test market to improve sales, is disastrous. It is expensive: it confounds the whole purpose of test marketing and confuses the results. For which product are the sales figures valid? The original one? Or the final tweaked product? After tweaking, the product may no longer resemble what the promotional material promised it to be and deliver. Customers are confused. If tweaking is necessary, a newly launched product should be pulled immediately.

A new product, a condiment sauce, much favoured by management, was launched into a test market. First, it lost colour in test market; artificial colours were added. This, of course, necessitated a label change. Then it was found through a flood of customer complaints that one of the artificial colours selectively broke down in the sauce medium leaving a most bizarrely coloured ring around the neck of the container. The project should have been aborted at the first sign of instability – actually it should never have been released for test market without a thorough and rigorous shelf stability test. However, the product is still on the market. Success or failure?

A distinct difference must be made between tweaking a product and product maintenance. By product maintenance I mean a systematic plan to make improvements in or variations of an established product that is perhaps in decline. It provides an extension of the life of a product. Product maintenance should be a critical part of the development process. It answers 'what will we do next with this product?'.

A product may be launched, as mentioned earlier, for purely tactical reasons. It is out in the marketplace to counter a similar competitor's established product. Whether the product is profitable is immaterial: it is a statement, a tactical move. It maintains brand awareness in that product category. Success or failure is determined by whether the overt challenge serves the company's tactical purposes in the marketplaces where the challenge was made.

### 19.2.2   Bumps and clanks along the road to marketing and test market

Any journey is filled with surprises. My first surprise came with 'slotting fees' in test marketing. Retailers demand fees for making space on their shelves for new products. If the retailer makes space for an untried new product, then some established, proven product must be displaced. This displaced, established product brought in a reliable profit per running foot of shelf that will be lost to the retailer. This lost income must be repaid to the retailer. Therefore, the company launching the new product pays a slotting fee to the retailer to make up for the loss that will occur for giving shelf space to an untried product. Slotting fees can be so high as to be restrictive for small companies undertaking the test market phase. The solution for many smaller companies is to get the same information that a test market would bring by using computer-generated simulated test marketing software or test market by other means.

Our company got a severe bump when we planned to introduce a salad-cooking oil blend of sunflower and olive oils hoping to capitalise on the value of unsaturated fatty acids in the diet that was in vogue at the time and the upscale value that olive oil would give. Both oils were expensive at that time: we had bought long on sunflower oil. When Russian sunflower oil flooded the market, sunflower oil prices fell. We were left holding large inventories of unsaturated, hence unstable, sunflower oil purchased at a high price. Another time we were interested in developing a pre-cooked bacon for the retail market. We partnered with a small company with a flow-through belt microwave oven for preparing the product and produced several trial runs of product. Product was excellent and shelf stability good for a refrigerated processed food. Costs estimates were very favourable until Japan bought heavily into the pork belly market and the price of pork bellies soared as they became scarce. When product cost estimates went through the roof the project was dropped. We had not anticipated the price sensitivity of the commodity markets for meat and oil, areas we were unfamiliar with.

Unfamiliarity with technologies is, with hindsight, apparent in Sorenson's chapter describing three case histories. The surimi affair was plagued with:

- communication difficulties between the principals
- ignorance of how seasonal variations influence the composition of fish
- lack of good methodology for quality control.

These all affected quality: only good advice saved the outcome. The bakery expansion problems brought back vivid memories I have of other scale-up problems from what was a craft production to a modern plant production. Murphy's Law takes hold: what can go wrong, will go wrong. The bakery problems and the sachet-filling problem also have elements of ignorance on the part of the baker with respect to baking technology (he was a craftsman, not a technologist) and the sauce manufacturer with microbiology of sanitation. However, in all three case studies the principals sought expert advice.

A startling government decision brought one flourishing export business to a close. Canada's province of Saskatchewan had been developing a saskatoon berry industry. This is a small fruit similar to but not of the same family as the blueberry. It has been a staple of indigenous Indian peoples of North America for centuries and for several decades has been used in both early and modern Canadian cuisine. Nevertheless, it has been declared a 'novel food' by the European Union (EU) and will require a pre-market safety approval before it can be exported into the EU. The saskatoon berry was not a part of the European diet prior to 1997 and so must undergo food safety and nutrition testing by a national authority in one of the EU member countries. Is this a real safety issue or an artificial trade barrier?

Edwardson and Best (Chapter 5) saw their successful product development work come to nought partly through government economic legislation which set prices topsy-turvy. Perhaps another factor for disappointment was that efforts and the products developed, e.g. cassava flour, were for the benefit of the agricultural community, i.e. making farmers into manufacturers, and not for targeted customers. A strong history of co-operatives is essential to make farmers into food manufacturers.

Naming new products can pose problems. A controversy over names is presently before the courts. It involves a small distillery in Glenville, Inverness county, on Cape Breton Island in Nova Scotia, Canada. This little distillery makes Canada's only single malt whiskey. It does not attempt to call their product scotch whiskey: Scotland would quite rightly take exception to this generic name. The name, Glen Breton, was chosen to take advantage of the name of the town and the island on which it is situated. Scotland claims the name is misleading and might lead customers to believe this is a single malt scotch whiskey. One might have some difficulty determining whether this was a deliberate attempt to ride on the coat-tails of true single malt scotches or not.

Such bumps and clanks, however, can often lead to ideas for new products for companies to pursue. For example, one of our subsidiary companies was hit badly when the US FDA (Food and Drug Agency) banned the use of an artificial sweetener that we had used successfully for several years in a line of very popular dietetic canned fruits. This ban forced us to explore fruits canned either in water or their own juices. There had been enough prior rumours surrounding the imminent ban to research alternatives which proved very successful in test market and our position in dietetic foods was maintained. Sometimes detours take one through interesting territory.

### 19.2.3    Intelligence gathering

A company can learn from the past experience of other companies in the development, marketing and selling, all processes involved in getting a product successfully into the hands of satisfied consumers. This learning process is a management technique called 'benchmarking' (Camp, 1989). It is useful. Certainly Currie (Chapter 17) used this learning process for their in-house brands.

It must be understood, however, that learning experiences gleaned through benchmarking occurred at other times, under other circumstances and, obviously, from conditions faced by other companies which may not be applicable to the times or the company making the comparisons. The comparisons are essentially historical data: that is, that's what that other company did under different circumstances that they faced back then. The data must be tempered by current world, national or local events, by results from on-going marketing research, by today's exciting discoveries in the agricultural and food sciences, and by today's developments in the business and economic communities.

To obviate the shortcomings of benchmarking, many companies, both large and small, enter into legitimate but shocking-sounding industrial espionage activities, euphemistically entitled 'intelligence gathering' or 'strategic business development' (Fuller, 2001). Boiled down, it simply means establishing a communication centre within the company to collate, assimilate and analyse all information about competitors, product launches, marketplaces and marketing technology, and be knowledgeable about personnel and executive changes within the industry for what these mean. It includes networking in trade associations, chambers of commerce and in fraternal associations. It includes such simple things as counting how many facings your competitor has on store shelves. It involves being acquainted with everything affecting your food business at all times but especially when developing new markets with new products. When one considers that the time spent from concept to the launch can be anywhere from a few weeks for me-too products to years for genetically modified fruit or vegetable products much can happen. This is time in which:

- markets have matured
- marketplaces have changed, for example e-tailing has come on stream
- customers and consumers on whom the original concepts were tested have matured or been exposed to a multitude of new stimuli or both
- time has passed during which your business development plans have been exposed intentionally or unintentionally to outsiders such as consultants, advertising agencies, technical sales representatives, etc.

Three personal anecdotes illustrate the latter: the first occurred on a long distance air flight. My seat companion instantly engaged me in conversation. He was a senior executive of a Canadian aircraft manufacturer. With little prompting – and over gin and tonics – he related internal politics and unrest within the senior ranks of the company. A few months later there was a major executive staff upheaval and their stock tumbled. Fortunately I had none but I could have moved on the information. The second was much more personal. At that time my wife worked for Bell Telephone and told me that a particular air surveying company were several months in arrears. I promptly sold my shares before the notice of bankruptcy appeared. The third incident is food related. A very eager salesman of pouch sealing equipment told me that our competitor (an Italian food manufacturer as we were) was experimenting with thermally processed pouch-packed foods and that we could borrow the sealer as our competitor was finished. Needless to say this information was relayed to the company president. What useful information can be harvested when all personnel in a company are encouraged to relate any information they come across in their daily activities to a central group within the company for analysis!

### 19.2.4    Analysis of failure

Failures were easy to review with the aid of hindsight to my students. The hard part was distinguishing between extrinsic and intrinsic factors in failures.

Extrinsic factors are those largely outside the control of the company's staff. For example, those described in the previous section were largely extrinsic factors but even here should we not have understood the oils and meat commodity markets better with expert advisors? Should we not have sought professional advice regarding the state of the markets and world supplies? Intrinsic factors for a product's failure are those within the control of the company's staff occurring because somebody 'dropped the ball'. But, shouldn't internal weaknesses have been recognised?

If during the launch the competition skews the company's test market results with false sales data by buying up the new product, that surely must be an extrinsic factor. Or the competition may withhold from a retailer one of its bell-ringer products if the retailer introduces the new product onto its shelves. The competitor simply tells the retailer 'there's a shortage'. The competition just happens to be out of stock of its popular product as long as the new product is on the shelves. This denies the newcomer shelf space. This would be an extrinsic factor. The marketing department of the launching company could not have foreseen such a move on the part of the competition, or could it have? Why hadn't someone within the company questioned either stunningly good sales figures – the competition buying up product – or abysmally bad sales figures? The launching company ought to have anticipated some retaliatory move on the part of the competition and been overseeing everything that was happening in the marketplace during the test launch. Why was some countermeasure not in place? Was this an extrinsic factor or an intrinsic factor?

An example of a company's disastrous foray into NPD was related to me some years ago by Robert Baker of Cornell University. A small food company introduced a canned fish chowder into a local marketing area. Sales were excellent, beyond expectation, and plans were made for expansion into new market areas and for an increase in production capabilities. Nobody within the company took into account the fact that the major competitor for this product was on strike during the launch period. Of course, when the strike ended the competitor entered aggressively into the marketplace with massive advertising and promotional campaigns which swamped the upstart company and destroyed all plans for expansion. Was the cause of failure an extrinsic one or an intrinsic one, the result of a company with its head in the sand concerning activities in the marketplace?

A reliance on a single supplier for a commodity or an essential ingredient can be disastrous if and when that grower-supplier or manufacturer is hit with a dreadfully poor harvest or the manufacturer suffers a strike. These were the very circumstances for one unfortunate company. In the preceding year they had introduced a hot sauce using a new hot pepper variety that was the basis for what was proving to be a very popular product. Now, in the second year, they had no raw material as heavy rains during harvest ruined the crop. One certainly cannot control the weather (extrinsic factor) but was reliance on a single grower-supplier a wise decision (intrinsic factor) on the part of the company? Identifying suitable suppliers and critical characteristics of ingredients are essential tasks of the development team.

Equally dangerous is over-reliance on a single or major customer. Often the customer dominates the food manufacturer: the customer dictates prices, delivery dates and demands new products to satisfy its needs, not the manufacturer's needs. The food company must toe the mark or face the loss of the customer. Building a company on the requirements of the military, or a large retail chain, or a private label customer, or a large food service giant, may be bad for a company's health should the major customer start making demands regarding price, delivery, quality and new products. I know of several poultry processors who have faced disaster because they relied on a single large retail customer. When the customer wanted a promotion, they dictated the price, size of birds, and delivery dates. If the processor cannot offer what the customer wants at the price the customer offers and when the customer wants it, the customer moves on to another supplier. This is largely an extrinsic factor in failure or is it?

Edwardson and Best (Chapter 5) could not have foreseen the impact of a government decision – an extrinsic factor. I met a very colourful man employed by a large soft drink manufacturer at a Codex Alimentarius meeting. He was one of a group of individuals whose jobs were to liaise with government food legislative personnel, attend Codex Alimentarius meetings and be observers at meetings wherever food matters were discussed. Admittedly only very large corporations can afford such intelligence gathering.

Every extrinsic factor in a product's failure or success may have an element of some intrinsic factor and vice versa. My confusion for classifying extrinsic and intrinsic factors can perhaps be understood.

### 19.2.5   Analysis of success

Reasons for success are rather harder to characterise and analyse. For failures one looks for errors in the process from concept development through to market launch and with hindsight pounces on the incorrect or incompetent action. In-depth discussion of successes and failures, not all of them on food products, can be found in Kraushar (1969) and Gersham (1990). I have come to believe there are few technical product failures; failures are primarily due to misunderstanding the customer/consumer interface, faulty market research and faulty promotion.

The benefits of hindsight in the analysis of successful new products merely tells us that everything was done correctly to make the products successful. The customer was targeted successfully and communicated with successfully. Marketing assessed the customer/consumer needs clearly, timed the entry into the market precisely and communicated the product's message effectively. The culinologists developed a highly desired product, which production was able to mass-produce at a price and quality customers would pay and so on. In developing and introducing the product, the company avoided errors of omission and commission.

Ooraikul's brilliant work (Chapter 9) describing their instant baked potato project presents an interesting mix of extrinsic and intrinsic elements involving both success and failure. That is, was the project successful? Yes, a process was developed but there were no buyers; a product was developed but there were no customers. Ooraikul described themselves as naïve but based solely on his narrative I suggest they did not have or use all the tools available – particularly intelligence gathering. They missed or ignored some signs. First, patent applications in the offing ought to have alerted the university's industrial liaison officer (ILO) to look for customers where such innovative vegetable film-making could be useful. Looking would not have hurt exclusivity for their sponsor especially in non-competing areas such as cigar wrappings, for one.

I find it hard to believe that both the retirement of the head of their principal and the sale of his company would not have been known to the development team. They and the principal's workers worked closely together for two or three years according to the narrative. Did no one over a beer, over lunch, at a coffee break, leak any information? Did no one ask how things were going in the principal's company? Was there no gossip in the work areas, in the industry or was such information ignored? The tool missing was industrial intelligence gathering, however informal and unrecognised as a tool this might be in an academic setting. Experience gained both as a hunter of acquisitions and the hunted is that there are signs and rumours that foretell something is up. When our Imasco Foods division of Imperial Tobacco Company was on the market we knew it before any official word was out.

Gwee (Chapter 13) describes a remarkably successful building of a new coconut beverage industry. Her steps were picture perfect from training herself and supplementing her skills with outside resources where needed. She established quality standards for her raw material and sought financial assistance. She did astonishingly little market and consumer research judging solely from her narrative. She saw the variety of product on the market, judged it inferior to what she knew could be made and did it. She selected new flavours without benefit of focus groups. One cannot argue with her success. I do wonder how successful she would have been had she developed this beverage within an established company with a range of products such as occurred with Russell's situation (Chapter 11). Within a company there are competing forces for money:

- finance departments can invest money in various financial vehicles which often earn more or have a greater profit margin than many food product categories
- production departments and the plants can always argue with excellent cost benefit proposals for more plant, newer, larger, faster production equipment or better in-line process control equipment to reduce waste
- engineering departments want better-equipped pilot plants
- marketing departments want to expand markets, perhaps develop export markets.

There is always competition for money within a company. Entrepreneurs striking out on their own do not have these internal competitive forces. Russell (Chapter 11) describes the competition within a company. He had to convince other departments (marketing, finance and production) of the merits of his team's innovation. Thus the routes to success of the entrepreneurs, (Ashworth) Bowie (Chapter 8) and Gwee (Chapter 13), are vastly different from that of Russell.

## 19.3    Steering the Product Development Process

A product development team brings together people with diverse skills from within and outside the company. Direction of the team is, in small companies, often at the executive level: in large companies, management is delegated to brand managers who direct a stable of products. There will be someone responsible for an accounting of expenses, for labour costs reflecting the manpower hours and for managing the budget: this individual is often the source of friction on the team, since this person is divorced from the drive and intensity or the team. Other internal resources are:

- The research and development department. This may be complemented with outside research companies, academic facilities, and partnering arrangements with other companies, etc.
- The production department. Their facilities are often necessary for test production runs – these often interfere with production schedules and production bonuses. There may be a need for a co-packer.
- The marketing and sales departments. These are closest to customers and consumers; they feed back market information to the team on what needs and niches in the marketplace can be exploited with new products. These internal resources are frequently supplemented by market research companies or advertising agencies or both.
- The engineering department. They modify existing production equipment where necessary (again, to the annoyance of the production schedules), write specifications for new equipment or redesign existing equipment according to the needs of the team.

The development team grows and diminishes, according to the skills required at a particular stage in development and launch. The lead changes according to the skill set required at that stage. But at all stages there must be control and direction provided by the appointed manager. The manager sees to it that direction of the project is not lost and rivalries do not arise between the skill sets. Human errors and frailties can enter here without the intervention of the manager. The manager is, in effect, a team coach, cajoling and encouraging his players to meet targets.

Markets and marketplaces, customers and consumers, and technology occurring 'outside' the often inward-looking team must be communicated 'in real time' to responsible people for sound decision making throughout development. Development plans will be drawn and redrawn based on real time information. The company may be changing too, undertaking new directions requiring new objectives as the economic and competitive climate surrounding it keeps changing.

Those who promise a sure method that, if followed, will ensure success in the development and launch of a new product are wrong – note Bertrand Russell's aphorism at the start of this chapter. There is no such methodology; New Product Development is a transformative process.

### 19.3.1    A critical path programme to audit the New Product Development process

Food manufacturing processes are subjected to a hazard analysis critical control point analysis (HACCP). This analysis reviews a product's entire manufacturing process from raw material purchasing and receiving to finished goods warehousing even through to distribution. It looks for weaknesses where the product might become contaminated, quality lost or safety respecting hazards of public health is compromised or economic hazards encountered. On the basis of this analysis a HACCP programme is established whereby weak points in the process are monitored: the hazard at that weak point is eliminated, minimised or controlled to safe or harmless levels.

Successful new products are essential for the growth and stability of any company, therefore the NPD process with its high cost and often a high failure rate should be subjected to an analysis for the weakest points in it similar to a HACCP programme. The major hazard involved in NPD is an economic one, i.e.:

- the product fails in the marketplace and a degree of brand image is lost. It is embarrassing. For this reason products are launched under a different brand name until a proven success is a *fait accompli*
- the development dollars, consultant fees and costs of labour time are lost.

This is the hazard. Therefore a critical path analysis programme is needed to minimise, prevent errors in, control direction of the development process and monitor the launch and by so doing reduce the possibilities of such losses.

For example, where promotional marketing skills are weak within a company, a critical path analysis would recommend use of an advertising agency. Where the company is weak in physical plant resources, it may have to seek a partner or a co-packer.

This critical path analysis programme for product development attempts to anticipate hazards in the new product process and strengthen weaknesses leading to potentially ill-considered decision making. It is as necessary as a HACCP programme is required in the manufacturing process. Such an analysis of critical path areas is presented in Table 19.1 as a series of questions requiring an answer.

The most important critical step begins with senior management with the seemingly silly question, 'Does the company know who or what it is?' Levitt (1960, 1975) describes some excellent examples of the results of 'not knowing'

**Table 19.1**   A critical path analysis programme for New Product Development

| Hazard | Remedial action for elimination, minimising or controlling of hazard |
| --- | --- |
| Does the company have clearly stated *raison d'être* with clearly stated goals and purpose? | • Company must have a clear statement of purpose and objectives.<br>• Company must know what services its products provide, why these please customers and consumers, and how their services are used.<br>• Company must know who its customers and consumers are. |
| Have company strengths and weaknesses been identified and objectively assessed? | • Strengths and weaknesses must be critically assessed.<br>• Weaknesses are to be buttressed by either partnership or other legal arrangements with individuals, companies with strengths in weak areas. This may require co-packer arrangements, partnerships, use of research and development contractors (private firms, universities, trade associations, etc.), market research companies, etc. |
| Has a business analysis and plan for the project been developed and approved by senior management? | • A business plan is required to set objectives for the project. This analysis and the objectives are not necessarily those for the development team. |
| Does the product concept promise a customer and/or consumer benefit? | • The technical and marketing benefits require identification so that these are not compromised during development by the team.<br>• Once identified, the benefits must be assessed for their uniqueness and competitiveness.<br>• Is the benefit, process or product patentable?<br>• Is it brand identifiable? Is it relevant to the brand? |
| Have the targeted customers and consumers been clearly identified? | • Market research must clearly identify targeted customers and consumers.<br>• Potential marketplaces where targeted customers and consumers are found must be identified and competitive activity (and products) in these markets identified.<br>• Competitive activity in the identified marketplaces must be assessed and strategic plans in place to counter any aggressive competitive action. |
| Are product concepts, once developed, clearly communicated? | • The product development team requires, on a need to know basis, clear communication of the product concept, product benefits and product uniqueness. Has this been provided? |

**Table 19.1**    A critical path analysis programme for New Product Development

| Hazard | Remedial action for elimination, minimising or controlling of hazard |
|---|---|
| | • Thereafter, the project manager needs to monitor the various skill sets to ensure that: (a) all understand their task, (b) all have the resources and capability to accomplish their tasks, and (c) progress is reported regularly.<br>• Have patentable ideas been identified for exploration? |
| Has a test market protocol been developed? | • Criteria for assessment of test market activities are established.<br>• A suitable test launch area has been chosen that contains the targeted customers and consumers.<br>• A watching brief is established to monitor all marketplace activities concerning test market, including competitive activities.<br>• Are promotional materials and advertising copy, suitable for marketplaces where customers and consumers are, in preparation or sourced?<br>• Has choice of media for advertising been considered? If so, does advertising take full advantage of the media under consideration?<br>• Have all promotional materials, advertising, label copy and claims been approved by the legal department?<br>• Does the advertising promise a basic customer/consumer advantage? Is it clear, credible, uncomplicated and sincere?<br>• Does the advertising demand action that will lead to the purchase? |
| Product maintenance | • Has a plan for future redesign of the product been developed? |

and these papers have become the cornerstone of many MBA marketing programmes. Levitt discusses the loss of business by the railways in the USA because they did not know what business they were in. They viewed themselves as being in the railroad business. They were not: they were in the transportation business. Transportation was the service, the values and assistance they provided their customers. They were not in rolling stock, i.e. material things. Consequently, the railroads lost out to other means of transportation. Similarly, the movie makers thought they were in the movie theatre business; they were not, they were in the entertainment business. The result of their narrow definition of who they were was their loss initially to the television industry whom they regarded initially as a competitive threat.

The first critical analysis assesses whether the company is too narrowly or too broadly defined or not defined at all. Management must know what services they provide their clients; what they are good at providing that pleases their customers and consumers. The nature of the services that the company provides and their objectives then define the PD Process.

Such an analysis as the above allows the company to evaluate its strengths and weaknesses in the services it provides. This review of internal resources allows the team to feel comfortable with those outside resources selected for complementing weaknesses. Any contracts with outside resources require review by legal teams.

Business plans contain a budget and objectives and provide management with one tool to abort the project if the attainment of objectives will exceed funds that were budgeted. The criteria for aborting a product development project will always require some internal clarification if for no other reasons than to assuage the pride of those who worked on the project; their self-confidence can be shaken and they may feel their security within the company challenged. Who stops a project, for what reasons is it stopped, and when is it to be stopped require documentation.

Product concepts must be clearly written down and understood by the development team: the development team must know what it is working on. Concepts require some flexibility to accommodate changing conditions in the marketplace that market research and competitive intelligence-gathering detects during development. An effective means of communicating all information impinging on a company's marketing thrust with a new product may require incorporation of changes to products-in-development to eliminate the hazard of being inert to marketplace changes.

It is assumed all steps in the technical aspects development process have been performed and signed off on by responsible supervisory persons, that is:

- Safety has been designed into the product. If heat penetration tests and thermal processes calculated, these must be signed off by a responsible certified thermal process engineer.
- Storage tests, abuse tests or both have been conducted, and shelf life calculated.
- A HACCP programme has been prepared and all critical stages in a production run are detailed with corrective action.
- Label and nutritional declarations have been approved by the legal department.
- Ingredients with specifications have been approved and alternate sources identified.

These requirements are verified, collated, approved and signed by a senior authority before the new product is ever turned over to production for a test launch.

The site of a test market, its timing and how success will be assessed, are important considerations. One is careful to launch a product where customers and consumers understand the product and its benefits and are not offended by it: in short, products are not launched in highly ethnic areas where the benefit of a product is not understood or appreciated or offends religious or cultural values.

The test launch market area must be protected from, or marketing must be aware of, untoward competitive market activity. The launching company must know what is happening in the marketplace. There are available simulated test market programs based on computer software programs. All software programs used in assisting in the analysis of any phase of the NPD process should be understood for the assumptions (for assumptions there are) used in these programs; these assumptions may hide, covertly or overtly, possibilities of misinterpretation of results.

Product maintenance has been briefly mentioned earlier. All products have a life cycle with growth, a steady state and then a decline in popularity as measured in sales. Product maintenance is a system of planning – not desperation moves to maintain flagging sales of a moribund product – designed to make a constant series of improvements or variations (new flavours, new forms, new uses, perhaps) to revitalise the product after maturity is reached.

## 19.4    Conclusions

It should be apparent by now that I consider markets, marketplaces, customers and consumers to be volatile and erratic. They, like any actual journey, are a far cry from what is depicted on a roadmap. They are subject to change so quickly that a good product launched with the wrong timing is a dead product. New products must ride the wave of customer and consumer enthusiasm generated in the various marketplaces available to them; timing is critical. Only if your product is the source of the enthusiasm can timing be off, compare if you will the Sony Walkman® and the Blackberry® technologies.

We now have the internet with its bloggers (and blogvertising), Youtube, Facebook and other social communication networks used not just by teenagers but by many adults (and the competition) to anonymously influence negatively or positively any product for the buying public in our marketplaces. These must be carefully scrutinised for adverse publicity or used effectively for the audience they can reach for effective communication.

### 19.4.1    Problems areas for the future
Future problems to product development will come, in my opinion, from three sources:

- first, legislation which has always accompanied the food industry will continue to be a source of problems for innovators;
- second, marketing research and its methodologies will come under closer government scrutiny with the result that accurate market research may become more difficult to pursue, costly to perform and morally repugnant to customers and consumers and that will lead to legislation; and

- finally, safety will become a greater concern for all in the food industry. Safety is ultimately legislated. Where there is doubt about safety, legislation will come in.

These three problem areas are so closely interconnected they could be considered one big roadblock – more legislation and regulation – on the road to successful new products.

### 19.4.2   Food legislation
The bad publicity trans-fats has received caused developers to reformulate products containing them. To my knowledge, only in New York City and the province of Ontario is there legislation to ban them.

Earlier in this chapter, examples of legislation impacting NPD were discussed:

- the banning of a previously permitted artificial sweetener by a government;
- the banning of products based on the saskatoon berry because it was new to the area it was being introduced into; and
- the controversy over nomenclature to stop the initial marketing of a malt whiskey, not yet a legislative problem but claimed to contravene 'country of origin' legislation.

Each could be both a legitimate cause for problems in marketing new products abroad but also be seen as a source of artificial trade barriers between nations. Edwardson and Best (Chapter 5) found changes in economic policy an impediment to successful agroindustry development.

Genetic engineering will engender legislation, partly because of the controversy between those in the public opposed to it for religious, ethical, safety, or environmental reasons and those, largely within the scientific and industrial communities, who stand to profit from it. Until those in favour of genetic modification can unequivocally prove to the satisfaction of the doubters and governments, legislation, public outcry or both will hamper the introduction of genetically modified novel products. One certain target will be cloned meat which has already received a bad press when none has been in the marketplaces. New genetically modified crops will possibly be legislated in ways to prevent ecological and environmental damage thus restricting their availability.

Nutrition has reached a new milestone with the discovery that non-nutritive substances in foods have protective action against certain diseases. These materials, the nutriceuticals (there are many other names and classifications for them), are being used as new food products themselves and to fortify food products. Their naturally occurring presence in many food products is heavily advertised. Problems arise when manufacturers want to advertise these health benefits for products to which extracts of nutriceuticals have been added – consider the promotional potential of anti-ageing nutriceuticals (Gruenwald, 2006). People on prescribed medications for medical conditions have had adverse reactions to extracts of nutriceuticals or to foods fortified with these materials. People self-medicate and manufacturers take advantage of this tendency to fortify foods with beneficial adjuncts to increase sales. Regulation can be expected with new products fortified with nutriceuticals, as neither their safety or efficacy has been established, e.g. kavakava has been condemned recently as a liver poison.

### 19.4.3   Marketing research

Market research is becoming increasingly expensive and difficult to undertake, especially with respect to understanding and motivating the customer and consumer. Wrong concepts and directions for development need to be eliminated early, before money is wasted pursuing them. Costs could be reduced if bad concepts, wrongly targeted customers and consumers, and products with no benefits were eliminated earlier in the development process. By knowing what pleased their targeted customers beforehand, developers might be more assured of success. That is the argument.

Marketing research has developed better ways to understand and manipulate customer and consumer likes and dislikes using advanced technology. Marketing research has grown from mail-in questionnaires, focus groups, analyses of courtesy and client card purchases, to neuromarketing (Carmichael, 2006). Studies of the likes and dislikes of customers and consumers have become sophisticated by neuromarketing techniques employing fMRI (functional magnetic resonance imaging) to determine *individual* reactions of customers and consumers to marketing stimuli (Biederman and Vessel, 2006). In short, neuromarketing and other less intrusive techniques such as ZMET (Zaltman metaphor elicitation technique) (Zaltman, 2000) permit the development of stimuli (advertising materials, mock-ups or products) to over-ride the brain's normal likes and dislikes. Shades of *1984*! There are groups opposed, often vehemently opposed, to neuromarketing technologies likening them to thought control (Carmichael, 2006).

Customers and consumers fear the potential loss of privacy that many marketing research techniques represent. Market researchers, through devices and techniques such as cookies, bots and phishing, can gain information about the interests and habits of internet users as they search the Web for shopping or personal interest. Even the use of in-store cameras to follow the movements and buying habits of customers is questioned as an invasion of privacy.

Client and courtesy cards do not elicit as much negative reaction, since their use is accompanied by rewards programmes, nevertheless an analysis of the data they provide is just as intrusive about customers' habits. Regulation (i.e., legislation) of the tools of market research is sure to come with a heavy hand. One need only recall the hand wringing and breast-beating when such information is carelessly handled or accessed by hackers. Market research will be forced to develop less invasive and less privacy-comprising technologies.

I have always received a chuckle from audiences when using a cartoon by Thach Bui and Geoff Johnson showing a customer reading a sign posted outside the door of a large retail outlet. The sign reads:

'Here at MegaFoodCorp, you're more than just a customer ... You're a completely predictable compilation of spending habits and product data.'

Quite apart from the sins inherent in marketing research technology are the problems market researchers face with virtual retailing, that is, e-tailing: this requires virtual media to promote and advertise the virtues of new products. The marketing department has the task of conceptualising a product – and eventually marketing and launching it – that appeals to virtual customers who, with the power of the internet, may be anywhere in the world. How does the marketing department target such virtual customers? Where is that customer and what are that customer's characteristics?

With the growth of virtual retailing there will come a lessening of the importance of bricks-and-mortar stores: distribution centres will be important and point-of-purchase advertising less important. Many real marketplaces will be unaffected, for example the farmer's market or farm gate stands, the coffee and doughnut stores as well as much of the food service and entertainment industry. The ubiquitous corner store will continue to serve where e-tailing cannot, for the spontaneous spur of the moment or impulse buying where one can run to for food items when unexpected company drops in. A growing segment of the population is finding e-tailing convenient and easy to use. If the real marketplace is diminished in the retail scene, then the market (a concept) emerges in company with virtual stores. With some marketplaces eliminated, market research is free to explore directly the use of markets as vehicles within which to retail worldwide without the middleman, i.e. the marketplace. Market research's tasks become more difficult. The challenge to develop new food products will become difficult; the chances of failure greater.

### 19.4.4   Safety

New food products, the novel ingredients they may contain (e.g., nutriceuticals and preparations such as hoodia extracts from the plant *Hoodia gordonii* for diet control and weight loss) but also the novel preservative systems established for products employing multiple minimally processed technology. New processing techniques, e.g. high-pressure processing, ohmic heating, pulsed electric fields, even irradiation are novel technologies with no long-term established history of safe use with respect to problems of public health or economic significance. Therefore, until safety is clearly established and both theoretical and practical knowledge gained on the predictability of safety, legislation will be enacted to protect the public and confound the developers.

Associated with safety and health issues, promotional health and disease-fighting claims of products fortified with nutriceuticals will be challenged. Oregano has been touted as a flu fighter. Therefore, the question becomes: if oregano as a herbal additive in cooking is safe, are we justified in saying that a concentrated extract of oregano (oil of oregano) is also safe as an anti-flu preventative? May promoters advertise these claims? We do not possess the knowledge based on experience or on theory of safe use that such claims are justified.

Foods preserved by novel techniques or fortified with novel preservative ingredients must provide the same degree of safety with respect to hazards of public health significance or to economic hazards as conventional foods. For example, in some jurisdictions *sous vide* products are banned in restaurants for fear of botulism and listeriosis. Some irradiated products have spoiled in unfamiliar ways, suggesting unexpected spoilage mechanisms permitting the growth of unsuspected microorganisms and not observed in these products conserved by conventional methods. The newer technologies present challenges yet to be overcome before safe new products emerge.

Foods genetically modified to produce greater quantities of, for example, some nutriceutical, may prove to have some deleterious effect on certain members of the population through the extra amounts of the desired nutriceutical. The safety of genetically altered or cloned products will be challenged.

Where there is concern respecting safety or reluctance to accept on the part of the buying public or both there will also be governance and regulation. How to market controversial, but safe, products in face of opposition by some segments of the population (cf. irradiated produce) will plague market researchers. Making claims for and promoting the enhanced value of novel products will engender regulation respecting advertising. Thus the cycle, legislation, marketing and safety, closes to present one or three major problems to be faced in the future for new product development.

Safety concerns, marketing technologies and government intervention all come back to one thing, more legislation to hamper the entrepreneur, the craftsperson, the small business person and the large multinational corporation.

## 19.5   Summary

Successful development must:

- be practical and profitable for the developer
- allow the developer to grow by resulting in repeat purchases and satisfy retailers in the various marketplaces the developer chooses
- satisfy the needs of customers and consumers at a price customers are willing to pay.

Examples of both successful and unsuccessful new product and project developments have been presented in the previous chapters and briefly discussed in this chapter.

## 19.6   References

BIEDERMAN, I. and VESSEL, E.A. (2006), Perceptual pleasure and the brain. *American Scientist*, 94, May–June, 247.

CAMP, R. (1989), Benchmarking: the search for best practices that lead to superior performance. *Quality Progress*, January–May.

CARMICHAEL, M. (2006), Neuromarketing: Is it coming to a lab near you? http://www.pbs.org/wgbh/pages/frontline/shows/persuaders/etc/neuro.html (accessed 26 December 2006).

FULLER, G. (1994), *New Food Product Development: From Concept to Marketplace*. Boca Raton, FL: CRC Press.

FULLER, G. (2001), *Food, Consumers, and the Food Industry: Catastrophe or Opportunity?* Boca Raton, FL: CRC Press.

FULLER, G. (2005), *New Food Product Development: From Concept to Marketplace*, 2nd edn. Boca Raton, FL: CRC Press.

GERSHAM, M. (1990), *Getting It Right the Second Time: How American Ingenuity Transformed Forty-nine Failures into Some of Our Most Successful Products*. Reading, MA: Addison-Wesley.

GIBSON, L.D. (1981), The psychology of food: why we eat what we eat when we eat it. *Food Technol.*, 35, 54.

GRUENWALD, J. (2006), Anti-ageing nutraceuticals, *Food Science and Technology*, 20 (3), 50.

KRAUSHAR, P.M. (1969), *New Products and Diversification*. London: Business Books.

LEVITT, T. (1960), Marketing myopia. *Harv. Bus. Rev.*, 38, 45.

LEVITT, T. (1975), Marketing myopia. *Harv. Bus. Rev.*, 53, 26.

MANSKY, M.H. (2006), New product startups: dealing with the unexpected. *Food Tech.*, 60, Sept., 40.

MATTSON, P. (1970), Eleven steps to low cost product development. *Food Prod. Dev.*, 6, 106.

SKARRA, L. (1998), Rollout roulette, *Prep. Foods*, 167, Aug.

ZALTMAN, G. (2000), The dimensions of brand equity for Nestlé Crunch Bar. A research case, Harvard Business School, N9-500-083, 27 January.

# Part VII

# Education for New Product Development

# 20

# Effective education for product development – building New Product Development courses

Vichai Haruthaithanasan, Penkwan Chompreeda, Hathairat Rimkeeree and Thongchai Suwonsichon, Thailand

*In Thailand a unique opportunity emerged to explore and establish a formal product development programme, multidisciplinary, and integrated into both an academic and an industrial environment. Against a backdrop of the phases of the successive 10-year Thai government social and economic development plans needed to build an essentially rural society into an integrated modern economy, there was the real need for many new innovative products. These could best come, with highest chances of success, from trained people. The universities could be the educators. So they were strengthened. They borrowed extensively from overseas. But the chance was also there to move beyond and into something new. Courses and programmes were built up over the years within the expanding agro-industry framework.*

*At Kasetsart University, it was recognised early that product development was both a necessary and a distinctive element if the future was to extend beyond the traditional in a society where tradition was strong. This area needed wide multidisciplinary strength ranging across a spectrum from engineering and the physical sciences, through biology and social and behavioural science, to business and management and marketing. It built from established areas, each studied by individual staff members trained to advanced levels through research scholarships overseas. Then it was brought together into courses, combined and unified in a way that was internationally novel, and moulded and strengthened by the special needs of Thailand.*

*The courses developed and improved. Hands-on practice and projects were encouraged and accomplished. Unfamiliar at first, the graduates became*

*welcomed in industry because they could perform. Prospective entrants saw success, and so able students were encouraged in substantial numbers to undertake demanding courses and the wide coverage necessary. Research and collaboration with industry were obvious out-growths and both grew extensively and into applied research institutes, set up in parallel to the more academic side. For all this to be established in a new field, even one so appropriate to the needs of an industrialising Thailand, was very adventurous and far-sighted. The success is a tribute to the vision and determination of the dedicated team who made it happen. It certainly demonstrates what can be done; and its results, 2000 graduates from Kasetsart University ranging to doctorates, a thriving department, two research institutes, and industrial collaboration, are impressive. Throughout Thailand, there are now many other similar successful university courses in product development.*

*The chapter particularly relates to Chapter 4, pages 149–193, 299–304, 287–304 in* Food Product Development *by Earle, Earle and Anderson.*

## 20.1   Introduction

Thailand is known as an agricultural country, with 65% of the total households farmers. The main crops grown by Thai farmers include rice, maize, cassava, sugar cane, soybean, mungbean, peanuts and coconuts. Besides these main crops, Thailand is known for the production of tropical fruits and vegetables as well as for the production of fish, poultry and pork. For the development of the country, the Thai government has established a national social and economic plan using the strengthening of the agricultural base to run and push the economic growth as well as increase the GDP of the country.

The development of the agro-industries in Thailand can be divided into six phases. From the first to the fourth phases, the development of food industries was emphasised. In the fifth and sixth phases, the food industries' effort was to produce high quality products to meet specified standards of either Thai food law or importing country or both. During this phase both food and non-food industries were readily established.

In the first phase (before 1960), the Thai food industries learned how to preserve food materials to prevent spoilage and to keep them for longer periods. In the second phase (1960–1970), the Thai food industries were established by importing 'know-how' technology, such as sweetened condensed milk, cooking oil from soybean, canned pineapple, canned fishery products and frozen foods. In the third phase (1970–1980), the food industries put their effort into producing varieties of food products both for local consumption and export markets. The fourth phase (1980–1990) was an era of food safety. Food industries emphasised quality assurance by adopting GMP and HACCP as the important tools to obtain food products safe for human consumption.

In the fifth phase (1990–2000), the food industries strongly emphasised the use of environmentally safe technology, environmentally safe packaging materials, setting quality assurance of the products, nutritional food products. Finally certified Good Manufacturing Practice (GMP) was required in the food factories to fit the world competitiveness for export of agro-industrial products, both food and non-food. Some technologies imported in the early phases were out of date, so some food industries were seeking new technologies or developing their own technologies by adapting overseas technology to fit their needs. Some food industries started research and development on their own products to satisfy customer needs.

In the sixth phase (2000–2010), the food market is becoming globalised, therefore the Thai agro-industries are meeting strong competition and need people with knowledge in both technical and marketing skills as well as management, especially e-commerce to fit the world today. So over the last forty years there has been a need for people educated in product development, and never more so than today.

In Thailand, education in food product development started in the Faculties of Agriculture and of Technology with Departments of Food Science and Technology, but with the development of the Faculties of Agro-Industry in the 1980s, separate departments of product development and specific degree courses in product development were started. During the next 20 years, these multidisciplinary undergraduate degrees were developed and later also Masters and PhD programmes.

## 20.2    Education for the agro-industry in Thailand

Because of the development of the agro-industries in Thailand, people were needed that were trained at university level. In 1962, the first programme of food science was established in the Faculty of Agriculture, Kasetsart University. It was a five-year programme; in 1963, the programme was changed to four years to fit the world academic standard for the food science curriculum. At the same time, the Faculty of Science, Chulalongkorn University offered a chemical technology programme in which food technology was a minor option. It was also a four-year programme, similar to the food science programme at Kasetsart University but this programme stressed technology rather than science. The food science programme was taught continuously at Kasetsart University for 18 years before it expanded into the Faculty of Agro-Industry.

The Faculty of Agro-Industry was established in 1980 because of the fourth national social and economic plan (1976–1980) of Thailand. There were two reasons for starting the new faculty at Kasetsart, firstly the need to strengthen and develop to large scale the production of many kinds of agricultural food materials, and secondly the need for university graduates from other fields of agro-industry as well as food science graduates. Therefore, the Faculty of Agro-Industry at Kasetsart University was composed of four departments: Food Science and Technology, Biotechnology, Packaging Technology, and Agro-Industrial Product Development. During the fifth national social and economic plan (1990–2000), there was more demand for graduates in the agro-industrial disciplines; therefore, Chiang Mai University and Prince of Songkla University established Faculties of Agro-Industry, and Khon Kaen University established a Faculty of Technology. Milestones of education for product development in Thailand are shown in Table 20.1.

Later on those universities offered higher education at graduate level, both MS and PhD to serve the needs of the country. The Product Development Department at Kasetsart University offered the masterate programme in 1987 and the doctorate programme in 1996. The Department of Product Development Technology at Chiang Mai University recently offered a masterate programme in Product Development in 2005. And in 2006, the Department of Product Development at Kasetsart University offered a special masterate programme in Product Development for students who are working during the week. They attend classes during the weekend.

These are the main universities teaching Product Development as major and minor parts both in foods and non-foods. Owing to the fact that the majority of

**Table 20.1**  Milestones of education for the agro-industry in Thailand

| Year | Department | Faculty | University | Degree |
|------|-----------|---------|-----------|--------|
| 1962 | Food Science | Agriculture | Kasetsart | Bachelor (5 years) |
| 1963 | Food Science | Agriculture | Kasetsart | Bachelor (4 years) |
|      | Chemical Technology (Option Food Technology) | Science | Chulalongkorn | Bachelor (4 years) |
| 1973 | Agri. Product | Agriculture | Khon Kaen | No degree |
| 1975 | Agro-Industry | Natural Resource | Prince of Songkla | Bachelor (4 years) |
| 1980 | Food Science and Technology<br>Biotechnology<br>Packaging Technology<br>Product Development | Agro-Industry | Kasetsart | Bachelor (4 years) |
| 1990 | Food Technology | Science | Chulalongkorn | Bachelor (4 years) |
| 1991 | Food Science and Technology<br>Food Engineering<br>Biotechnology<br>Packaging Technology<br>Product Development Technology | Agro-Industry | Chiang Mai | Bachelor (4 years) |
| 1991 | Food Science and Technology<br>Biotechnology<br>Material Product Technology | Agro-Industry | Prince of Songkla | Bachelor (4 years) |
| 1991 | Food Science and Technology<br>Biotechnology | Technology | Khon Kaen | Bachelor (4 years) |

lecturers were educated as food scientists and food technologists, the major teaching and research in product development is oriented mostly to food products.

There were numbers of universities that established Faculties of Agro-Industry or Agro-Industrial Technology during the period of the end of the fifth and the sixth national social and economic plan (1990/2000–2010). There were few universities that had a curriculum in product development and a PD department. However, the Departments of either Food Science and Technology or Food Technology and Biotechnology offered students an option in product development or packaging technology. Just recently (2005), there were more than 100 universities in Thailand and all are now under the Ministry of Education. Some of the new public universities were formed by changing the status from Teacher College or Institute of Technology to University and restructuring and reforming the curriculum. For example, a group of Teacher Colleges changed their name to Rajabhat University (40 universities as the cluster). The other group of Institutes of Technology also renamed their Institutes to Rajamangala University of Technology (9 Universities as the cluster). Both Rajabhat University and Rajamangala University of Technology offered at least one option in the following disciplines: Food Science, Food Science and Nutrition, Food Science and Technology, Food Technology, Biotechnology and Agro-Industrial Product Development.

## 20.3    History of education for New Product Development: the Kasetsart University (KU) story

'Kasetsart' is the Thai word meaning agriculture. Therefore, Kasetsart University (KU) had a mission to provide education concerning all agriculture programmes since 1943. In early days (before 1970), almost all graduates from Kasetsart University worked as researchers in the Ministry of Agriculture. Students who graduated with honours in any fields were invited to join the university as teaching staff and later on received scholarships for higher education abroad. After graduation, they came back and worked in Kasetsart University. Not many graduates worked in their own business or in commercial food companies.

The starting point of education in product development was in 1971–1975. At that period, the Thai government received a soft loan from the World Bank for education development at Kasetsart University. Part of this budget gave an opportunity for the Department of Food Science in the Faculty of Agriculture to reconsider and restructure its curriculum that had been taught for almost 20 years. In 1973, Professor Narudom Boon-Long, the head of Department of Food Science invited Professor Richard Earle, head of the Department of Biotechnology, Massey University, New Zealand to be the academic advisor for three months at Kasetsart University. That mission resulted in building up the new direction of academic development in other branches of technology and also met the requirements of the food industry development plan. The proposal to set up a Faculty of Agro-Industry at KU was submitted to the government during the fourth national social and economic plan (1976–1980). Then in 1980, the Faculty of Agro-Industry was established at Kasetsart University and it was the first in Thailand. The Faculty was composed of four departments; Food Science and Technology, Biotechnology, Packaging Technology, and Product Development.

The first batch of students in Product Development graduated in 1984 and they were PD 1. Managing education in product development at the start was not easy. There were three big problems and their solutions could be a complete chapter and a case study in developing a PD degree. They were curriculum, teaching and staff matters which were all solved eventually.

### 20.3.1   Curriculum

At the starting point of setting up the Department of Product Development, there were some problems that usually occur for all new departments. For Product Development, it was worse than the other departments since no one know what 'Product Development' was. It was explained that product development is a special tool used for developing a new product. A series of questions came one after another of what, when and how, it could be done. What was it and how to use it? Many answers were advanced, trying to explain, but it still was not clear. It took time – until the fifth batch of students graduated in 1988!

At that time, Thailand's economic situation was stable; food industries expanded their business for export. Therefore, they needed creative persons to make new products, new value, new styles and new fashions from the agricultural base for export markets. That was a very good opportunity for PD graduates; they were more welcome in the food industries than before. At that time PD graduates could get good jobs and showed their high ability in developing new products. It also made clear to the public that product development is a tool for creating a food product that meets the requirements of a group of consumers under consideration by a company.

### 20.3.2   Staff

In 1980, there were only three lecturers in the Department of Product Development; Professor Narudom Boon-long, Professor Vichai Haruthaithanasan, and Dr Chintana Oupadissakoon, and with two technicians. Then in 1982, Dr Prasert Saisitti was transferred from the Food Research Institute. The department recruited two young lecturers; Professor Sombat Khotavivattana in 1982 and Professor Dr Penkwan Chompreeda in 1983, just in time to teach students, especially third- and fourth-year students for which almost all the classes were taught by the Department. There were 15 students each year and at the end of year 4 there were 60 students in the department. Therefore, with five teaching staff and two technicians, the whole Product Development programme with 60 students was taught until 1987. But gradually, especially in the 1990s, the staff increased up to 15 by 2007.

The majority of the staff were educated in Food Science and Technology, but there was a variety of other disciplines – quality control, chemical technology, product development, agricultural chemistry, industrial engineering and in later years Agro-Industrial Product Development.

In 1987, Dr Siriluk Sinthavalai transferred from Food Research and Development Institute to the Department and decided to pursue a doctorate degree in PD at Massey University, New Zealand. Unfortunately, after working for three years, she had to retire due to ill health. In 1988, Dr Rungnaphar Pongsawatmanit, a first degree honour student of PD 1 transferred from King Mongkut's University of Technology Ladkrabang to replace Dr Siriluk's position. In 1989, Professor Narudom Boon-long retired, the department had two positions to replace him, and both of them had MS in PD, Dr Anuvat Jangchud and Dr Kamolwan Jangchud. Later on both of them received PhDs from University of Georgia, USA. Dr Anuvat Jangchud was supported by USAID-Peanut CRSP and Dr Kamolwan Jangchud was funded by a King's Scholarship. In 1989, Dr Hathairat Rimkeeree received a government scholarship to study PhD in PD at Massey University, New Zealand. In 1991, Professor Paiboon Thamaratwasik joined the department and left in 1995 then Dr Paisan Wuttijumnong transferred from Prince of Songkla University to replace his position. Dr Saowanee Lertworasirikul received a King's Scholarship to study at North Carolina State University, USA and graduated in 2004. During 1992–1996, the Department was allocated four government scholarships which were given to Dr Thongchai Suwonsichon, and Dr Withida Chantrapornchai, both of them graduated from University of Massachusetts, USA; Dr. Walairut Chantarapanont, who graduated from University of Georgia, USA; and Dr Nantawan Therdthai, who graduated from University of Western Sydney, Australia. Professor Vichai Haruthaithanasan retired in 2004, his position was replaced by Dr Pisit Dhamvithee, PhD from Kasetsart University. In 2005, the department had recruited new staff; Miss Rudeepan Wattanapat unfortunately resigned after working for only two years. Later on in 2007, her position was replaced by Dr Thepkunya Harnsilawat, PhD from Kasetsart University. In 2007, Dr Paisan Wuttijumnong left the department to be the Dean at Faculty of Agro-Industry at Prince of Songkla University.

The department had built staff from 5 lecturers in 1983 to 15 lecturers in 2007, and each of them has unique expertise. As now they are very active and work together to educate graduates at BS, MS and PhD levels to serve the Thai community. The total of students numbers 450 in 2007. There are more than 1,000 PD alumni working in universities, government research agencies, food and non-food industries and their own businesses.

Strong human resource development at the Department of Product Development at Kasetsart University could not have happened without the wide vision of the five pioneer lecturers as well as some support from the Professors Earle and USAID-Peanut CRSP.

### 20.3.3   Teaching

At the early stages, none of the lecturers had a background in product development by training, the first three, Narudom, Vichai and Chintana transferred from the Department of Food Science. Sombat had a background in chemical technology and Penkwan a background in food science and nutrition. They all realised that product development is industrial research and should be run with a good concept and correct direction. There are several steps to run to make a new product or improve an existing product. It is not just a cooking procedure as done in street food. It is not just a cooking by order from plate to plate, from time to time. It is not just marketing research. The problems facing them were how to teach students to understand the system of product development. All lecturers worked very hard to make themselves first understand.

It was a very hard work to push students to learn and to understand the desk research. To learn by practising in a group of 3–5 persons was easy to control and made them understand more quickly. For example to learn how to form a product concept, it started from outside the product itself. Firstly, to find out the product idea, who wants/needs such a consumer product? Secondly to know in which market we need to place the product, and which market segment might need and want the product. We should make it clear from the starting point; no evidence is kept hidden. If not, it will fail from the starting point. Whenever the information from the market research is completed and made clear, then the development work can run forward in the laboratory. However, our students were too young and had no experience in working with their hands. That was the big problem in teaching product development at that time. The staff had to input more time and to work with them in the laboratory. They had to learn how to handle the product. The textbook of Professor Mary Earle provided a very good guide of the product development system for staff to follow in teaching students at the department.

### 20.3.4   Research

In the storm of hard work by staff, the department was lucky to have collaborative research on utilisation of peanuts with the Department of Food Science and Technology, University of Georgia, USA funding by USAID Peanut-CRSP started in 1982 and continued until 1995. The collaborative work provided an opportunity for staff to run research on peanuts for more than 10 years.

## 20.4   Curriculum creation and implementation at Kasetsart University

Product development programmes focus on product innovation, technology innovation, process development, quality assurance, marketing and business management. The aim is to teach how to design and manage the development of agro-industrial products through the systematic application of the Product

Development Process. Product development is defined as the translation of new ideas into marketable new products, processes or services. The area of study involves the multi-disciplinary skills of consumer and market research, product design, product development, process development, quality assurance, technological innovation, marketing and business management. There is the need for students going into the food industry to have the ability to interconnect various areas of food science in problem solving and particularly in creation of new products. There is an emphasis on quantitative skills, as it is important that product development decisions in industry are made on data and analysis and not just descriptive 'feelings'.

A small group of people designed the Product Development courses for Bachelor and Masterate degrees at Kasetsart University, amongst the earliest in Thailand and overseas. The numbers of students in these courses increased rapidly and the Department of Product Development grew in knowledge and skills. Combining a group of people, with a wide variety of skills including market and consumer research, experimental design, quality assurance, food engineering, in advanced academic product development is not an easy task.

### 20.4.1   Undergraduate course in product development
The major aim of Kasetsart University's Bachelor of Product Development in Agro-Industrial Products is to produce graduates who are professionals in new agro-industrial product development, processing, and quality assurance at industrial scale, food safety and quality management. The four-year degree also gives graduates an understanding of consumer research, nutrition, packaging and project management. Graduates can work developing new foods and non-food products, improving manufacturing processes, developing packaging, helping companies make quality products, providing support and expertise to management, and doing scientific research.

The courses are the combination of three areas: product development, processing and quality measurement. Product development includes: principle of agro-industrial product development, techniques of product development. Processing includes processing of agro-industrial products, principles of process development. Quality measurement includes: sensory evaluation of quality, chemical and physical quality measurement, biological quality measurement and quality management system. Apart from this, the minor elective subjects include marketing subjects: principle of marketing, consumer behaviour, and marketing research, which fulfil the skills needed in product development.

In the Bachelor degree curriculum, students have the following courses:

- First year: introduction to agro-industry, calculus, chemistry, biology, abridged physics, standards and regulations of agro-industrial products, organic chemistry, and art of living.
- Second year: packaging materials for product development, processing of agro-industrial products, principles of management, quantitative chemical analysis, principles of statistics, processing of agro-industrial products,

biological quality measurement, statistics for product development, biochemistry, general psychology, man and society.

- Third year: fundamental process engineering, chemical and physical quality measurement, principles of agro-industrial product development, sensory evaluation of quality, principles of marketing, introduction to economics, principles of process development, techniques of product development and consumer behaviour.
- Fourth year: research methods in agro-industrial product development, research and development for agro-industry, quality management system in agro-industry.

They must work during the third year summer vacation for a minimum of 200 hours in a relevant industry and are required to submit a report on their experience.

### 20.4.2   Teaching methods for undergraduate courses
Product Development courses at Bachelor degree level have been traditionally theory-based, supported by laboratory work, but there is now a worldwide trend towards project-based learning. In product development education, project-based learning is essential in order to integrate the disciplines of design, marketing and manufacturing towards the common goal of creating a new product. The process of product development consists of a set of activities, tools and methods that are themselves evolving and improving with practice.

### 20.4.3   Masterate courses
The Department of Product Development offers two graduate programmes: a masters degree and a doctoral degree in Product Development. The programmes were established to meet the great demand for qualified product developers and to supply much needed R&D activities for the fast-growing food and agro-processing industries. The programmes are multidisciplinary courses which offer graduate degrees majoring in product development and minoring in either marketing or other related areas such as biotechnology. The programmes provide important product development knowledge to students with no product development background and also the advanced product development tools are taught. These people will be an important resource for developing the Thai agro-industries in the future.

The Masters programme offers two plans. Plan A is for regular students: comprising a minimum of 24 credits of both core and elective courses, and a minimum of 12 thesis credits. Plan B, which is suitable for people working in the industry, comprises a minimum of 35 credits of both core and elective courses, and 6 credits of an independent study course.

Problem-based learning (PBL) activities are incorporated into the graduate course – Development of Agro-Industrial Products. This can aid in student understanding of basic product development principles while developing students' problem-solving and critical thinking skills. Students investigate the problems from the industries and solve the problems using product development tools. Integrated problem-based learning aids students in developing communication, problem-solving, self-directed learning, and other desired skills and demonstrates the potential to be an enjoyable and challenging classroom experience for both students and teachers.

### 20.4.4   PhD courses

The PhD programme comprises a minimum of 16 credits of both core and elective course and a minumum of 36 thesis credits. The programmes are established to meet the great demand for qualified product development teachers for universities and also researchers for research institutes in Thailand. The programmes provide advanced product development tools and product development management courses are taught. These people will be an important resource for universities in Thailand in the area of product development.

The department has joined the Agro-Industry PhD Programme Consortium which consists of seven premier Thai universities at the start (Chiang Mai University, Chulalongkorn University, Kasetsart University, Khon Kaen University, Prince of Songkla University, KMUTT and Suranaree University of Technology) and eleven overseas partner universities (University of Alberta, ENSIA, University of Georgia, University of Guelph, University of Massachusetts, Massey University, University of Nottingham, Purdue University, Michigan State University, Wageningen Agricultural University, and University of Reading). Students enrolled in the PhD Programme through this consortium will spend approximately one year doing their research at one of the overseas partner universities.

## 20.5   University co-operation

The Department of Product Development, Kasetsart University has co-operated with universities both within Thailand and overseas. The types of co-operation were research and human resource development.

### 20.5.1   Co-operation with Thai Universities

Under the Agro-Industry PhD Programme Consortium Programme, the Department of Product Development has built PhD graduates for various universities in Thailand since 1999. Up to 2007, the department has accepted 11 PhD candidates under the consortium programme. They are from Chiang Mai University (3), Prince of Songkla University (1), Khon Kaen University (1), Mahidol University (1), Thammasart University (1), Burapha University (1), Mae Fah Luang University (1), King Mongkut's Institute of Technology Ladkrabang (1) and King Mongkut's Institute of Technology North of Bangkok (1). Five of them had graduated and are working at their universities such as Chiang Mai University, Prince of Songkla Univeristy, Burapha University and Mahidol University. They are very active and important manpower for their universities.

The staff of Department of Product Development co-operate with staff of several universities in conducting research especially in area of food product development dealing with specific raw materials and nationwide consumers.

Within Kasetsart University, the Department also built some doctorates/ masterates in PD for Kasetsart University Agriculture and Agro-Industrial Product Improvement Institute (KAPI) and Department of Fishery Product, Faculty of Fishery and Institute of Food Research and Product Development.

### 20.5.2    Co-operation with overseas universities

The Department of Product Development has long relationships with Massey University, New Zealand since Prof. Mary D. Earle helped establish the four-year Bachelor degree in Product Development in 1980. A number of international universities hold a Memorandum of Understanding (MOU) with the Faculty of Agro-Industry for education and research purposes. There are student exchanges with students from University of Tennessee, Institute of Agriculture, USA, and National Pingtung University of Science and Technology, Taiwan.

The PhD candidates under the Consortium programme at the Department of Product Development will spend approximately one year doing their research at one of the overseas partner universities. Therefore, the department has co-operated with overseas universities to build up human resources for Thai universities including Kasetsart University. The overseas universities through the consortium are: USA – University of Georgia, Purdue University, and University of Massachusetts; UK – University of Reading and the University of Nottingham; and New Zealand – Massey University.

In terms of research, the Department of Product Development received funding from the United States Agency for International Development Peanut Collaborative Research Support Programme (USAID Peanut CRSP) through the University of Georgia for 13 years (1982–1995). The project conducted research on appropriate technology for peanut storage and utilisation. In addition to research, the department in co-operation with the University of Georgia had trained a number of staff from developing countries such as Cambodia, Laos, Myanmar, Indonesia, Bhutan, Bangladesh, Philippines, Sri Lanka, Vietnam and a few countries from Africa.

## 20.6    Industrial co-operation: industrial-based projects

A problem-based interdisciplinary capstone learning experience is designed to enhance career skills (critical thinking, decision making, team work, communication, etc.) in the context of industry's approach to developing new and improved food products. The Department of Product Development has co-operated with various scales of industry; medium, small and cottage (village). The types of co-operation are varied depending on needs of each industry.

### 20.6.1    Co-operation with medium and small industries

There are two types of co-operation with medium and small industry, product research and development and training. Numbers of R&D projects were conducted according to the needs of industry, mostly on formulation and process development/improvement. The training of industry people was conducted both at the department and in-house. The disciplines were Techniques in New Product Development, Sensory Evaluation and Consumer Testing, Product Shelf-life Determination, Quality Assurance – GMP, HACCP, Total Product Quality Measurement and Statistical Tools for Product Development.

### 20.6.2    Co-operation with small and cottage industries

There are also two types of co-operation with small and cottage industries but in different activities. These industries need ready-to-operate technology since they have fewer technical staff. Technology transfer in various types of products has been often done to the cottage industry. In-house training was also conducted, mostly related to safety matters. Technical consultations are constantly given to those industries either by telephone, letter or visit.

## 20.7    Alumni

The Department of Product Development is dedicated to the education of undergraduate, graduate students and industrial employees in the field of research and development of agro-industrial products. The knowledge applications of science and technology, quality assurance as well as marketing and business management are integrated for graduates to teach and do research. Our department has quality measurements laboratories and a processing plant to serve for hundreds of students each year. Every year, at least a hundred undergraduate students and 15 graduate students graduate in our programmes. Recently, the number of our alumni reached more than 1,000 people. They work in either government or private sectors. These following are some outstanding alumni playing important roles in their careers.

- Professor Witoon Prinyawiwatkul is a professor in the Department of Food Science, Louisiana State University. His expertise is sensory quality/ evaluation and consumer acceptance of foods and beverages; new food product development from concept development to pre-market test; functional/physicochemical characteristics of food ingredients. He contributes his knowledge to Thai food science and technology research. Every year he comes to Thailand and provides short training courses in sensory evaluation, product development and statistical analysis.
- Associate Professor Rungnaphar Pongsawamanit works at the Department of

Product Development, Kasetsart University. She was the first generation of undergraduate students. She received BS (Hons) in Product Development. Her expertise in hydrocolloids, shelf life and food engineering is well-known in Thai food industries. She has published many journal articles, three food science and technology books. In 2005, she gained the first rank Kasetsart University researcher publishing many articles in international journals.

- Mrs Oratai Sillapanapaporn graduated MS in 1991. She is Director, Office of Commodity and System Standards, National Bureau of Agricultural Commodity and Food Standards (ACFS), Ministry of Agriculture and Co-operatives. She has evolved the ACFS organisation to control agricultural products, food, and processed agricultural products by certifying and enforcing standards from food producers to consumers, to negotiate with international partners in order to reduce technical barrier to trade (TBT) and to improve and enhance competitiveness of Thai agricultural and food standards. Her dedication is invaluable for the food safety programme in Thailand.

- Mr Vitoon Niyhatiwatchanchai and Mrs Mali Niyhatiwatchanchai graduated BS in 1987. This couple are very successful in food business and are a great model for our recent students to establish their own businesses in the future. Since 1988 Mrs Niyahatiwatchanchai had decided to pursue her MS in Product Development and worked with Assoc. Prof. Chompreeda. Her thesis was the development of reduced calorie salad cream. It had a highly strong impact on this couple to ignite their own food business in salad dressing, mayonnaise, Thai sauce and fruit fillings. Recently, they have two food companies. One is the Pure Food Co., Ltd, and the other one is the Win-Win Food Co., Ltd. Every year during the summer session, at least two of our junior students have a good chance to be apprenticed in those companies.

- Mr Boonanake Chaisam is another successful alumnus. After he graduated in 1986, he has been in the business of food flavours and ingredients as the sales manager. With his experience and talent, he established his own business named the Flavour Force Company Limited.

- Mrs Sirin Pusayapaiboon graduated MS in 1993. She is the daughter-in-law of the founders of the Useful Food Co., Ltd. It is one of Thailand's leading snack manufacturers. She is the research and development manager and has been very successful in developing new product lines, especially in producing new breakfast cereal, and launching into the market.

## 20.8    Staff in 2007

Between the late 1980s and early 1990s, the department had only six academic staff and provided teaching in three programmes: Bachelor degree, Master degree and diploma. Recently, this grew to five programmes and 15 academic staff as follows:

- Assistant Prof. Walairut Chantarapanont: Food microbiology, Food safety, Food sanitation.

- Assistant Prof. Withida Chantrapornchai: Colour and colour evaluation, Optical properties of emulsions, Natural colorants.
- Associate Prof. Penkwan Chompreeda: Nutritional product development, Sensory evaluation and consumer study, Food safety, Functional foods product development, Utilisation of peanut and rice.
- Pisit Dhamvithee: Product development management, Multivariate analysis, Applied statistical techniques for agro-industry research.
- Associate Prof. Anuvat Jangchud: Modelling and experimental design in product development, Development and characterisation of edible films and coatings, fruit leather/vegetable leather, Prolonging shelf life of fruits.
- Associate Prof. Kamolwan Jangchud: Extrusion technology, Effect of ingredients on qualities of food products, Development of Thai dessert and snack products, Product development from rice and sweet potato.
- Associate Prof. Sombat Khotavivattana: Agro-industrial processing, Sugar technology, Drying and processing development.
- Assistant Prof. Saowanee Lertworasirikul: Data envelopment analysis, Production and inventory control, Logistics management, Product and process optimisation and prediction with fuzzy logic and neural networks.
- Associate Prof. Chintana Oupadissakoon: Sensory science, Food quality and factors affecting quality of foods, Fruit paste/candies.
- Associate Prof. Rungnaphar Pongsawatmanit: Food process engineering, Snack food technology, Hydrocolloid applications, Rheological properties for process and product development, Shelf life evaluation of agricultural products.
- Associate Prof. Hathairat Rimkeeree: Cosmetic product development, Consumer product development, Consumer research, Consumer sensory evaluation, Non-food product development.
- Associate Prof. Thongchai Suwonsichon: Texture and texture evaluation, Sensory evaluation and Consumer testing, Non-destructive measurement using near infrared spectroscopy (NIR).
- Assistant Prof. Nantawan Therdthai: Process modeling and optimisation, Baking technology.
- Thepkunya Harnsilawat: Food process engineering, Food Emulsion and Hydrocolloid applications in food products.

The Department provides support for academic staff for continuing education, attending conferences and seminars, and co-operating with local and oversea specialists. Our academic staff are dedicated to transfer their knowledge in the form of academic training, workshops and consulting to both governmental and private organisations.

## 20.9    Research and centres of excellence

### 20.9.1    Research activities
The department has established several research activities. These are:

- application of hydrocolloids for product development
- measuring shelf life and extending shelf life
- Thai snack research
- texture and colour evaluation
- cosmetic research
- Thai tropical fruit and vegetable products
- modelling and simulation processes
- Thai fermented products.

Funding of research ranges over internal sources, government agencies and domestic and international corporations. Our staff are very active in their own research. Every year, they receive research grants from many sources including The National Research Council of Thailand (NRCT), The Thailand Research Fund (TRF), The Royal Jubilee PhD grant, the National Science and Technology Development Agency (NSTDA), and National Bureau of Agricultural Commodity and Food Standards (ACFS).

### 20.9.2    Centres of excellence

The popular research areas needed in the Thai food industry are product development and sensory evaluation. Our department is the pioneer in teaching and research in those two areas. In 2005, Kasetsart University funded the budget to our department in order to establish two special research units (SRU), one in sensory and consumer research and the other in product innovation technology and management.

*Special research unit in sensory analysis and consumer research (SRU–SCR)*
The Department has been a recognised leader in sensory evaluation and consumer testing. Dr Oupadissakoon and Dr Chompreeda have been comprehensively teaching and training in sensory evaluation since the late 1980s. From their pioneering work, this SRU–SCR was established in 2005 with funds from Kasetsart University. It is a teaching, research, and service unit. The main activities are education, research and services in a full range of sensory evaluation, perception and acceptance. Every year, this SRU provides at least two complete theoretical and practical short courses to meet the industrial needs. The SRU also provides consultancy services to industry in all aspects of sensory science such as trained panels to assess product quality and consumer and focus group research. The relationships between sensory, chemical and physical properties are also studied. The collaborative research with other universities such as Kansas State University, University of Georgia, and Louisiana State University is very active.

*Special research unit in product innovation and technology management*
This research unit works collaboratively with all centres and research units in the Faculty of Agro-Industry, especially with the Sensory Evaluation and Consumer Research Unit, to develop concepts, design and manage new products. The unit

uses multidisciplinary approaches to integrate science and technology with marketing and management, in order to develop, implement, manufacture and sustain new products. This group provides expertise in information systems management for scientific data and related information. The specialisations for the food industry and other agro-industries of this SRU are the following:

- new product and package design
- marketing of food textile products
- information management system for food industry
- material handling, distribution system, and supply chain management
- manufacturing system modelling and optimisation
- production planning and inventory control
- quality function development in product innovation
- artificial neural network and fuzzy logic applications
- systems analysis for sustainability of food production systems.

The group uses their expertise to educate students and industrial employees in the applications of the sciences and technology to agro-industrial systems, and to understand the necessary problem-solving skills.

## 20.10    Problems met and lessons learnt

Our problems and the lessons we had to learn in building up the department could be elaborated under the categories: facilities, staff, curriculum and students. Problems concerning facilities were equipment for the laboratory and teaching environment. At the start, the facilities were very limited and not sufficient to serve all students. This problem resulted from the limitation of the budget from the government. The problems were gradually solved when the department received some non-government budgets such as research grants from USAID-Peanut CRSP, industries and special BS programme. Problems related to staff were number, performance and experience. As mentioned earlier, at the start the department had a very limited number of staff (5) which was not really enough to handle 60 students. New staff for the department were recruited on the basis of their high performance, for example they held honours degrees, or had obtained the government's scholarship or the King's Scholarship. Even though the department's staff were capable of handling a hundred students, some were too young (newly graduated) to teach PD. Only after 4–5 years, had they built up sufficient experience. Therefore, the lessons learned about staff building were that it is not only number that is important, but that they should be high performing and experienced as well. Curriculum was not much of a problem because there was a pattern from Massey University to follow. The last problem was teaching the students. By culture, most university students in Thailand are not mature. Therefore, teaching them how to think is so difficult it is sometimes impossible. We had to make them all realise that PD involves creative thinking and then put them to work.

## 20.11   The future of New Product Development education in Thailand

Many universities in Thailand will be offering BS programmes in PD and some MS programmes in the future. Within five years, PhD candidates in PD will graduate and return to their universities. Then they will be PD resources for their universities.

The Department of Product Development, Kasetsart University will provide an opportunity to Thais as well as foreign students to enrol in higher degrees (MS and PhD) especially in research or non-coursework programmes. Teaching techniques will be adapted to fit marketing trends as consumer behaviour is changing in the world.

As to the future of PD education in Thailand, it could be said that there is a worldwide trend towards the development of health-care products both in the food and non-food sectors. The main research will focus on using organic-agricultural materials such as tropical fruits and vegetables, spices and medical herbs to develop new products. There is a trend to use modified agricultural materials such as germinated rice grains to increase gamma amino butyric acid (GABA). This improves nutritive value for better health and also increases the value of rice grains. There is also a trend to use vegetable oils such as coconut oil and rice-bran oil to replace mineral oil in cosmetic product development.

## 20.12   Summary

This chapter demonstrates the effective education for product development at Kasetsart University. It can be used as a case study for educators/administrators in developing countries to learn how to organise and create a high standard academic department. Over two decades the Department of Product Development has developed slowly in term of curriculum, infrastructure, manpower and trust. In 2007, the Department is the most outstanding centre for product development in the country in terms of staff, and curriculum, which covers BS, MS and PhD programmes. This chapter also provides many good lessons in how to study PD and how to plan and solve problems relating to curriculum, staff, and teaching in a PD department. It also demonstrated how hard the staff of the Department of Product Development worked to develop the human resources in this area. The Department produced more than 1,000 PD alumni working all over the country to create new products for local consumption as well as for export. It is a good and unique programme for those worldwide to study. The BS programme is based around a structure totalling 138 credits, covering basic courses, core courses, and main courses of Product Development including Marketing and Quality Measurement. For MS, and PhD programmes, there are 36 credits and 52 credits, respectively. The programmes stress agro-industrial research by focussing on the use agricultural produce as raw materials to develop new products. Teaching methods include project-based learning as well as problem-based learning. Therefore, it is very interesting for everyone to follow.

## 20.13   Bibliography

BOON-LONG, NARUDOM. 1989. New hope: Agro-Industry. The Book for Memory of Professor Narudom Boon-Long's Retirement. Bangkok, Thailand.

DEPARTMENT OF FOOD SCIENCE AND TECHNOLOGY. 2003. Food Science and Technology. Faculty of Agro-Industry, Kasetsart University. Bangkok, Thailand.

FACULTY OF AGRO-INDUSTRY. 2007. Faculty of Agro-Industry Information Book. Kasetsart University, Bangkok, Thailand.

FOOD SCIENCE AND TECHNOLOGY ASSOCIATION OF THAILAND. 2006. FoSTAT 30th Anniversary Profile Book. Bangkok, Thailand.

HARUTHAITHANASAN, VICHAI. 2004. *Vichai Haruthaithanasan: 60 Years of His Life and Works. The Book for a Memory of Professor Vichai Haruthaithanasan's Retirement.* Bangkok, Thailand.

KASETSART UNIVERSITY. 2006. KU Directory. Bangkok, Thailand.

THAI GOVERNMENT ORGANIZATION DIRECTORY 2003–2004. Office of Higher Education Commission, Ministry of Education.

# Index